Calculus of One Variable:
An Eclectic Approach

Michel Helfgott

ii

Michel Helfgott
Department of Mathematics and Statistics
East Tennessee State University
Johnson City, Tennessee
USA

ISBN 978-1477633878

Cover Photograph: Puente de los Suspiros, Lima, Peru [Photo Credit: Federico Helfgott]

Calculus of One Real Variable

Contents

Preface

This textbook is intended for a two-semester course on calculus of one variable. The target audience is comprised of first-year students in biology, chemistry, physics and other related disciplines. It is an extension of *Introductory Calculus for the Natural Sciences* (Helfgott and Moore, 2011), a book that grew out of the Symbiosis Project at East Tennessee State University.

The title of the book reflects the fact that it is not limited to one single approach to calculus. Rather, we use technology or applications whenever they are necessary to introduce certain topics. Nonetheless, as expected, a conceptual framework permeates the whole book. It has been my purpose to reach a balance between theory and applications, relying on graphing calculators when the need arises.

In agreement with established pedagogical principles, some propositions are proven but others are relegated to enrichment notes or to an appendix, or sometimes a proper reference is provided for further study. Let us not forget that students can learn calculus, and how to apply calculus to solve real-life problems, even if they have not fully assimilated the formal definition of limit. An intuitive understanding of limits is often all that is needed to work with derivatives and integrals.

A distinctive characteristic of the book is the early introduction of sequences and geometric series, and a gradual development of simple differential equations, as well as the use of linear regression to analyze data. The core of the book is to be found in the first three chapters, in which examples from biology, chemistry and physics are analyzed with care, emphasizing the close links between calculus and the natural sciences. The last two chapters, or sections thereof, can be used as a sort of capstone in order to show how mathematics helps in the understanding of enzyme kinetics and transport across cell membranes.

I wish to thank Robert Beeler and Harald Andrés Helfgott for reading the manuscript with care and making many useful suggestions. My gratitude also goes to Edith Seier for being the driving force behind section 4.8. Last but not least, I should add that in matters of grammar and style I received invaluable advice from Federico Helfgott. To all of them I am deeply grateful.

Chapter 1

Differential Calculus

1.1 Introduction

Integral Calculus, the calculus of areas, goes back to early Hellenistic times, some 2,300 years ago. Archimedes (287-212 BCE) was able to calculate the area of circles or segments of parabolas to a high degree of accuracy. Differential Calculus, the calculus of tangents, is also an ancient subject. We might remember that at the time of Euclid (323-285 BCE) mathematicians knew how to construct the tangent at any point of a circle.

The development of symbolic algebra in the 16th century played a crucial role in the history of mathematics because it led to the geometry with coordinates in the first half of the 17th century and the works of Isaac Newton (1642-1727) and Gottfried Leibniz (1646-1716) in the second half of the 17th century. Newton and Leibniz, independently of each other, were among the first to realize that the problem of tangents and the problem of areas are intimately linked. Thus, Integral Calculus and Differential Calculus were recognized not as two separate disciplines but rather as parts of an organic whole called simply calculus.

Calculus was successfully applied to physics, especially in the 18th and 19th centuries, and later to chemistry and biology, as well as many areas of engineering. However, it took almost 200 years, roughly from 1684 when Leibniz published his first paper on calculus until about 1880, to put the subject under a solid foundation. It was a long process that culminated with the works of Karl Weierstrass (1815-1897) and his disciples.

A review of Cartesian geometry

Geometry with coordinates, also called Cartesian geometry, was developed by René Descartes (1596-1650), whose name in Latin was Cartesius, and his contemporary Pierre de Fermat (1601-1665). In 1637 Descartes published 'Discourse on the Method of Rightly Conducting the Reason in the Search for Truth in the Sciences,' wherein one of the appendices dealt with his new approach to the solution of geometrical problems. This appendix, called 'The Geometry,' was one of the most notable advances in geometry since Greek classical times, some 2,000 years before. Descartes' main idea was to use symbolic algebra to solve problems in geometry. With the benefit of hindsight, we can assert that Cartesian geometry created the necessary tools for the development of calculus.

We will start the review by analyzing lines in Cartesian geometry, the simplest geometrical object. The equation of any vertical line is $x = h$, where $(h, 0)$ is the intersection with the x-axis, while the equation of any horizontal line is $y = k$, where $(0, k)$ is the intersection with the y-axis. Given any two points (x_1, y_1) and (x_2, y_2), $x_2 \neq x_1$, we define m:

$$m = \frac{y_2 - y_1}{x_2 - x_1}$$

and call it the 'slope' of the segment with endpoints (x_1, y_1) and (x_2, y_2). Any non-vertical line L is characterized by the fact that the slope of the segment between any two points on L is a constant, which is usually denoted by the letter m. Thus, if (x_1, y_1) is any point on L, and m is the slope, any other point (x, y) on L must satisfy the equality

$$\frac{y - y_1}{x - x_1} = m.$$

That is to say, $y - y_1 = m(x - x_1)$. This is called the 'point-slope' equation of a line, which can be written as $y = mx + b$. The latter is called the 'slope-intercept' equation of a non-vertical line because $(0, b)$ happens to be the intersection of the line with the y-axis. Of course, if (x_1, y_1) and (x_2, y_2) are any two points on a non-vertical line, we can define $m = \frac{y_2 - y_1}{x_2 - x_1}$. Then $y - y_1 = m(x - x_1)$, or $y - y_2 = m(x - x_2)$, becomes the equation of the given line. Both equations are equivalent, in the sense that they define the same line. For instance, the points $(1, 2)$ and $(3, 7)$ determine the line defined by the equation

$$y - 7 = \tfrac{7-2}{3-1}(x - 3)$$

or, what is the same,

$$y - 2 = \tfrac{5}{2}(x - 1).$$

Its intersections with the axes are $(0, -1/2)$ and $(1/5, 0)$, which are found by solving $y - 2 = \tfrac{5}{2}(0 - 1)$ and $0 - 2 = \tfrac{5}{2}(x - 1)$, respectively.

What about distance between points on the plane? Applying the Pythagorean proposition we obtain a formula for the distance d between any two points (x_1, y_1) and (x_2, y_2), namely

$$d = \sqrt{(x_1 - x_2)^2 + (y_2 - y_1)^2}.$$

This important formula opens the way to discuss vertical parabolas. Given a point $(0, p)$, the 'focus,' and the horizontal line $y = -p$, the 'directrix,' we are interested in the set of all points (x, y) whose distance to the focus is the same as the distance to the directrix. Thus,

$$\sqrt{x^2 + (y - p)^2} = \sqrt{(y + p)^2}. \tag{1.1}$$

Consequently $x^2 + y^2 - 2py + p^2 = y^2 + 2py + p^2$, which yields

$$y = \frac{1}{4p}x^2. \tag{1.2}$$

It can be shown that any point (x, y) that satisfies (1.2) will also satisfy (1.1). Hence, the equation (1.2) completely characterizes the vertical parabola with focus $(0, p)$ and directrix $y = -p$. The point $(0, 0)$ is called the 'vertex' of the parabola.

It is to be noted that, given any equation $y = ax^2$, we can conclude that it defines a parabola with focus at $(0, \frac{1}{4a})$ and directrix $y = -\frac{1}{4a}$. All we have to do is set $\frac{1}{4p} = a$ and solve for p. For instance, $y = x^2$ defines the parabola with focus at $(0, 1/4)$ and directrix $y = -1/4$. When $a > 0$ the parabola opens upwards, while for $a < 0$ the parabola opens downwards. Moreover, if we move a parabola, defined by $y = ax^2$, from the vertex $(0,0)$ to a vertex (h, k), a simple translation with no rotation, the new vertical parabola will be defined by the equation

$$y - k = a(x - h)^2.$$

This parabola is an 'undeformed' copy of the original parabola. The idea of translation, without deformation or rotation, is important for what comes next in our review.

Given any quadratic expression $y = ax^2 + bx + c$, what curve does it represent? The method of 'completion of squares' leads to

$$y = ax^2 + bx + c = a(x^2 + \frac{b}{a}x + \frac{c}{a}) = a(x^2 + \frac{b}{a}x + \frac{b^2}{4a^2} - \frac{b^2}{4a^2} + \frac{c}{a}) =$$

$$= a((x + \frac{b}{2a})^2 + (\frac{c}{a} - \frac{b^2}{4a^2})) = a(x + \frac{b}{2a})^2 + (c - \frac{b^2}{4a}).$$

Thus, $y = ax^2 + bx + c$ defines a parabola with vertex at $(-\frac{b}{2a}, c - \frac{b^2}{4a})$. As stated before, it opens upwards if $a > 0$, it opens downwards if $a < 0$. Letting $h = -\frac{b}{2a}$, $k = c - \frac{b^2}{4a}$ we get $y - k = a(x - h)^2$. Defining $y' = y - k$ and $x' = x - h$, we can assert that the parabola $y' = ax'^2$ has its focus at $(0, \frac{1}{4a})'$, where the apostrophe emphasizes the fact that we are working with the translated axes.

The translated parabola, defined by $y = ax^2 + bx + c$, crosses the x-axis depending on whether $\Delta = b^2 - 4ac \geq 0$ (Δ is called the 'discriminant' of the quadratic equation

$ax^2 + bx + c = 0$). Why is this so? If $a > 0$, the parabola opens upwards. It will cross the x-axis provided that the second coordinate of its vertex is ≤ 0; that is to say, $c - \frac{b^2}{4a} \leq 0$, which is equivalent to

$$b^2 - 4ac \geq 0.$$

If $a < 0$, the parabola opens downwards and, in order to cross the x-axis, the second coordinate of its vertex has to be ≥ 0, that is to say, $c - \frac{b^2}{4a} \geq 0$. But $\frac{b^2}{4a} \leq c$ is equivalent to $b^2 - 4ac \geq 0$ since $a < 0$. Thus, in either case $\Delta \geq 0$. We could have reached the same answer if we were to remember that the solutions of $ax^2 + bx + c = 0$ are given by the quadratic formula

$$x = \frac{-b \pm \sqrt{b^2 - 4ac}}{2a}.$$

It is worth mentioning here that the method of 'completion of squares' lies behind the justification of the quadratic formula.

As expected, parabolas play an important role in the solution of quadratic inequalities. Suppose we wish to solve $x^2 - 3x + 2 < 0$. The parabola $y = x^2 - 3x + 2$ has its vertex at $(3/2, -1/4)$ and opens upwards, crossing the x-axis at $x = 1$ and $x = 2$. Therefore (Figure 1.1) the parabola adopts negative values whenever $1 < x < 2$. On the other hand, we can see that if $x < 1$ or $x > 2$ the parabola adopts positive values; consequently, any number smaller than 1 or bigger than 2 is a solution of the inequality $x^2 - 3x + 2 > 0$.

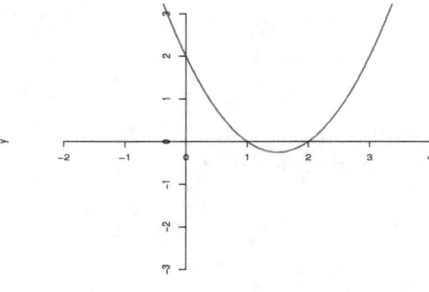

Figure 1.1: The parabola defined by $y = x^2 - 3x + 2$

In general, suppose that we wish to solve the inequality

$$(x - a)(x - b) < 0.$$

Let $y = (x - a)(x - b) = x^2 - (a + b)x + ab$. This expression defines a parabola that opens upwards, intersects the x-axis at $x = a$ and $x = b$, and has its vertex at

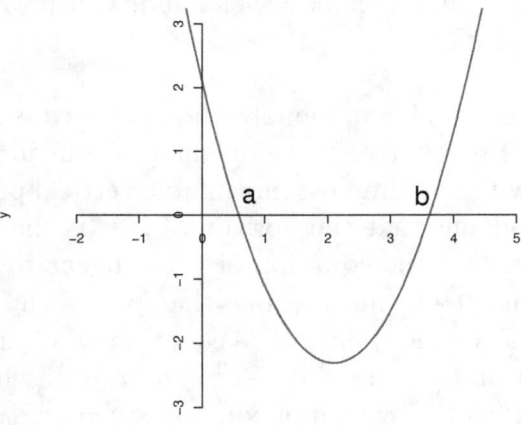

Figure 1.2: The parabola defined by $y = x^2 - (a+b)x + ab$

$(\frac{a+b}{2}, -\frac{(a+b)^2}{4})$ (Figure 1.2). Therefore, any x that satisfies the inequality $a < x < b$ (if $a < b$) or $b < x < a$ (if $b < a$) is a solution to the problem.

To dispel any doubts about the importance of lines and parabolas, let us discuss two problems of considerable interest.

A problem of optimization and a problem about tangents

Suppose that we wish to build a rectangular pen, next to a river, with 3000 ft of wire. The river serves as one side of the pen. What dimensions should the pen have in order to maximize its area? Let x and y be the unknowns and A denote the area. We have $2x + y = 3000$ and $A = xy$. Therefore $A = x(3000 - 2x) = -2x^2 - 3000x$. But the parabola $y = -2x^2 + 3000x$ opens downwards and adopts its maximum at $x = \frac{-3000}{2(-2)} = 750$ (let us recall that the vertex of any parabola, defined by $y = ax^2 + bx + c$, is attained at $x = -\frac{b}{2a}$). The best we can do is to select $x = 750$ and $y = 3000 - 2 \times 750 = 1500$.

Is it possible to solve all optimization problems using only Cartesian geometry? Unfortunately not. For example, suppose we have 100 cm^2 of tin and wish to build a right cylinder, without a top, of maximum volume. Let r be the radius of the base and h the height. Then $100 = \pi r^2 + 2\pi rh$, and therefore $h = \frac{100 - \pi r^2}{2\pi r}$. Since $V = \pi r^2 h$, we can conclude that

$$V = 50r - \frac{\pi}{2}r^3$$

This is a cubic polynomial, and there is no easy algebraic method comparable to 'completion of squares.' Calculus, on the other hand, will provide us the necessary tools to solve this and similar challenges.

Let us consider a different problem, namely finding the equation of the tangent to a parabola, which happens to be a very important problem in mathematics and its applications. Let us deal with the simplest imaginable vertical parabola, the parabola defined by $y = x^2$. It is evident that the horizontal axis is the tangent to the given parabola at $(0,0)$. Can we find the equation of the tangent to $y = x^2$ at any point (c, c^2)? Of course, by 'tangent' we mean a line that passes through (c, c^2) and does not touch the parabola at any other point. The equation of a line that passes through (c, c^2) and any other point on the curve is $y - c^2 = m(x - c)$, where m is the slope of the line. Then the solution to the system of equations $y = x^2$, $y - c^2 = m(x - c)$ will provide the points of intersection between the parabola and the above-mentioned line. To solve this system, we need to pay close attention to the equation $x^2 - c^2 = m(x - c)$, which is equivalent to $x^2 - mx + (mc - c^2) = 0$. If we wish to find the tangent, we have to choose m in such a way that the last equation has only one root. In order for this to happen the discriminant has to be zero; that is to say, $m^2 - 4(mc - c^2) = 0$, which in turn is equivalent to $(m - 2c)^2 = 0$. Thus, $m = 2c$ will be the slope of the tangent to the parabola at (c, c^2). In summary, the equation of the tangent is:

$$y - c^2 = 2c(x - c).$$

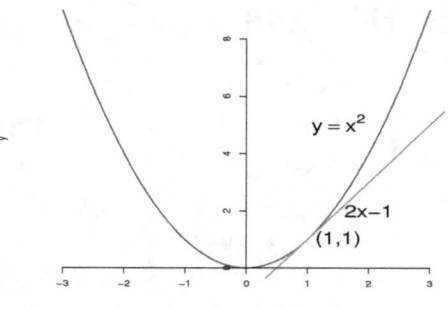

Figure 1.3: Tangent to a parabola

Although the above-mentioned approach works for any conic (that is, parabolas, ellipses, circles, and hyperbolas), it fails for simple curves such as $y = x^3$. Calculus provides a systematic method for calculating the tangents to a wide variety of curves.

Modeling

Let us start by recalling that, if an object is moving at constant speed v, the distance traveled during a period of time t is $d = vt$. Hence, if we draw the coordinate axes with t on the horizontal axis and v on the vertical axis, the area of the corresponding rectangle will provide the required distance. What happens if the velocity is not constant? Suppose a metal ball starts from the top of an inclined plane and slides down under the influence of gravity (we do not take into consideration friction or air resistance). We observe that the velocity of the ball increases with time; thus, we might put forward the hypothesis that, for $t \geq 0$

$$v = kt \qquad (1.3)$$

where k is an unknown constant. This is the simplest imaginable formula for the velocity of the sliding ball. Our hypothesis will be supported if the mathematical consequences of it are in agreement with experiments. In the first place, we need to calculate the distance traveled by the ball up to time t. We divide $[0, t]$ in n subintervals $[t_{i-1}, t_i]$ of small and equal widths, n being a large natural number. Since the width of each subinterval is very small, we may assume that the velocity is practically constant on each $[t_{i-1}, t_i]$. Hence the distance traveled by the ball during this subinterval of time is approximately the area of the rectangle of width $t_i - t_{i-1}$ and height v_i (the value of v at t_i). Then, the total distance that the ball travels will be, approximately, the area of the triangle with base t and height kt, that is to say,

$$d = \frac{1}{2}kt^2. \qquad (1.4)$$

We feel confident that this formula makes sense because, through an effort of imagination, the number of subintervals can be increased at will. In other words, we have

$$\frac{d}{t^2} = \text{constant}.$$

Consequently, if we are able to measure distances each second and let d_i be the total distance traveled by the ball between $t = 0$ and $t = i$, the numbers in the third column of table 1.1 should remain practically constant (of course, errors of measurement will intervene). If this is so, we may accept the 'model' $v = kt$ and proceed to estimate k as twice the constant value found on the third column of the table 1.1. The constant k, unknown at the beginning, receives the name of **parameter** of the model. If there is concordance between the consequences of the model and data, we can proceed to estimate k. Thus, a parameter is an unknown constant whose value is determined using data, once the model is accepted.

We might observe that $d_2 - d_1 = \frac{k}{2} \cdot 4 - \frac{k}{2} = \frac{k}{2} \cdot 3 = 3d_1$, $d_3 - d_2 = \frac{k}{2} \cdot 9 - \frac{k}{2} \cdot 4 = \frac{k}{2} \cdot 5 = 5d_1$,..., $d_{i+1} - d_i = d_1(2i + 1)$. Thus, if we measure d_1 and put marks on the

Table 1.1: Prediction

t	d	$\frac{d}{t^2}$
1	d_1	$d_1/1$
2	d_2	$d_2/4$
3	d_3	$d_3/9$
4	d_4	$d_4/16$
5	d_5	$d_5/25$
-	-	-
-	-	-
-	-	-

inclined plane at d_1, $3d_1$, $5d_1$, $7d_1$,... the model $v = kt$ predicts that the ball will pass through these marks after each second. Galileo Galilei (1564-1642), the founder of modern mechanics, was the first scientist to perform experiments with inclined planes.

Everything seems to work well, but what might we do if, in a different context, the model $v = kt$ were not in agreement with data? We could try to work with $v = kt^2$. Since we wish to find out how distance varies with time under these circumstances, we would have to calculate the area under the parabola $v = kt^2$. This is a much harder task than calculating the area of a triangle! Calculus techniques provide a fast and convincing solution to the problem of areas bounded by curves of different types.

Quite often, especially in chapter 2, we will discuss diverse models of the natural sciences. An important activity will be to test the validity of a model by contrasting, with data, the mathematical consequences of it. The idea of a model is intimately linked to what we understand nowadays as the **scientific method**. Scientists often work in three stages (Frank, 1957):

1. *Setting up principles.*

2. *Making logical conclusions from these principles in order to derive observable facts about them.*

3. *Experimental checking of these observable facts.*

That is to say, a scientist often builds a model of a particular aspect of the real world. After obtaining observable facts from it, through a process that frequently involves mathematics or statistics, we have to check these facts with data obtained from experiments. If there is concordance we accept the model, otherwise we discard it and build another model. As expected, we will study with special care those models that involve calculus. However, sometimes we will not engage in 'modeling' but we will use well-established laws of physics in order to solve applied problems.

Exercises for Section 1.1

1. Find the equation of the line that passes through the points $(3, 1)$ and $(5, 2)$. Where does it cross the axes?

2. Find the focus and vertex of the parabola defined by $y = x^2 + 3x + 1$. Do the same for the parabola defined by $y = -2x^2 + x - 1$.

3. Solve the inequality $x^2 - 3x - 4 < 0$.

4. What is the solution of $x(x - b) < 0$? What is the solution of $x(x - b) > 0$?

5. Solve the inequality $x^2 + x - 20 > 0$.

6. Suppose that you have 500 m of wire and wish to build a rectangular pen of maximum area, next to a river and with no fence on the side adjacent to the river. What dimensions should you choose?

7. Use only Cartesian geometry to find an equation of the tangent to the parabola $y = x^2 + x$ at $(1, 2)$.

8. Use the method of completion of squares to deduce the formula for an arbitrary quadratic equation.

9. Recall, from Pre-Calculus, that two non-vertical lines are perpendicular if and only if the product of their slopes is -1. Find an equation of the line that passes through $(1, 2)$ and is perpendicular to the line defined by $y = 3x - 5$.

10. Which point on the line $y = 2x + 3$ is closest to $(1, 1)$?

11. A runner is 10 m. behind a bus. At $t = 0$ she starts to run at the constant speed of 2 m/s and the bus starts from rest with an acceleration of 1 m/s^2. At what time will the runner be closest to the bus? (Hint: One needs to find where the parabola $y = 10 + \frac{1}{2}t^2 - 2t$ adopts its minimum.)

1.2 Functions

Examples of relationships between variables are abundant in nature. For instance, we measure the height of a growing plant over a certain period of time or the concentration of a substance as it is being used in an enzymatic reaction. Some relationships are 'functions'; thus, we need to give a precise mathematical meaning to the intuitive concept of *function*, a word that often appears in the English language.

Let A and B be two sets. By a function f from A to B, to be denoted $f : A \to B$, we mean a rule or correspondence by which every element a in A is assigned to an element b in B (we will write $b = f(a)$). The set A is called the **domain** of the function ($dom(f)$), while the set of all images of A under f is called the **range** of the function ($ran(f)$). Some authors denote the range of f by the symbol $f(A)$.

The sets under discussion can be quite arbitrary, as the rule itself, but some of the most interesting and useful functions arise when A and B are subsets of $\Re$ (the set of real numbers) and the rule is defined through a formula. For example, let $f : \Re \to \Re$ be defined by $f(x) = x + 3$. This is the function that takes any number x and sends it into $x + 3$. Its domain is $\Re$ and its range is also $\Re$ because for any real number y we are able to find a number, namely $y - 3$, such that $f(y - 3) = y$.

In calculus, usually the domain of a function will be an interval of the real line. That is to say, a subset of $\Re$ defined by one or two inequalities, as well as $\Re$: $(a, b) = \{x : a < x < b\}$, $[a, b] = \{x : a \le x \le b\}$, $[a, b) = \{x : a \le x < b\}$, $(a, b] = \{x : a < x \le b\}$, $(a, \infty) = \{x : x > a\}$, $[a, \infty) = \{x : x \ge a\}$, $(-\infty, b) = \{x : x < b\}$, $(-\infty, b] = \{x : x \le b\}$. Let us agree to call a, b the endpoints of $[a, b]$, similarly a is the endpoint of $[a, b)$, b is the endpoint of $(a, b]$, a the endpoint of $[a, \infty)$, and b the endpoint of $(-\infty, b]$. A point x_o of an interval I is an **interior point** if it is not an endpoint, while the **interior** of an interval I ($int(I)$) is the collection of all the interior points. That is, $int(I)$ is the interval itself with the exclusion of the endpoints; for instance[1] $int[a, b] = (a, b)$, $int[a, b) = (a, b)$. An interval I is said to be **open** if and only if $I = int(I)$, that is, all the points of I are interior points. Thus, (a, b), (a, ∞), $(-\infty, b)$, and the real line itself, are open intervals. Actually, they are the only types of open intervals. A **neighborhood** of x_o is any open interval that contains x_o. For instance, $(-2, 1)$ and $(-1, 3)$ are neighborhoods of 0 while $(0, 5)$ is a neighborhood of 2. A **deleted neighborhood** of x_o is a neighborhood of x_o where x_o has been eliminated.

The **graph** of $f : A \to B$ is the set of all ordered pairs $(a, f(a))$, where a is an arbitrary element in the domain. For example, if f has domain $\Re$ and is defined by the rule $f(x) = x + 3$, its graph is the set of all ordered pairs (x, y) such that $y = x + 3$. This is the well-known line $y = x + 3$. If $g : \Re \to \Re$ is defined by $g(x) = x^2$ we will be dealing with a function whose graph is a vertical parabola. From now on we will relax our notation and just write $y = f(x)$ to denote both the function and its graph; from the context it will be understood what we mean. Thus, when we write $y = x^2$ we have in mind the function that squares every real number and whose graph is a parabola. **Sometimes we will just write** $x \to f(x)$ **to denote the function**

---

[1] The symbol (a, b) may denote an interval or an ordered pair. From the context under discussion we have to determine which meaning is intended.

$y = f(x)$, **or simply** $f(x)$. Thus, the symbol $x \rightarrow x^2$ will sometimes denote the function f defined by the rule $f(x) = x^2$, whose domain is the whole real line and its range is the interval $[0, \infty)$. Moreover, often we will identify the graph of a function with the **curve** that it describes.

The function $y = x^2 - 3x + 4$ has a parabola as its graph too. Indeed

$$y = x^2 - 3x + (\tfrac{3}{2})^2 - (\tfrac{3}{2})^2 + 4 = (x - \tfrac{3}{2})^2 + \tfrac{7}{4}.$$

We recall that the expression $y = a(x - h)^2 + k$ represents a parabola with vertex at (h, k). It opens upwards if $a > 0$, downwards if $a < 0$. Thus, the graph of $y = x^2 - 3x + 4$ is a parabola with vertex at $(3/2, 7/4)$, which opens upwards. It should come as no surprise that this parabola does not cross the x-axis because the equation $x^2 - 3x + 4 = 0$ has no real roots. In section 1.1 we saw that any function $y = ax^2 + bx + c$ has as its graph a vertical parabola with vertex at $(-\frac{b}{2a}, c - \frac{b^2}{4a})$. Whether it opens upwards or downwards depends on the sign of a.

Expressions of the type $y^2 = x$ or $x^2 + y^2 = 1$ do not define a function. Rather, each of them defines what is called a **relation**. In the first case, it is a horizontal parabola. In the second case, we are dealing with a circle centered at the origin and a radius of 1. Both define 'implicitly' a pair of functions, namely $y = \pm\sqrt{x}$ and $y = \pm\sqrt{1 - x^2}$. To be a function, a relation has to pass the **vertical test**: every vertical line can touch the graph of the relation at most at one point.

We can add, subtract, multiply or divide functions in a natural way. Let $f : A \rightarrow \Re$ and $g : A \rightarrow \Re$. Then $f + g$ is the function with domain A such that $(f + g)(x) = f(x) + g(x)$. Similarly, $f - g$ is the function with domain A such that $(f - g)(x) = f(x) - g(x)$ and fg, $\frac{f}{g}$ are the functions with domain A such that $(fg)(x) = f(x)g(x)$, $(\frac{f}{g})(x) = \frac{f(x)}{g(x)}$ respectively. In the case of quotient of functions we are assuming that $g(x) \neq 0$ for every x in A. There is another operation between functions that we have to keep in mind; namely, 'composition' of two functions. Suppose $f : A \rightarrow B$ and $g : C \rightarrow D$, where B is a subset of C. Then $g \circ f : A \rightarrow D$ is defined by the rule $(g \circ f)(x) = g(f(x))$. The function $g \circ f$ is called 'the composition of g and f.' For instance, $h(x) = \sqrt{x^2 + 1}$ is the composition of $x \rightarrow \sqrt{x}$ and $x \rightarrow x^2 + 1$. Similarly, $\cos(x^2)$ is the composition of $x \rightarrow \cos x$ and $x \rightarrow x^2$.

It should be noted that the operation of composition is not commutative; that is to say, in general $g \circ f \neq f \circ g$ even if both are defined. The last example is a clear indication of this fact because $(\cos x)^2$ (Figure 1.4) is quite different from $\cos(x^2)$ (Figure 1.5).

A function $f : A \rightarrow B$ is said to be **one-to-one** if $x_1 \neq x_2$ implies $f(x_1) \neq f(x_2)$. Graphically, the property of being one-to-one means that any horizontal line can touch the graph of f at most at one point. For instance, $f : \Re \rightarrow \Re$, $f(x) = x^2$, is not one-to-one. However, $g : [0, \infty) \rightarrow \Re$, $g(x) = x^2$, is one-to-one. Obviously, any

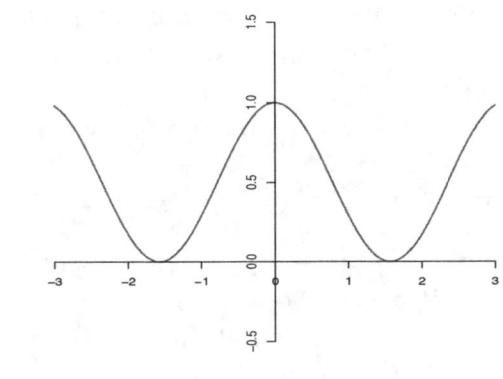

Figure 1.4: The function $(\cos x)^2$

linear function $x \rightarrow ax + b$ $(a \neq 0)$ is one-to-one. A function $f : A \rightarrow B$ is said to be **onto** if for any b in B there exists a in A such that $f(a) = b$. That is to say, f is onto if and only if $ran(f) = B$. For instance, $h : \Re \rightarrow \Re$, $h(x) = \sin x$, is not onto because $ran(h) = [-1, 1]$, but $h : \Re \rightarrow [-1, 1]$, $h(x) = \cos x$, is onto. Evidently, any function $f : A \rightarrow ran(f)$ is onto. The definitions of one-to-one and onto functions will be particularly useful in the next section and later on when we will have to deal with inverse trigonometric functions as well as the inverse of the natural logarithm function.

If a function $f : A \rightarrow B$ is one-to-one we can define its **inverse**

$$f^{-1} : ran(f) \rightarrow A,$$

namely $f^{-1}(y) = x$, where x is the unique element of A such that $f(x) = y$. That is to say, $f^{-1}(y) = x$ if and only if $f(x) = y$ for any x in A and any y in $ran(f)$. We might also observe that

$$(f^{-1} \circ f)(x) = x \qquad (1.5)$$

and

$$(f \circ f^{-1})(y) = y \qquad (1.6)$$

for any x in A and any y in $ran(f)$. As a matter of fact, if there exists a function $g : ran(f) \rightarrow A$ such that

$$(g \circ f)(x) = x \text{ for any } x \text{ in } A$$

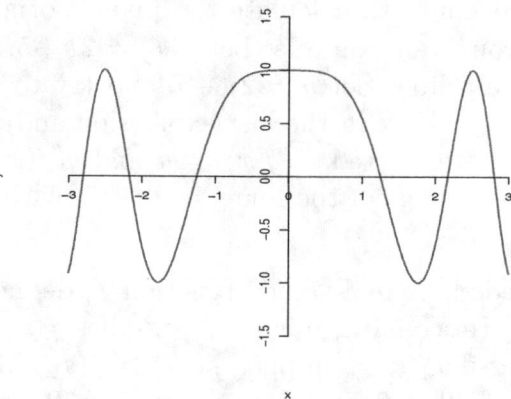

Figure 1.5: The function $\cos(x^2)$

and

$$(f \circ g)(y) = y \text{ for any } y \text{ in } ran(f)$$

then $g = f^{-1}$. For instance, let $f(x) = x^2$ for any x in $[0, \infty)$. This is a one-to-one function with domain $[0, \infty)$ and range $[0, \infty)$. Evidently $g(x) = \sqrt{x}$ is its inverse because $\sqrt{x^2} = x$ and $(\sqrt{x})^2 = x$ for any $x \geq 0$.

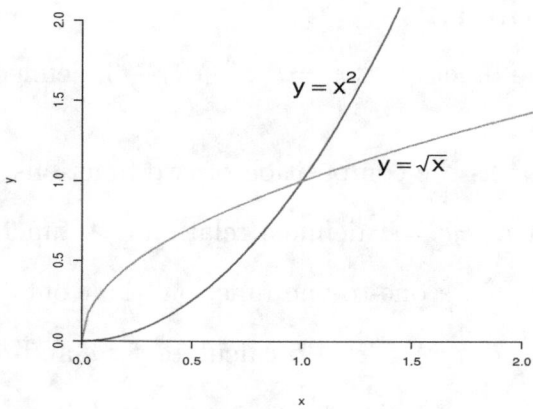

Figure 1.6: The square root function and its inverse

Paying close attention to Figure 1.6 we realize that x^2 and $\sqrt{x}$, defined on $[0, \infty)$, are 'mirror images' of each other, with the line $y = x$ acting as the mirror. This is a property shared by any function and its inverse. We should mention that there is a very useful algebraic procedure to calculate the inverse of a one-to-one function, which we will illustrate through an example. Let $f(x) = 2x + 5$, a one to-one function whose graph happens to be a line. Set $y = 2x + 5$ and try to express x in terms of y: $x = \frac{1}{2}y - \frac{2}{5}$. Then change y by x in the last expression and define $g(x) = \frac{1}{2}x - \frac{2}{5}$. The reader can check that, as expected, $f(g(x)) = x$ and $g(f(x)) = x$ for any x. The line $y = \frac{1}{2}x - \frac{2}{5}$ is the mirror image of the line $y = 2x + 5$, with the line $y = x$ acting as the mirror.

Another concept is needed. A real-valued function f, defined on an interval I of the real line, is said to be **increasing** if $x_1 < x_2$ implies $f(x_1) < f(x_2)$. Similarly, f is said to be **decreasing** if $x_1 < x_2$ implies $f(x_1) > f(x_2)$. For example, the sine function is increasing on $[-\pi/2, \pi/2]$ and decreasing on $[\pi/2, 3\pi/2]$. Obviously, every increasing or decreasing function is one-to-one.

To bring this section to an end, let us recall the notion of absolute value: For any real number a, $|a| = a$ if $a \geq 0$, $|a| = -a$ if $a < 0$. For instance, $|5| = 5$ while $|-7| = 7$. The reader must have seen the five main properties linked to the absolute value of a number: $|a| \geq 0$, $a \leq |a|$, $|a|^2 = a^2$, $|ab| = |a||b|$, $|a + b| \leq |a| + |b|$. Besides, let us keep in mind that $||a| - |b|| \leq |a - b|$, and $|a| < b$ if and only if $-b < a < b$. We should emphasize that the absolute value function, namely $f(x) = |x|$ for all x, will play an important role in the sections to come. Its graph is to be found in Figure 1.24.

Exercises for Section 1.2

1. Find the range of the function $f(x) = -x^2 + 2x - 3$, defined over the whole real line.

2. Write $h(x) = (\cos x)^2$ as the composition of two functions.

3. Does the expression $x + y^2 = 1$ define a relation or a function?

4. Is $f(x) = x^5$, $f : \Re \to \Re$, a one-to-one function? Is it onto?

5. Given $f(x) = \sin x$ and $g(x) = x^2 - 5$ calculate $f \circ g$ and $g \circ f$.

6. Let $f : I \to \Re$ be an increasing function, where I is an interval. Accepting that $ran(f)$ is an interval too, show that its inverse $g : ran(f) \to \Re$ is also increasing.

7. Show that, in the preceding exercise, the word 'increasing' can be replaced by 'decreasing.'

8. Find the inverse of the function $f(x) = x^2$, $x \leq 0$, and sketch a graph of it.

9. Given the function $f(x) = \sqrt{-x}$, $x \leq 0$, find its inverse and sketch a graph of it.

10. Although the sine function is not one-to-one, if we restrict its domain to $[-\pi/2, \pi/2]$ it becomes one-to-one. Sketch the inverse of this restriction of the sine function, keeping in mind that the inverse is the mirror image with respect to to the line $y = x$.

1.3 The Common Logarithm Function

Let us recall that $10^n = 10 \times 10 \times \ldots \times 10$ (n factors), for any natural number n. Besides, $10^{1/n}$ is the unique positive number b such that $b^n = 10$. Moreover, $10^{m/n} = (10^{1/n})^m$ for any natural number m. Thus 10^r is defined for any positive rational number r. If s is a negative rational number, say $s = -r$, we define $10^s = 10^{-r} = 1/10^r$. We are also aware, from our High School experience, that $(10^r)^s = 10^{rs}$ and $10^r 10^s = 10^{r+s}$, for any rational numbers r, s. As a matter of fact, $(a^r)^s = a^{rs}$ and $a^r a^s = a^{r+s}$ provided that $a > 0$.

How could we define $10^{\sqrt{2}}$? The difficulty stems from the fact that $\sqrt{2}$ is an irrational number. However, being an irrational number it can be approximated by a rational number with any degree of accuracy, say 1.414213. We expect that $10^{\sqrt{2}}$ is very close to $10^{1.414213}$; actually, both agree up to the fourth decimal. Hence, it seems plausible to define a function $f : \Re \to \Re$, $f(x) = 10^x$, and accept the fact that $10^x 10^y = 10^{x+y}$ and $(10^x)^y = 10^{xy}$ for any real numbers x, y. The same reasoning leads to the definition of a^x, where $a > 0$ and x is any real number.

In chapter 2 (section 2.10) we will revisit this function and provide a precise definition of a^x, where a is a fixed positive number and x is an arbitrary real number. In the meantime we can rely on the graphical representation of 10^x, observing that it is an increasing function whose range is $(0, \infty)$. Since $f(x) = 10^x$ is increasing, we can conclude that it is one-to-one. Thus, its inverse $f^{-1} : (0, \infty) \to \Re$ exists. This inverse function has a particular name, it is called the 'common logarithm' or log for short (Figure 1.7). Hence $\log(1) = 0$ because $10^0 = 1$ and $\log(10) = 1$ because $10^1 = 10$. Moreover, (1.5) and (1.6) lead to

$$10^{\log x} = x \text{ for every } x > 0,$$

$$\log 10^x = x \text{ for any } x.$$

Right away we can prove a well-known property, namely

$$\log x_1 x_2 = \log x_1 + \log x_2 \text{ for any } x_1, x_2 > 0.$$

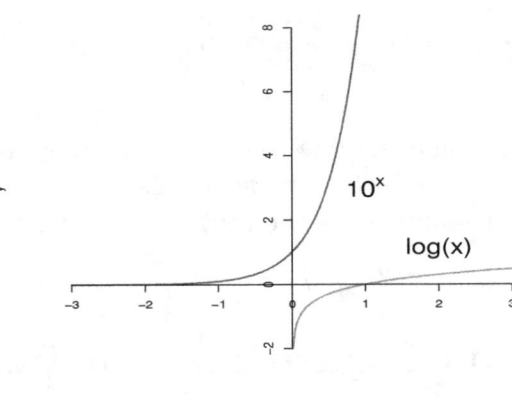

Figure 1.7: Common logarithm function and its inverse

Indeed, let $\log x_1 = y_1$, $\log x_2 = y_2$. Hence $x_1 = 10^{y_1}$, $x_2 = 10^{y_2}$, which in turn leads to $x_1 x_2 = 10^{y_1} 10^{y_2} = 10^{y_1 + y_2}$. Therefore $\log x_1 x_2 = y_1 + y_2 = \log x_1 + \log x_2$. The fact that the logarithm of a product is the sum of the logarithms of the factors allowed scientists, since the early 17th century, to 'transform' products into sums through tables of logarithms, thus avoiding complicated calculations.

There are other properties of logarithms that can be proven with ease: $\log x^n = n \log x$, $\log x^{-1} = -\log x$, $\log \frac{x_1}{x_2} = \log x_1 - \log x_2$. Furthermore, for the time being let us accept that $\log x^r = r \log x$ for any $x > 0$ and any real number r; a full justification will be provided in section 2.10.

We used 10^x to define log but we could have chosen 2^x to define $\log_2$ or for that matter a^x, $a > 1$, to define $\log_a$. The base $a = 10$ was the preferred one, right after logarithms were invented by John Napier (1550-1617) at the beginning of the 17th century, because it is related to the decimal system. Nowadays, mathematicians and scientists often work with the base e, where e is the irrational number to be discussed extensively in sections 2.1, 2.8, and 2.9 ($e \approx 2.71828$). The inverse of the function $\exp : \Re \to (0, \infty)$, where $\exp(x) = e^x$, is denoted ln and is called the 'natural logarithm.' That is to say, $\ln : (0, \infty) \to \Re$ and $\ln = \log_e$. The natural logarithm is the logarithm of choice in higher mathematics because, as we will see in section 1.11, the slope of the tangent line to the curve $y = \ln x$ at any point $(x, \ln x)$ is $1/x$ while the slope of the tangent line to the curve $y = \log_a x$ at $(x, \log_a x)$ is the more complicated expression $\frac{1}{x \ln a}$. However, the common logarithm is still used in many applications; one of them is the pH scale, a measure of acidity of a solution, and the other is the decibel scale, a measure of the relative intensity of sound.

The pH scale

In order to measure the acidity of a solution, chemists define $pH = -\log[H^+]$, where $[H^+]$ is the concentration (in moles/liter) of hydrogen ions. $[H^+]$ goes from 1 (extremely acidic) to 10^{-14} (extremely alkaline); a concentration level of 10^{-7} is considered neutral. Thus, the pH scale goes from 0 to 14. Due to the negative sign in front of $\log[H^+]$, the higher the pH the lower the acidity. A solution with pH=13 is very alkaline, while a solution with pH=14 is extremely alkaline (10 times more alkaline). In a similar fashion, a solution with pH=3 is quite acidic, 100 times more acidic than a solution with pH=5. Note that a neutral solution has pH=7.

Consider one solution having $pH = 5$, and another solution with $pH = 4.2$. How many times more acidic is the second solution? Let $5 = -\log[H^+]_1$, $4.2 = -\log[H^+]_2$. Hence $0.8 = \log \frac{[H^+]_2}{[H^+]_1}$, which in turn leads to $10^{0.8} = \frac{[H^+]_2}{[H^+]_1}$. Therefore $[H^+]_2 = 10^{0.8}[H^+]_1$, almost 6.31 times more acidic.

The decibel scale

To measure sound, physicists define

$$D = 10 \log \frac{W}{10^{-12}}$$

where W is the sound intensity in watts/m^2. The letter D stands for 'decibels' and is a measure of relative sound intensity (10^{-12} watts/m^2 is the lowest sound intensity that the human ear can hear). Note that when $W = 10^{-12}$ we have $D = 10 \log \frac{10^{-12}}{10^{-12}} = 0$ decibels. A normal conversation has a sound intensity of 10^{-6} watts/m^2, thus, under these circumstances

$$D = 10 \log 10^6 = 60 \text{ decibels.}$$

On the other hand, if the relative intensity in a coffee shop is 80 decibels, we might like to know the sound intensity. Indeed,

$$80 = 10 \log \frac{W}{10^{-12}}.$$

Hence $10^8 = \frac{W}{10^{-12}}$, which in turn leads to $W = 10^{-4}$ watts/m^2.

Suppose that the relative sound intensity in the above-mentioned coffee house increases from 80 to 85 decibels. How many times more intense sound is? We have

$$85 = 10 \log \frac{W_1}{10^{-12}}$$

and

$$80 = 10 \log \frac{W_2}{10^{-12}}.$$

Thus

$$0.5 = \log \tfrac{W_1}{10^{-12}} - \log \tfrac{W_2}{10^{-12}}.$$

Therefore $0.5 = \log \tfrac{W_1}{W_2}$. But, 10^x and log are inverse functions, consequently

$$10^{0.5} = \tfrac{W_1}{W_2}.$$

In summary, $W_1 = \sqrt{10}W_2$. Although the number of decibels increases by only 5 units (from 80 to 85 decibels), the intensity of the sound increases more than threefold!

Exercises for Section 1.3

1. Use the logarithm function to solve the equation $3^{x+2} = 5$.

2. Consider one solution having $pH = 7.1$ and another solution with $pH = 3.6$. How many times more acidic is the second solution?

3. Suppose that the relative sound intensity in a room decreases from 70 to 64 decibels. How many times less intense sound is?

4. Show that $\log x^{-1} = -\log x$.

5. Suppose that you wish to multiply a = 365,737 by b = 1,215,011. Is it correct to obtain the answer by calculating $10^{\log a + \log b}$? Please justify your answer.

6. What sound intensity corresponds to 66 decibels?

7. What is the hydrogen ion concentration of a solution with $pH = 11.8$?

8. Assume that the relative sound intensity in a coffee house increases from 72 decibels to 80 decibels. How many times more intense sound will become?

9. Could you solve the equation $2^x = 5$ by hand, or is a graphing calculator your only alternative? What about $x2^x = 5$?

10. Can you find a way of solving the equation $\log x + 3x = 1$, or is technology your only option? Either way, find the solution of the given equation.

1.4   Limits of Functions at a Point

The height of Japanese boys, between the ages of 1 and 13 years of age, is known (Altman and Dittmer, 1964). From table 1.2 we can find the average rate of growth between any two years. For instance, the average rate of growth between 6 and 8 years of age is

Table 1.2: Height of Japanese Boys

t (years)	1	2	3	4	5	6	7	8
height (cm)	77.9	85.7	93.4	99.6	104.7	110.8	116.7	121.5

t (years)	9	10	11	12	13
height (cm)	126.6	130.8	135.9	141	147.6

$$\frac{121.5-110.8}{8-6} = 5.35 \text{ cm/year.}$$

Similarly, the average rate of growth between 6 and 7 years of age is

$$\frac{116.7-110.8}{7-6} = 5.9 \text{ cm/year.}$$

Suppose that we would like to know the 'instantaneous' rate of growth at $t = 5$. It would be necessary to compute

$$\frac{H(t) - H(5)}{t - 5} \tag{1.7}$$

for values of t closer and closer to 5 ($H(t)$ denotes the height as a function of time). We cannot obtain this information from the table, but we can plot the points $(t, H(t))$ and draw a curve that passes, as close as possible, through these points (see Figure 1.8). Then we would have to carefully measure $H(t)$ for values of t very close to 5 (either to the left or right of 5) and thereafter calculate the quotient (1.7) or, what is equivalent, we should draw with great care the tangent to the curve at $(5, H(5))$ and then estimate its slope. A graphical approach is needed because we do not know the equation of the curve. The next example will illustrate what to do when the equation of the curve is known.

Suppose that $s(t) = t^2$, $t \geq 0$, is a function of time that describes the space traveled by an object (t is measured in seconds, s in meters). What is its 'instantaneous velocity' at $t = 3$ seconds? Average velocity is not a problem; for instance, the average velocity between $t = 3$ and $t = 4$ is

$$\frac{s(4)-s(3)}{4-3} = 16 - 9 = 7 \text{ m/sec.}$$

Since we wish to find the velocity at $t = 3$, we calculate the quotient $\frac{s(t)-s(3)}{t-3}$ for values of t closer and closer to 3; say

$$\frac{s(3.01)-s(3)}{3.01-3} = \frac{3.01^2-9}{0.01} = 6.01,$$

$$\frac{s(3.001)-s(3)}{3.001-3} = \frac{3.001^2-9}{0.001} = 6.001,$$

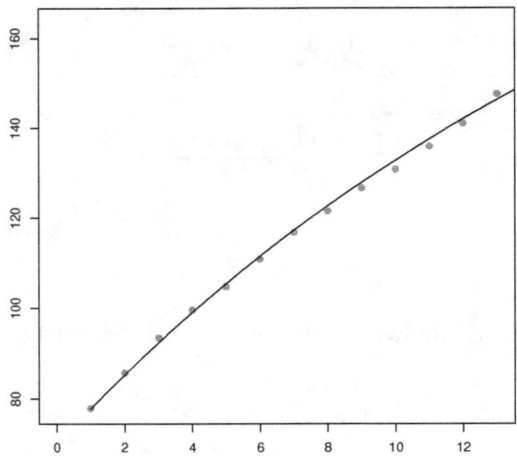

Figure 1.8: Height of Japanese Boys (1-13 years of age)

$$\frac{s(2.999)-s(3)}{2.999-3} = \frac{2.999^2-9}{-0.001} = 5.999.$$

Approaching 3 either from the right or the left we seem to approach the value 6. Indeed, for $t \neq 3$ we have

$$\frac{s(t)-s(3)}{t-3} = \frac{t^2-9}{t-3} = \frac{(t+3)(t-3)}{t-3} = t + 3.$$

Thus, the functions $t \rightarrow \frac{s(t)-s(3)}{t-3}$ and $t \rightarrow t + 3$ are identical for any $t \neq 3$ (the first function is not even defined at $t = 3$).

When we approach 3, the quotient $\frac{s(t)-s(3)}{t-3}$ approaches 6 because the function $t \rightarrow t + 3$ certainly does so! We will then write

$$\lim_{t \rightarrow 3} \frac{s(t)-s(3)}{t-3} = 6.$$

Let us emphasize that although the function $t \rightarrow \frac{s(t)-s(3)}{t-3}$ is not defined at $t = 3$, we can analyze what happens when t gets closer and closer to 3. The reader may have noticed that we found the slope of the tangent to the curve, defined by the function $s(t) = t^2$, at $t = 3$.

Next let us try to find the slope of the tangent to the curve $y = \sin x$ at $x = 0$. We need to calculate

$$\lim_{x \rightarrow 0} \frac{\sin x - \sin 0}{x - 0}.$$

That is to say, our task is to calculate $\lim_{x\to 0} \frac{\sin x}{x}$. First of all, we note that the function $x \to \frac{\sin x}{x}$ is not defined at $x = 0$. Making sure that we work with radians, a graphing calculator (with an 8-digits decimal approximation) provides the following numbers:

$$\tfrac{\sin 0.1}{0.1} = 0.99833417,$$

$$\tfrac{\sin 0.01}{0.01} = 0.99998333,$$

$$\tfrac{\sin 0.001}{0.001} = 0.99999983.$$

Approaching zero from the left we would also obtain numbers very close to 1 because $\frac{\sin(-x)}{-x} = \frac{-\sin x}{-x} = \frac{\sin x}{x}$. There is then compelling numerical evidence pointing towards the assertion that

$$\lim_{x\to 0} \frac{\sin x}{x} = 1. \tag{1.8}$$

A graph of $x \to \frac{\sin x}{x}$ close to the origin (Figure 1.9) would help to corroborate the fact that the limit (1.8) holds.

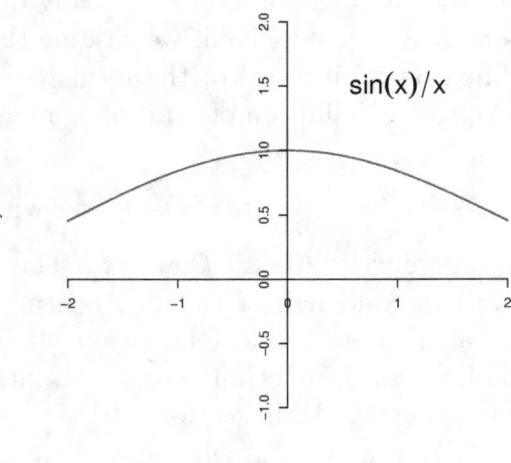

Figure 1.9: An important limit

Thus, the equation of the tangent to the curve $y = \sin x$ at $(0,0)$ is $y - \sin 0 = 1(x - 0)$, that is, $y = x$ (Figure 1.10). There is no simple algebraic approach to reach this result. We will have to wait until section 1.8 to learn about a geometric argument that leads to (1.8).

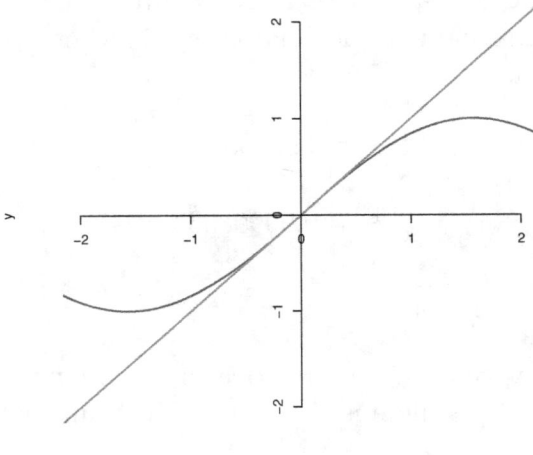

Figure 1.10: The tangent to the sine function at the origin

The two preceding examples suggest that the problem of tangents to curves involves finding the limit of a function defined in a neighborhood of a point c but not necessarily at c itself (recall that a neighborhood of c is any open interval that contains c). Actually, there is no loss of generality if we assume that there exists $\delta_o > 0$ such that f is at least defined on an interval of the form $(c - \delta_o, c + \delta_o)$ where the point c has been deleted. A precise definition of limit of $f(x)$ as x approaches c is as follows:

For every $\epsilon > 0$ there exists $\delta > 0$ such that $|f(x) - L| < \epsilon$ whenever $0 < |x - c| < \delta$.

Under these circumstances we write $\lim_{x \to c} f(x) = L$. The above-mentioned definition came into being in the second half of the 19th century after an almost two hundred year effort to put calculus on a solid foundation. It may well happen that $L = f(c)$, in which case the function is 'continuous' at c. Continuity at a point, and on an interval, will be discussed at length in section 1.6.

As expected, if $\lim_{x \to c} f(x)$ and $\lim_{x \to c} g(x)$ exist, the following properties hold:

$$(i) \lim_{x \to c}(f + g)(x) = \lim_{x \to c} f(x) + \lim_{x \to c} g(x)$$

$$(ii) \lim_{x \to c}(f - g)(x) = \lim_{x \to c} f(x) - \lim_{x \to c} g(x)$$

$$(iii) \lim_{x \to c} f(x)g(x) = \lim_{x \to c} f(x) \times \lim_{x \to c} g(x)$$

$$(iv) \lim_{x \to c} \frac{f(x)}{g(x)} = \frac{\lim_{x \to c} f(x)}{\lim_{\to c} g(x)}$$

where in (iv) we are assuming that $\lim_{x \to c} g(x) \neq 0$.

We could prove any of them using the definition of limit. For instance, let us provide a proof for (i). Define $L_1 = \lim_{x \to c} f(x)$ and $L_2 = \lim_{x \to c} g(x)$. Given any $\epsilon > 0$ there exists $\delta_1 > 0$ such that if $0 < |x - c| < \delta_1$ then $|f(x) - L_1| < \epsilon/2$, and there exists $\delta_2 > 0$ such that if $0 < |x - c| < \delta_2$ then $|g(x) - L_2| < \epsilon/2$. Let $\delta = $ minimum $\{\delta_1, \delta_2\}$ and assume that $0 < |x - c| < \delta$. Then

$$|(f+g)(x) - (L_1 + L_2)| = |f(x) - L_1 + g(x) - L_2| \leq |f(x) - L_1| + |g(x) - L_2| < \tfrac{\epsilon}{2} + \tfrac{\epsilon}{2} = \epsilon.$$

Thus $\lim_{x \to c}(f + g)(x) = \lim_{x \to c} f(x) + \lim_{x \to c} g(x)$. A proof of the other three properties can be found in appendix B.2.

Right away we observe that

$$\lim_{x \to c} x^n = c^n,$$

n an arbitrary natural number, because $\lim_{x \to c} x = c$ and the fact that property (iii) can be applied n times.

Question: Given $c > 0$, is it true that $\lim_{x \to c} \sqrt{x} = \sqrt{c}$? Let us use the ϵ, δ definition of limit to provide a proof. First of all, we observe that for any $x > 0$:

$$|\sqrt{x} - \sqrt{c}| = \frac{|\sqrt{x} - \sqrt{c}||\sqrt{x} + \sqrt{c}|}{\sqrt{x} + \sqrt{c}} = \frac{|x - c|}{\sqrt{x} + \sqrt{c}} < \frac{1}{\sqrt{c}}|x - c|.$$

Given any $\epsilon > 0$ we choose $\delta = \epsilon\sqrt{c}$. It follows that if $|x - c| < \delta$ then $\frac{1}{\sqrt{c}}|x - c| < \epsilon$, which in turn implies

$$|\sqrt{x} - \sqrt{c}| < \epsilon.$$

We have have been able to show that $\lim_{x \to c} \sqrt{x} = \sqrt{c}$, $c > 0$.

The good news is that we can learn many techniques of calculus without being forced to follow a level of rigor demanded by the definition of limit. The intuitive idea of limit will often be a useful guide in the future and seldom will lead us astray.

Let us illustrate the use of the above-mentioned properties. We have:

$$\lim_{x \to 0}(x^3 + x + 1) = \lim_{x \to 0} x^3 + \lim_{x \to 0} x + \lim_{x \to 0} 1 = 0 + 0 + 1 = 1,$$

$$\lim_{x \to 2} \frac{x^3 - 2^3}{x - 2} = \lim_{x \to 2} \frac{(x-2)(x^2 + 2x + 2^2)}{x - 2} = \lim_{x \to 2}(x^2 + 2x + 4) = 12,$$

$$\lim_{x \to 1} \frac{\sqrt{x} - 1}{x - 1} = \lim_{x \to 1} \frac{(\sqrt{x} - 1)(\sqrt{x} + 1)}{(x - 1)(\sqrt{x} + 1)} = \lim_{x \to 1} \frac{x - 1}{(x - 1)(\sqrt{x} + 1)} =$$

$$= \lim_{x \to 1} \frac{1}{\sqrt{x} + 1} = \frac{1}{\lim_{x \to 1} \sqrt{x} + 1} = \frac{1}{2}.$$

In the last two examples we could not apply the fourth property right away because the limit of the denominator was zero. A little bit of algebra was needed to cancel equal terms in the numerator and the denominator. By the way, we have been able to calculate the slope of the tangent line to the function $x \to x^3$ at $x = 2$ and the slope of the tangent line to the function $x \to \sqrt{x}$ at $x = 1$.

The squeeze property and the dragging property

We must get acquainted with two more properties of limits of functions. The first one is the 'squeeze property.' Suppose that f, g, h are functions defined on a neighborhood of c, not necessarily at c, and $f(x) \leq h(x) \leq g(x)$ for all x in their shared domain. Moreover, $f(x) \to L$ and $g(x) \to L$ as $x \to c$; then $\lim_{x \to c} h(x)$ exists and the value of this limit is also L. For instance, let $f(x)$ be a certain function defined on a neighborhood of the origin, but not at 0 itself, and $x^2 \leq f(x) \leq x^4$, $x \neq 0$. Since $x^4 \to 0$ and $x^2 \to 0$ as $x \to 0$, we can conclude that $\lim_{x \to 0} f(x)$ exists, and its value is 0.

Let us prove the squeeze property: Given any $\epsilon > 0$, since $\lim_{x \to c} g(x) = L$ there exists $\delta_1 > 0$ such that $g(x) < L + \epsilon$ whenever $0 < |x - c| < \delta_1$. Similarly, since $\lim_{x \to c} f(x) = L$ there exists $\delta_2 > 0$ such that $L - \epsilon < f(x)$ provided $0 < |x - c| < \delta_2$. Let $\delta = \text{minimum} \{\delta_1, \delta_2\}$ and assume that $0 < |x - c| < \delta$. Then $L - \epsilon < h(x) < L + \epsilon$ due to the fact that $f(x) \leq h(x) \leq g(x)$. Therefore $|h(x) - L| < \epsilon$ if $0 < |x - c| < \delta$, thus bringing the proof to an end.

The 'dragging property' asserts that if $\lim_{x \to c} g(x) = L > 0$, there exists a neighborhood $N = (c - \delta, c + \delta)$ such that $g(x) > 0$ for every x in N, $x \neq c$. Similarly, if $\lim_{x \to c} g(x) = L < 0$ there exists a neighborhood $N = (c - \delta, c + \delta)$ such that $g(x) < 0$ for every x in N, $x \neq c$. That is to say, a limit 'drags' a function forcing it to have the same sign of the limit in a deleted neighborhood of c.

The proof of the dragging property is straightforward. Suppose that $\lim_{x \to c} g(x) = L > 0$. Then there exists $\delta > 0$ such that $|g(x) - L| < \frac{L}{2}$ for every x in $(c - \delta, c + \delta)$, $x \neq c$. In particular $-\frac{L}{2} < g(x) - L$, i.e. $g(x) > \frac{L}{2} > 0$ for every x in $(c - \delta, c + \delta)$, $x \neq c$. The proof of the case $\lim_{x \to c} g(x) = L < 0$ is entirely similar.

Lateral limits

We can also deal with lateral limits, to be denoted $\lim_{x \to c+} f(x)$ (limit from the right) and $\lim_{x \to c-} f(x)$ (limit from the left). For instance, if $f(x) = 1$ when $x > 0$ and $f(x) = -1$ when $x < 0$, it is evident that $\lim_{x \to 0+} f(x) = 1$ and $\lim_{x \to 0-} f(x) = 1$. By the same token, if $g(x) = x^2 + 1$ when $x < 0$ and $g(x) = x^2$ when $x > 0$ (Figure 1.11), we can conclude that $\lim_{x \to 0-} g(x) = 1$ and $\lim_{x \to 0+} g(x) = 0$. We just have to keep in mind that when calculating a right limit we are approaching the point only from the right, and when calculating a left limit we are approaching the point only from the left. Undoubtedly, $\lim_{x \to c} f(x) = L$ if and only if the corresponding lateral limits at c exist and adopt the value L.

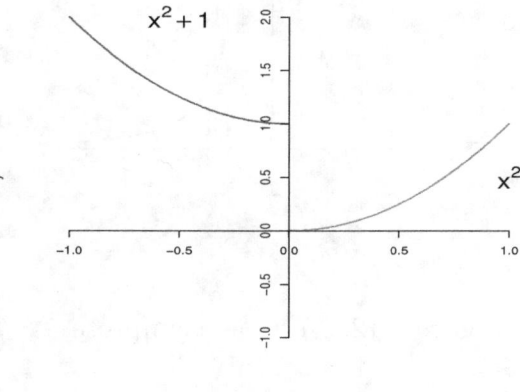

Figure 1.11: Lateral limits

It should be noted that both the squeeze property and the dragging property are valid for lateral limits, as well as properties (i) through (iv) with regard to the limit of a sum, product, etc. For instance, suppose we wish to calculate $\lim_{x \to 0+} x \cos \frac{1}{x}$. We cannot use property (iii) because we do not know whether $\lim_{x \to 0+} \cos \frac{1}{x}$ exists. Actually, it does not exist because $\cos 1/x$ oscillates wildly as one approaches zero from the right. However, $-1 \le \cos \frac{1}{x} \le 1$ for any $x > 0$. Hence $-x \le x \cos \frac{1}{x} \le x$ for any $x > 0$. Since $\lim_{x \to 0+} x = 0 = \lim_{x \to 0+}(-x)$, the squeeze property leads to the conclusion that $\lim_{x \to 0+} x \cos \frac{1}{x} = 0$. Moreover,

$$\lim_{x \to 0-} x \cos \frac{1}{x} = \lim_{x \to 0+}(-x) \cos \frac{1}{-x} = -\lim_{x \to 0+} x \cos \frac{1}{x} = -0 = 0$$

Thus, $\lim_{x \to 0} x \cos \frac{1}{x} = 0$.

Vertical asymptotes

Next let us consider the function $1/x^2$ (Figure 1.12). When x approaches 0 either from the left or the right, the function adopts bigger and bigger values. To denote this behavior, we write $\lim_{x \to 0} f(x) = \infty$. Strictly speaking, if $f(x)$ is defined in a neighborhood of c, but not at c, $\lim_{x \to c} f(x) = \infty$ if and only if for every $M > 0$ there exists $\delta > 0$ such that $f(x) > M$ whenever x lies on $(c - \delta, c + \delta)$, $x \ne c$. For instance, given $M > 0$, choose $\delta = 1/\sqrt{M}$. If $0 < |x| < 1/\sqrt{M}$ then $0 < x^2 < 1/M$, consequently $1/x^2 > M$. We have just confirmed what our intuition suggested, namely $\lim_{x \to 0} \frac{1}{x^2} = \infty$.

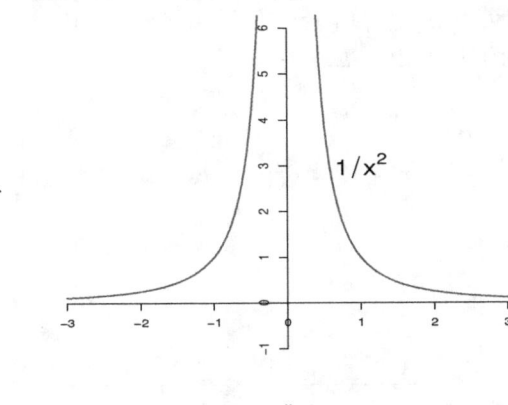

Figure 1.12: Vertical asymptote

Analogously, $\lim_{x \to c} f(x) = -\infty$ if and only if, given any $M > 0$, the inequality $f(x) < -M$ holds 'eventually.' That is, there exists $\delta > 0$ such that $f(x) < -M$ whenever $0 < |x - c| < \delta$. For instance, $\lim_{x \to 0} -1/x^2 = -\infty$ (Figure 1.13).

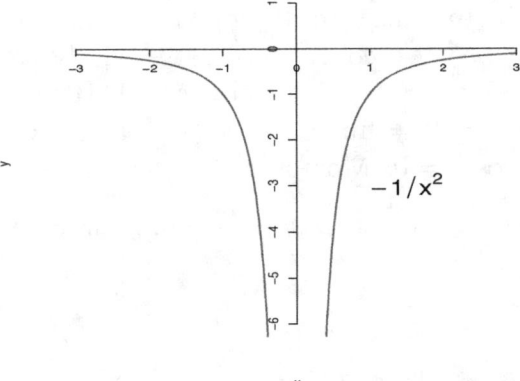

Figure 1.13: Another asymptotic behavior

These two definitions can be adapted to lateral limits. For instance, $\lim_{x \to 0+} \frac{1}{x} = \infty$, $\lim_{x \to 0-} \frac{1}{x} = -\infty$ (Figure 1.14). If either $\lim_{x \to c+} f(x) = \pm\infty$ or $\lim_{x \to c-} f(x) = \pm\infty$, or $\lim_{x \to c} f(x) = \pm\infty$, we will say that f has a vertical asymptote at c. Thus, if $f(x) = \frac{1}{x-2}$ for $x > 2$, and $f(x) = 1$ whenever $x \le 2$, we can conclude that f has a vertical asymptote at 2 because $\lim_{x \to 2+} f(x) = \infty$.

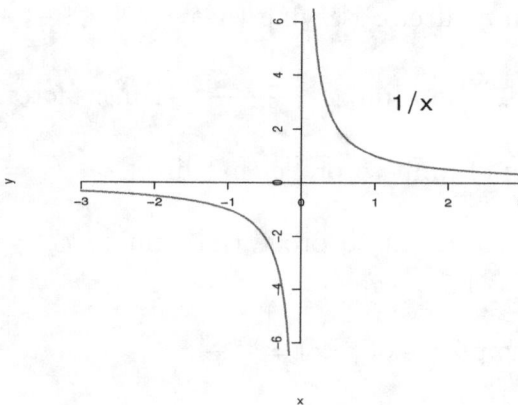

Figure 1.14: The function $1/x$

Exercises for Section 1.4

1. Calculate $\lim_{x\to 1} \frac{x^2-1}{x-1}$ and write the equation of the tangent line to the curve $y = x^2$ at $(1, 1)$.

2. What is the slope of the tangent line to the curve $y = \sqrt{x}$ at $(4, 2)$?

3. Accepting that $\lim_{h\to 0} \frac{\sin h}{h} = 1$, calculate $\lim_{h\to 0} \frac{\cos h - 1}{h}$ (Hint: Note that

$$\frac{\cos h - 1}{h} = \frac{(\cos h - 1)(\cos h + 1)}{h(\cos h + 1)}$$

for any $h \neq 0$).

4. Find $\lim_{x\to 0} \frac{\sin 5x}{x}$ and $\lim_{x\to 0} \frac{x}{\sin 4x}$.

5. Calculate $\lim_{x\to 0^-} \frac{|x|}{x}$ and $\lim_{x\to 0^+} \frac{|x|}{x}$.

6. Let a function f be defined over the whole real line as follows: $f(x) = -x^2 + 1$ whenever $x < 0$, and $f(x) = x$ whenever $x \geq 0$. Calculate $\lim_{x\to 0^-} f(x)$ and $\lim_{x\to 0^+} f(x)$.

7. We have a function g, defined over the whole real line (except the origin) in the following way: $g(x) = \cos x$ whenever $x < 0$, $g(x) = 2x$ whenever $x > 0$. Calculate $\lim_{x\to 0^-} g(x)$ and $\lim_{x\to 0^+} g(x)$.

8. What is the slope of the tangent line to the curve $y = \cos x$ at $(0, 1)$?

9. Given $f(x) = \frac{x+6}{x-4}$, $x \neq 4$, find the vertical asymptote of the function and sketch the corresponding curve after observing that $f(x) = 1 + \frac{10}{x-4}$.

10. Calculate $\lim_{x \to 2} \frac{x^2-x-2}{x-2}$ and $\lim_{x \to -4} \frac{x^2+x-12}{x+4}$ (Hint: factorize the numerator.)

11. Use the ϵ, δ definition of limit to prove that $\lim_{x \to 0+} \sqrt{x} = 0$.

12. Use the ϵ, δ definition of limit to prove that $\lim_{x \to c} x^2 = c^2$ (Hint: Note that $|x^2 - c^2| = |x + c||x - c| \leq (|x| + |c|)|x - c|$).

13. Find the vertical asymptotes of $f(x) = \frac{x+3}{x^2-2}$.

1.5   Limits of Functions at Infinity

There is another type of limit that we would like to analyze, namely the limit of a function $f(x)$, defined on an interval (a, ∞), as x becomes bigger and bigger. Or, as one might say, the limit of the function as x tends to ∞ ('infinity'). Probably everyone would agree that $\lim_{x \to \infty} \frac{1}{x} = 0$, and $\lim_{x \to \infty}(1 + \sin\frac{1}{x}) = 1$ (Figure 1.15). Furthermore, $\lim_{x \to \infty} \frac{1}{x^n} = 0$ $(n = 1, 2, 3, 4, ...)$.

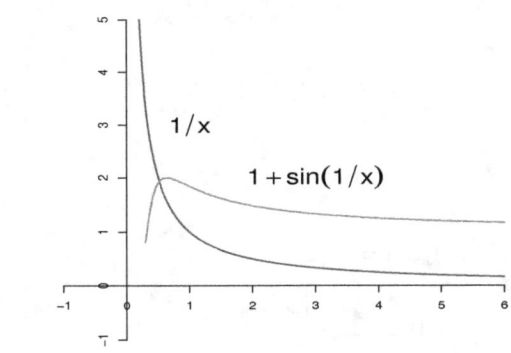

Figure 1.15: Limit at 'infinity'

Strictly speaking,

$\lim_{x \to \infty} f(x) = L$ if and only if for every $\epsilon > 0$ there exists p such that $x > p$ implies $|f(x) - L| < \epsilon$.

The usual limit laws are valid; that is to say, $\lim_{x\to\infty}(f(x)\pm g(x)) = \lim_{x\to\infty} f(x)\pm$ $\lim_{x\to\infty} g(x)$, $\lim_{x\to\infty} f(x)g(x) = (\lim_{x\to\infty} f(x))(\lim_{x\to\infty} g(x))$, $\lim_{x\to\infty} \frac{f(x)}{g(x)} = \frac{\lim_{x\to\infty} f(x)}{\lim_{x\to\infty} g(x)}$ provided that $\lim_{x\to\infty} f(x)$ and $\lim_{x\to\infty} g(x)$ exist and, furthermore, $\lim_{x\to\infty} g(x) \neq 0$ when dealing with a quotient. For instance

$$\lim_{x\to\infty} \frac{x+7}{x^2+x+1} = \lim_{x\to\infty} \frac{\frac{1}{x^2}+\frac{7}{x^2}}{1+\frac{1}{x}+\frac{1}{x^2}} = \frac{0}{1} = 0.$$

It could also happen that a function $f(x)$, defined on an interval (a, ∞), eventually surpasses any positive number. We will say that $\lim_{x\to\infty} f(x) = \infty$ if and only if for every $M > 0$ there exists p such that $f(x) > M$ provided that $x > p$. Very many functions have this kind of behavior. For example, $\lim_{x\to\infty} x^2 = \infty$ (given any $M > 0$ choose $p = \sqrt{M}$); actually, $\lim_{x\to\infty} x^n = \infty$ for every positive natural number n. However, $\lim_{x\to\infty} \sin x$ does not exist because the sine function simply oscillates between -1 and 1. We should note that if $\lim_{x\to\infty} f(x) = \infty$ then $\lim_{x\to\infty} \frac{1}{f(x)} = 0$.

Horizontal asymptotes

The function $f(x) = \frac{ax}{b+x}$ has particular interest in biology (a and b are positive parameters; see Figure 1.16). The Michaelis-Menten equation, namely

$$v = \frac{V_{max}[S]}{K_m+[S]}$$

is of this type ($[S]$ denotes the concentration of substrate and v is the rate of the enzymatic reaction). It will be discussed in detail in chapter 3. For the time being let us note that

$$\lim_{x\to\infty} f(x) = \lim_{x\to\infty} \frac{a}{\frac{b}{x}+1} = \frac{a}{\lim_{x\to\infty}(\frac{b}{x}+1)} = \frac{a}{0+1} = a.$$

A mathematician might think about a limit, but a scientist would rather ask: What happens when x, the concentration, becomes bigger and bigger? The answer is that it approaches the parameter a 'asymptotically.' Or we might say that $y = a$ is a 'horizontal asymptote.' Another function that appears in biological applications is

$$g(x) = \frac{ax^2}{b+x^2}.$$

We observe (see Figure 1.17) that

$$\lim_{x\to\infty} g(x) = \lim_{x\to\infty} \frac{a}{\frac{b}{x^2}+1} = \frac{a}{\lim_{x\to\infty}(\frac{b}{x^2}+1)} = \frac{a}{0+1} = a.$$

We might as well discuss the behavior at 'infinity' of the function $h(x) = \frac{ax}{b+x^2}$, also of interest in biology (see Figure 1.18). We note that

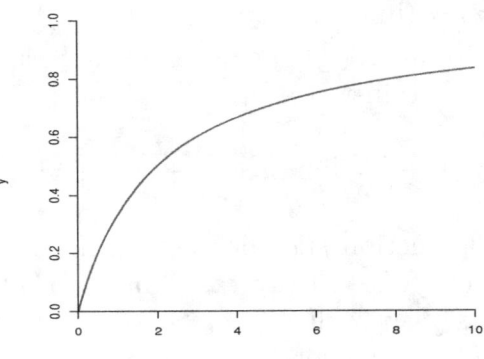

Figure 1.16: graph of $f(x) = \frac{x}{2+x}$

$$\lim_{x \to \infty} h(x) = \lim_{x \to \infty} \frac{a}{\frac{b}{x}+x}.$$

But $\lim_{x \to \infty} \frac{b}{x} = 0$ and $\lim_{x \to \infty} x = \infty$. Therefore

$$\lim_{x \to \infty} h(x) = \frac{a}{\lim_{x \to \infty} (\frac{b}{x}+x)} = 0.$$

Thus, $y = 0$ is a horizontal asymptote.

If a function f is defined on an interval $(-\infty, a)$, we could inquire whether $\lim_{x \to -\infty} f(x)$ exists; that is to say, it makes sense to find out what is the behavior of the function as we go farther and farther to the left. Based on our experience with limits, we would say that $\lim_{x \to -\infty} f(x) = L$ if and only if for every $\epsilon > 0$ there exists p such that $x < p$ implies $|f(x) - L| < \epsilon$. For instance, $\lim_{x \to -\infty} \frac{1}{x} = 0$ while $\lim_{x \to -\infty} (2 + \frac{1}{x-1}) = 2$. In a similar fashion we could define $\lim_{x \to \infty} f(x) = \infty$ or $\lim_{x \to \infty} f(x) = -\infty$.

Let $f(x) = \frac{x+9}{x-1}$. Using long division we get

$$\frac{x+9}{x-1} = 1 + \frac{10}{x-1}.$$

Therefore $\lim_{x \to 1^-} f(x) = -\infty$ and $\lim_{x \to 1^+} f(x) = \infty$. Thus, $x = 1$ is a vertical asymptote. On the other hand, $\lim_{x \to \pm\infty} \frac{10}{x-1} = 0$. Consequently $\lim_{x \to \pm\infty} f(x) = 1$, that is, $y = 1$ is a horizontal asymptote. Evidently, the function crosses the x-axis at $(-9, 0)$ and crosses the y-axis at $(0, -9)$. We have all the information needed to sketch a graph (Figure 1.19).

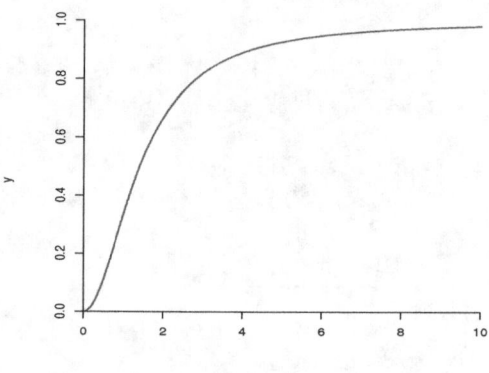

Figure 1.17: graph of $f(x) = \frac{x^2}{2+x^2}$

Oblique asymptotes

A function can have a vertical as well as an oblique asymptote. Consider for instance the function $f(x) = \frac{x^2+1}{x+1}$. Right away we see that $x = -1$ is a vertical asymptote since $\lim_{x \to -1+} f(x) = \infty$ and $\lim_{x \to -1-} f(x) = -\infty$. But

$$\frac{x^2+1}{x+1} = x - 1 + \frac{2}{x+1}.$$

Since $\lim_{x \to \pm\infty} \frac{2}{x+1} = 0$ we can conclude that $f(x)$ will get closer and closer to the line $y = x - 1$ as $x \to \pm\infty$. That is to say, the line $y = x - 1$ becomes an oblique asymptote of f (Figure 1.20).

Exercises for Section 1.5

1. Calculate $\lim_{x \to \infty} \frac{x+8}{x^3+x+2}$.

2. Let $f(x) = \frac{x-5}{x-2}$. Find its vertical asymptote. What happens when $x \to \pm\infty$? Sketch a graph of the corresponding curve.

3. Find the vertical and horizontal asymptote of the function $f(x) = \frac{x+3}{x-2}$. Sketch a graph.

4. Calculate the oblique asymptote of the function $f(x) = \frac{x^3+1}{x^2+x+1}$. Sketch a graph of the corresponding curve.

5. Calculate the vertical and oblique asymptote of $f(x) = \frac{x^2+3}{x-1}$. Sketch a graph.

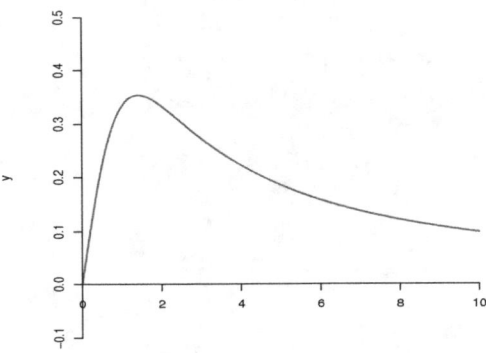

Figure 1.18: graph of $f(x) = \frac{x}{2+x^2}$

6. Sketch the graphs of $f(x) = \frac{x+5}{x^2+2x+1}$ and $g(x) = \frac{x^2+2x+1}{x+5}$, clearly specifying any vertical, horizontal, or oblique asymptote.

7. Calculate $\lim_{x \to -2} \frac{x^2+3x-10}{x^2-4}$ and $\lim_{x \to \pm\infty} \frac{x^2+3x-10}{x^2-4}$. Sketch a graph of the function under consideration.

8. Calculate $\lim_{x \to \infty} \frac{3x^2-1}{x^3+5}$ and sketch a graph of the function.

9. Define two functions f and g on $(1, \infty)$ such that $f(x) < g(x)$, $x > 1$, but $\lim_{x \to \infty} f(x) = \lim_{x \to \infty} g(x)$.

10. Solve the previous exercise if we demand that the functions are defined on $(0, \infty)$ instead of $(1, \infty)$.

1.6   Continuity

We all have an intuitive idea about the continuity of a function f defined on an interval I of the real line. We expect it not to have jumps; in other words, we expect to be able to draw its graph without having to raise the pencil from the paper. In this sense, all the polynomial functions as well as the sine and cosine function are continuous over the real line, while the log and ln functions are continuous on $(0, \infty)$, and a^x is continuous over the whole real line (a is any fixed positive number). The tangent and secant function are also continuous over their domain of definition, which happens to be the real line with the exception of numbers of the form $(2n+1)\frac{\pi}{2}$ (n being any integer). However, the function

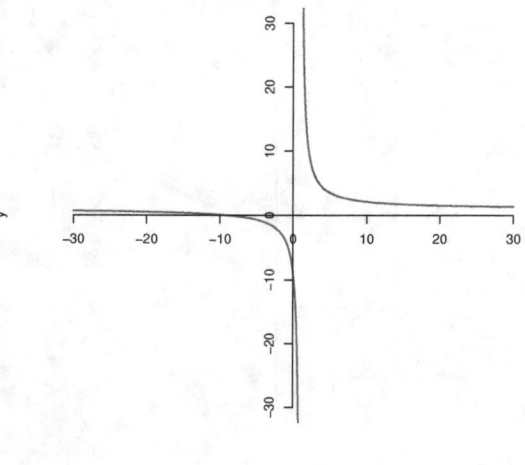

Figure 1.19: graph of $f(x) = \frac{x+9}{x-1}$

$$f(x) = \frac{\sin x}{x}, \; x \neq 0$$

$$f(0) = 2$$

is not continuous at the origin; it has a jump at $x = 0$ due to the fact that $\lim_{x \to 0} f(x) = 1$ but $f(0) \neq 1$. This example, and similar ones, suggests the definition of continuity at a point. Suppose that f is defined on an interval I and c is an interior point of I. We will say that f is continuous at c if $\lim_{x \to c} f(x)$ exists and $\lim_{x \to c} f(x) = f(c)$. According to this definition, all the polynomial functions as well as the trigonometric, exponential, and logarithmic functions are continuous at every point on their domain of definition. An equivalent definition of continuity at c, which will be used in section 1.8, demands that $\lim_{h \to 0} f(c + h) = f(c)$.

If c is an endpoint, continuity at c means continuity from the right or the left. For instance, let $f(x) = \sqrt{x}$, $x \geq 0$ (Figure 1.21). It is continuous everywhere on its domain because $\lim_{x \to c} \sqrt{x} = \sqrt{c}$ whenever $c > 0$ and $\lim_{x \to 0+} \sqrt{x} = \sqrt{0} = 0$. Thus, we have continuity from the right at the origin. Or take the function $h(x) = x^2$, $0 \leq x \leq 1$. It is continuous on $[0, 1]$, but let us not forget that at 0 we have right continuity and at 1 we have left continuity.

Let us analyze the function $g(x)$ defined by the rule $g(x) = 1$ (for $x > 0$), $g(x) = -1$ (for $x < 0$), and $g(0) = 0$. It is defined at $x = 0$ but $\lim_{x \to 0} g(x)$ does not exist. Thus, it is not continuous at $x = 0$. It has an 'essential discontinuity' there because there is no way we could redefine $g(0)$ in order to have continuity at $x = 0$. A different

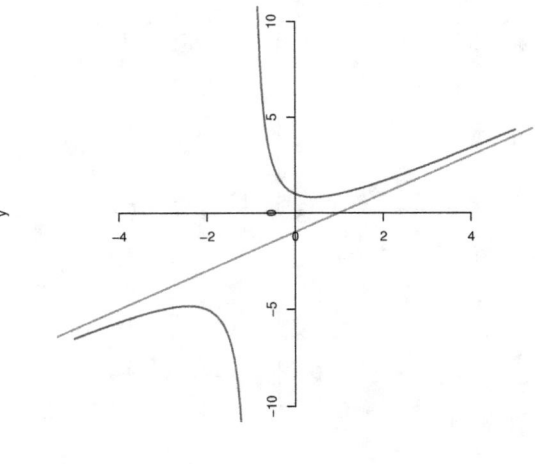

Figure 1.20: The function $\frac{x^2+1}{x+1}$ and its oblique asymptote

scenario takes place when analyzing the function $f(x) = \frac{\sin x}{x}$, $x \neq 0$, $f(0) = 2$: we could redefine $f(0) = 1$ and reach continuity at the origin because then

$$\lim_{x \to 0} f(x) = \lim_{x \to 0} \frac{\sin x}{x} = 1 = f(0).$$

This type of discontinuity is appropriately called 'removable.'

The algebra of continuous functions

The sum, product, difference and quotient of continuous functions at a point are also continuous at the point due to the fact that $\lim_{x \to c} f(x) = f(c)$ and $\lim_{x \to c} g(x) = g(c)$ imply

$$\lim_{x \to c}(f + g)(x) = f(c) + g(c),$$

$$\lim_{x \to c}(f - g)(x) = f(c) - g(c),$$

$$\lim_{x \to c}(fg)(x) = f(c)g(c),$$

$$\lim_{x \to c}(f/g)(x) = f(c)/g(c).$$

Of course, if we are dealing with quotients of functions, we are assuming that $g(c) \neq 0$. For instance, $x^2 + \sin x$ and $\frac{\cos x}{x^2+1}$ are continuous everywhere, as well as x^n (n an arbitrary natural number).

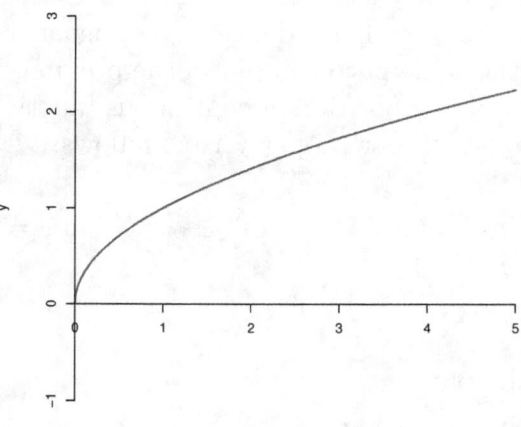

Figure 1.21: The square root function

We might add that the composition of continuous functions is also continuous. Indeed, if f is continuous at c and g is continuous at $f(c)$ then $g \circ f$ also is continuous at c. We can confirm this by observing that given any $\epsilon > 0$ there exists $\mu > 0$ such that $|y - f(c)| < \mu$ implies $g(y) - g(f(c))| < \epsilon$. Moreover, there exists $\delta > 0$ such that $|f(x) - f(c)| < \mu$ provided $|x - c| < \delta$. Hence $|g(f(x)) - g(f(c))| < \epsilon$ if $|x - c| < \delta$. Examples of composition of continuous functions are easy to find: $\sin(x^3)$, $(\cos x)^4$, $\sqrt{x^2 + 7}$, $\ln(x^2 + 1)$, $e^{\sin x}$, etc.

Two properties of continuous functions

We will study two remarkable properties of continuous functions.

1. **The gatekeeper property**

   A continuous function acts as a 'gatekeeper' with regard to limit processes. In other words, if $\lim_{x \to c} f(x) = L$, g is continuous at L, and the range of f lies inside the domain of g, it follows that $\lim_{x \to c} g(f(x)) = g(L)$. This can be written as

   $$\lim_{x \to c} g(f(x)) = g(\lim_{x \to c} f(x)). \tag{1.9}$$

   For instance,

   $$\lim_{x \to 0} \left(\tfrac{\sin x}{x}\right)^3 = \left(\lim_{x \to 0} \tfrac{\sin x}{x}\right)^3 = 1^3 = 1$$

because $p(x) = x^3$ is continuous at 1 (actually, it is continuous everywhere). Expressed in a rather literary fashion we might say that, under the presence of continuity, g and lim are interchanged after 'greeting each other.' The justification of (1.9) follows the same pattern used in the previous subsection, where we proved the continuity of composition of continuous functions. Another example about the use of the gatekeeper property is as follows:

$$\lim_{x \to 1} \ln(x^2) = \ln(\lim_{x \to 1} x^2) = \ln 1 = 0$$

2. **The dragging property**

Let us keep in mind that if a function g is defined on an interval I and is continuous at an interior point c, and $g(c) > 0$, then g 'drags' $g(x)$ values from a neighborhood of c making them positive, that is, there exists $\delta > 0$ such that $g(x) > 0$ for every x in $(c-\delta, c+\delta)$. It is a consequence of the dragging property of limits, which was discussed in section 1.4. As expected, the second property is also valid when $g(c) < 0$.

Two theorems

Two very important theorems are true for continuous functions defined on an interval of the form $[a, b]$:

Max-Min Theorem (MMT) Let $f : [a, b] \to \Re$ be continuous. Then the function attains its maximum and its minimum in $[a, b]$, that is, there exist points α and β in $[a, b]$ such that $f(\alpha) \geq f(x)$ for every x in $[a, b]$, and $f(\beta) \leq f(x)$ for every x in $[a, b]$.

Intermediate Value Theorem (IVT) Let $f : [a, b] \to \Re$ be continuous, $f(a) \neq f(b)$. If $f(a) < C < f(b)$ then there exists c in (a, b) such that $f(c) = C$. Similarly, if $f(a) > C > f(b)$ then there exists d in (a, b) such that $f(d) = C$. In other words, any horizontal line between $f(a)$ and $f(b)$ has to cross the curve $y = f(x)$ (Figure 1.22).

There is compelling graphical evidence about the validity of MMT (sometimes called the **Extreme Value Theorem**) and IVT; thus, we will not worry about a proof. Anyhow, a rigorous proof would take us far afield into the realm of mathematical analysis (Trench, 2002). Both theorems will be used later, for instance when we will need to discover a way of finding either a maximum or a minimum of a function or to show that the natural logarithm is an onto function.

It is important to keep in mind that the hypotheses of either MMT or IVT cannot be weakened. For instance, $f : [0, 1] \to \Re$ defined by the rule $f(x) = 0$ $(0 \leq x < 1/2)$,

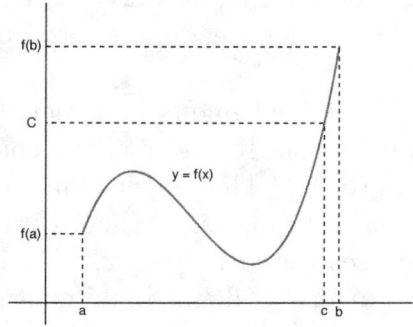

Figure 1.22: A pictorial representation of IVT

$f(x) = x$ $(1/2 \leq x \leq 1)$ is not continuous at $x = 1/2$. The horizontal line $y = 1/4$ does not cross the graph of f, thus the thesis of IVT does not hold.

Let $g : [0, 1] \to \Re$ be defined by $g(0) = 0$, $g(x) = 1/x$ $(0 < x \leq 1)$. This function is continuous on $(0, 1]$ but is not continuous at $x = 0$. Neither the Max-Min or IVT theorem hold in this example: g does not have a maximum and $y = 1/2$ does not cross the graph of g despite the fact that $g(0) < 1/2 < g(1)$.

The continuity of the inverse

Firstly, we should point out that if f is continuous and non-constant on an interval I then the range of f is an interval. This fact seems plausible. Indeed, we need to show that given any y_1, y_2 in $f(I)$, $y_1 < y_2$, it follows[2] that $[y_1, y_2]$ is a subset of $f(I)$. Since y_1, y_2 belong to $f(I)$ we can assert that there exist x_1, x_2 in I such that $f(x_1) = y_1$ and $f(x_2) = y_2$. Let us suppose that $x_1 < x_2$ (the case $x_2 < x_1$ can be dealt in a similar fashion). Since I is an interval we will have that $[x_1, x_2]$ is contained in I.

Let y be any element of $[y_1, y_2]$. Then $y_1 \leq y \leq y_2$, that is, $f(x_1) \leq y \leq f(x_2)$. If $y = f(x_1)$ or $y = f(x_2)$ we can assert that y belongs to $f(I)$ and we are done. Assume that $f(x_1) < y < f(x_2)$. Then IVT, applied to the restriction of f to $[x_1, x_2]$, implies that there exists x in (x_1, x_2) such that $y = f(x)$, consequently y is in $f(I)$,

---

[2] We must keep in mind that given any set A of real numbers, not containing just one element, A is an interval if and only if $[r, s]$ is a subset of A for any r, s in A, $r < s$ (Bartle and Ionescu-Tulcea 1970, p.850)

as we wished to show.

Moreover, if f is increasing (decreasing) and continuous, its inverse function is also increasing (decreasing) and continuous on the range of f. The challenge is not to show that the inverse is increasing (decreasing)[3] but to prove the continuity of the inverse (see appendix B4).

For instance, $f(x) = 10^x$ is continuous and increasing on the whole real line and its range is the interval $(0, \infty)$. Its inverse, the common logarithm function, is continuous and increasing on $(0, \infty)$. The tangent function, restricted to $(-\pi/2, \pi/2)$, is increasing and continuous. Its range is $\Re$, hence its inverse is continuous on $\Re$. Similarly, the sine function, restricted to $[-\pi/2, \pi/2]$, is increasing and continuous with range $[-1, 1]$. Then its inverse is defined and continuous on $[-1, 1]$.

Exercises for Section 1.6

1. Does the function $h(x) = \frac{\cos x - 1}{x}$, $x \neq 0$, and $h(0) = 3$, have a removable or essential discontinuity at $x = 0$? Please justify your answer.

2. Calculate $\lim_{x \to 0}(\frac{\cos x - 1}{x})^2$, $\lim_{x \to 1}(\frac{x+5}{x^2+1})^3$, $\lim_{x \to 0} \ln(\cos x)$.

3. Write $h(x) = (\sin x)^2$ as the composition of two continuous functions. Sketch a graph of $h(x)$. Do the same for $e^{\cos x}$.

4. Does the function $f : (0, 1] \to \Re$, $f(x) = x^2$, attain a minimum in $(0, 1]$?

5. Let $f(x) = x$, $0 \leq x \leq 1$, $f(x) = 2$, $1 < x \leq 3$. It is clearly discontinuous at $x = 1$. Define a horizontal line $y = k$, which does not cross the curve despite the fact that $f(0) < k < f(3)$.

6. Show that f is continuous at c if and only if $\lim_{h \to 0} f(c + h) = f(c)$.

7. Find the image of the interval $[1, \infty)$ under the continuous function $1/x$.

8. Let $f : [a, b] \to \Re$ be continuous and increasing. Show that the range of f is the interval $[f(a), f(b)]$ (Hint: Use the Intermediate Value Theorem).

9. What happens if, in the previous exercise, f is decreasing instead of increasing? In other words, on which interval is the inverse of f defined, continuous, and decreasing?

10. Let us consider the function $g : \Re \to \Re$, $g(x) = x$ for $x < 2$, and $g(x) = ax^3$ for $x \geq 2$. For what value of a will the function g be continuous at $x = 2$?

---

[3]See exercises 6 and 7 from section 1.2.

11. Define two functions such that they are continuous everywhere except at the origin but such that their sum is continuous everywhere (including the origin).

12. Define two functions such that they are continuous everywhere except at the origin but such that their product is continuous everywhere (including the origin).

Enrichment Note: The Range of a Continuous Function

We have seen that if I is an interval and $f : I \to \Re$ is continuous, and non-constant, then $ran(f)$ is an interval. A natural question to ask is whether $ran(f)$ is an open interval when I is an open interval. The answer is negative because, for instance, $f(x) = x^2$, defined on $(-1, 1)$, has its range equal to $[0, 1)$, which is not an open interval. Can we add an extra condition to f in order to make sure that its range is open too? Suppose that f, defined on the open interval I, is continuous and *increasing*. We wish to show that, under these circumstances, the range of f is open. With this purpose in mind, assume that y is an arbitrary element of $ran(f)$. Then there exists x in I such that $y = f(x)$. Since I is open, there exist x_1, x_2 in I, $x_1 < x < x_2$, such that the interval (x_1, x_2) is entirely contained in I. But f is increasing, therefore

$$f(x_1) < f(x) < f(x_2).$$

That is, $f(x_1) < y < f(x_2)$ or, what is the same, y belongs to the interval $(f(x_1), f(x_2))$. It is our purpose to show that $(f(x_1), f(x_2))$ is entirely contained in the range of f. Indeed, suppose that z is an arbitrary element of the interval $(f(x_1), f(x_2))$. Thus,

$$f(x_1) < z < f(x_2).$$

The Intermediate Value Theorem implies that there exists u in (x_1, x_2) such that $f(u) = z$. Consequently, z belongs to $ran(f)$. We have succeeded in showing that the range of f is an open interval. If we were to assume that f is decreasing instead of increasing, a similar argument would lead to the conclusion that the range is an open interval. Thus, we should not be surprised that the range of 10^x is the open interval $(0, \infty)$ while the range of the tangent function, restricted to $(-\pi/2, \pi/2)$, is the whole real line (an open interval too).

1.7 Derivatives

As we saw in sections 1.1 and 1.4, an important problem is to determine the slope of the tangent line to a curve $y = f(x)$ at a certain given point $(c, f(c))$. We define the tangent at $(c, f(c))$ through the equation $y - f(c) = m(x - c)$, where

$$m = \lim_{x \to c} \frac{f(x) - f(c)}{x - c}. \tag{1.10}$$

This definition of the slope of the tangent line makes a lot of sense because it emphasizes the fact that a tangent at a point is really found through secants that become closer and closer to the geometric idea of a tangent (Figure 1.23).

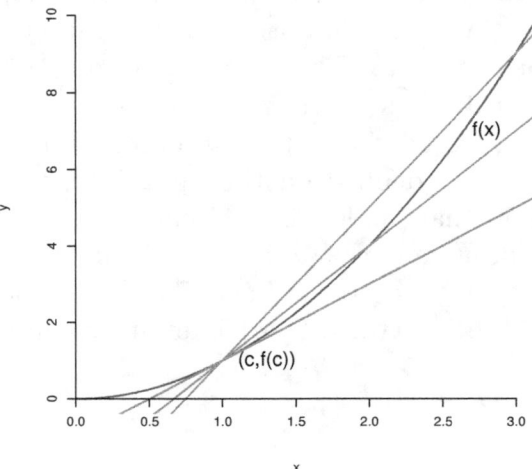

Figure 1.23: Finding the tangent at a point through secant lines

For instance, let $f(x) = x^2$ and let c be any fixed real number. Our task is to find the equation of the tangent line at (c, c^2), namely $y - c^2 = m(x - c)$ where m is given by (1.10). First of all, a simple algebraic identity shows that

$$\frac{x^2 - c^2}{x - c} = \frac{(x-c)(x+c)}{x-c} = x + c$$

whenever $x \neq c$. Therefore:

$$m = \lim_{x \to c} \frac{x^2 - c^2}{x - c} = \lim_{x \to c}(x + c) = c + c = 2c.$$

Consequently, the equation of the tangent to the curve $y = x^2$ at (c, c^2) is $y - c^2 = 2c(x - c)$. The reader may compare this approach to the one we employed in section 1.1.

Next let us consider the function $f(x) = x^3$. We wish to find the equation of the tangent line to the curve $y = x^3$ at an arbitrary point (c, c^3). The only challenge is to calculate the slope of the tangent. To accomplish this, we need to calculate $\lim_{x \to c} \frac{x^3 - c^3}{x - c}$. We note that

$$\frac{x^3 - c^3}{x-c} = \frac{(x-c)(x^2+cx+c^2)}{x-c} = x^2 + cx + c^2$$

for any $x \neq c$. Consequently

$$m = \lim_{x \to c} \frac{x^3 - c^3}{x-c} = \lim_{x \to c}(x^2 + cx + c^2) = 3c^2.$$

Now we are confident that the equation of the tangent to the curve $y = x^3$ at (c, c^3) is $y - c^3 = 3c^2(x - c)$.

A pattern seems to be emerging. For any natural number n, maybe the slope of the tangent to the curve $y = x^n$ at (c, c^n) is nc^{n-1} ? Indeed it is, as we will soon find out. For $x \neq c$ we have

$$\frac{x^n - c^n}{x-c} = \frac{(x-c)(x^{n-1}+cx^{n-2}+...+c^{n-1})}{x-c} = x^{n-1} + cx^{n-2} + ... + c^{n-1}.$$

Therefore

$$\lim_{x \to c} \frac{x^n - c^n}{x-c} = \lim_{x \to c}(x^{n-1} + cx^{n-2} + ... + c^{n-1}) = nc^{n-1}. \tag{1.11}$$

Thus, the equation of the tangent to the curve $y = x^n$ at any point (c, c^n) is $y - c^n = nc^{n-1}(x - c)$. Next let us try to find the equation of the tangent to the curve $y = \sqrt{x}$ at $(c, \sqrt{c})$, $c > 0$. We need to employ an algebraic 'trick.' For $x \neq c$ we have

$$\frac{\sqrt{x} - \sqrt{c}}{x-c} = \frac{(\sqrt{x} - \sqrt{c})(\sqrt{x} + \sqrt{c})}{(x-c)(\sqrt{x} + \sqrt{c})} = \frac{1}{\sqrt{x} + \sqrt{c}}.$$

Hence $m = \lim_{x \to c} \frac{\sqrt{x} - \sqrt{c}}{x-c} = \lim_{x \to c} \frac{1}{\sqrt{x} + \sqrt{c}} = \frac{1}{2\sqrt{c}}$. Thus, the equation of the tangent line to the curve $y = \sqrt{x}$ at $(c, \sqrt{c})$ is $y - \sqrt{c} = \frac{1}{2\sqrt{c}}(x - c)$.

In general, whenever $\lim_{x \to c} \frac{f(x) - f(c)}{x-c}$ exists, we will say that the function $f(x)$ is **differentiable** at c. This concept is extremely important, so much so that the slope of the tangent line to the curve $y = f(x)$ has a special name: it is called the **derivative** of $f(x)$ at c, and is usually denoted $f'(c)$ (a slight variation from **Newton's original symbol**). We should mention at this stage that an equivalent definition of derivative is as follows:

$$f'(c) = \lim_{h \to 0} \frac{f(c + h) - f(c)}{h}. \tag{1.12}$$

Let us note that differentiability at a point implies continuity at the point. Indeed

$$f(x) - f(c) = \frac{f(x) - f(c)}{x-c}(x - c).$$

Therefore

$$\lim_{x \to c} f(x) - f(c) = \lim_{x \to c} \frac{f(x) - f(c)}{x-c} \times \lim_{x \to c}(x - c) = f'(c) \times 0 = 0.$$

Consequently $\lim_{x \to c} f(x) = f(c)$. The converse is not true; in other words, a function can be continuous at a point without being differentiable there. Indeed, let $f(x) = |x|$ (the well-known absolute value function). It is certainly continuous at the origin, but

$$\lim_{x \to 0+} \frac{|x| - |0|}{x - 0} = \lim_{x \to 0+} \frac{x - 0}{x - 0} = 1,$$

$$\lim_{x \to 0-} \frac{|x| - |0|}{x - 0} = \lim_{x \to 0-} \frac{-x - 0}{x - 0} = -1.$$

Thus, $f'(0)$ is not defined because for the limit (1.10) to exist it has to adopt the same value whether we approach it from the left or the right. From a geometrical point of view we realize that there is no way of drawing a tangent of the absolute value function at the origin (Figure 1.24).

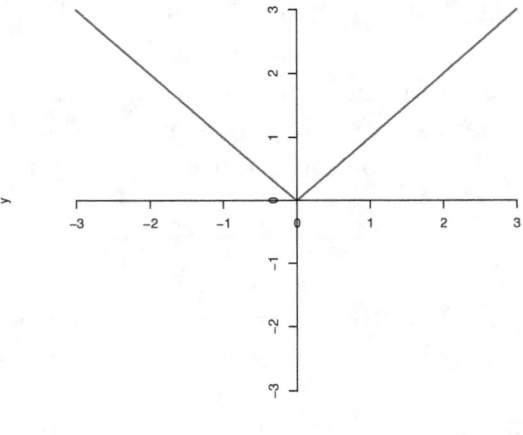

Figure 1.24: The absolute value function

The function $g(x) = x^{1/3}$ is continuous at the origin (Figure 1.25), but it is not differentiable there. Indeed

$$\frac{g(x) - g(0)}{x - 0} = \frac{x^{1/3}}{x} = \frac{1}{x^{2/3}}.$$

But $\lim_{x \to 0} 1/x^{2/3} = \infty$. Similarly, $h(x) = x^{2/3}$ is continuous at the origin (Figure 1.26). We have:

$$\frac{h(x) - h(0)}{x - 0} = \frac{x^{2/3}}{x} = \frac{1}{x^{1/3}}.$$

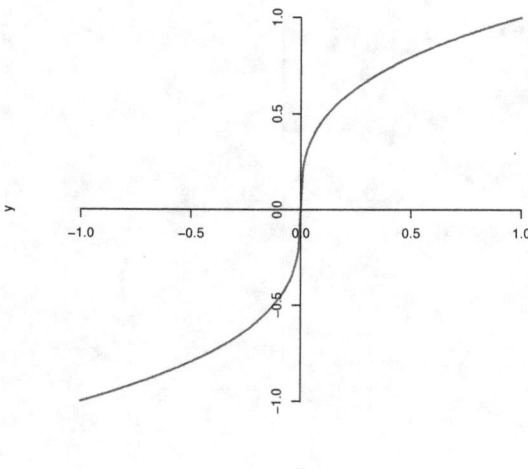

Figure 1.25: The function $x^{1/3}$

However, $\lim_{x \to 0^+} \frac{1}{x^{1/3}} = \infty$, and $\lim_{x \to 0^-} \frac{1}{x^{1/3}} = -\infty$. Thus, neither $g(x)$ nor $h(x)$ are differentiable at the origin despite the fact that they are continuous there.

Luckily, we will not have to use the definition of derivative as a limit each time we need to calculate $f'(c)$. As we will see next, there are six basic rules that we can follow in the future. Suppose that f and g are defined on a neighborhood of c and are differentiable at c. Then $f + g$, $f - g$, λf, fg, $1/g$, and f/g are differentiable at c. Moreover

$$(f + g)'(c) = f'(c) + g'(c) \text{ (The Sum Rule)},$$

$$(f - g)'(c) = f'(c) - g'(c) \text{ (The Difference Rule)},$$

$$(\lambda f)'(c) = \lambda f'(c) \text{ (The Constant Multiple Rule)},$$

$$(fg)'(c) = f'(c)g(c) + f(c)g'(c) \text{ (The Product Rule)},$$

$$\left(\tfrac{1}{g}\right)'(c) = \frac{-g'(c)}{g(c)^2} \text{ (The Reciprocal Rule)},$$

$$\left(\tfrac{f}{g}\right)'(c) = \frac{f'(c)g(c) - g'(c)f(c)}{g(c)^2} \text{ (The Quotient Rule)},$$

where it is to be understood that the last two rules are valid insofar as $g(c) \neq 0$.

The first three rules are to be expected and can be proved easily. But the next three rules are unexpected and rather surprising, so we might as well provide a proof. Let us start with the rule about the derivative of a product: We have

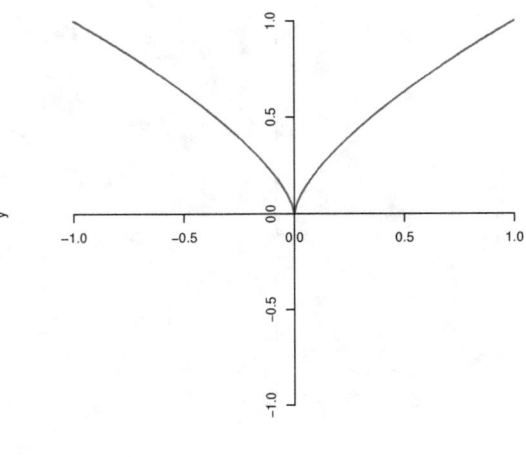

Figure 1.26: The function $x^{2/3}$

$$\frac{f(x)g(x)-f(c)g(c)}{x-c} = \frac{f(x)g(x)-f(c)g(x)+f(c)g(x)-f(c)g(c)}{x-c} =$$

$$= \frac{f(x)-f(c)}{x-c}g(x) + f(c)\frac{g(x)-g(c)}{x-c}.$$

Since g has to be continuous at c, we can see that $\lim_{x \to c} g(x) = g(c)$. Then

$$\lim_{x \to c}\left[\frac{f(x)-f(c)}{x-c}g(x) + f(c)\frac{g(x)-g(c)}{x-c}\right] = f'(c)g(c) + f(c)g'(c).$$

Consequently $(fg)'(c) = f'(c)g(c) + f(c)g'(c)$. We have just shown the Product Rule.

A proof of the rule about the derivative of $1/g$ is even simpler[4] Indeed

$$\frac{\frac{1}{g(x)}-\frac{1}{g(c)}}{x-c} = \frac{-\frac{g(x)-g(c)}{g(x)g(c)}}{x-c} = -\frac{g(x)-g(c)}{x-c}\frac{1}{g(x)g(c)}.$$

Therefore

$$\lim_{x \to c}\frac{\frac{1}{g(x)}-\frac{1}{g(c)}}{x-c} = -g'(c)\frac{1}{g(c)^2}.$$

We have succeeded in proving that $\left(\frac{1}{g}\right)'(c) = \frac{-g'(c)}{g(c)^2}$.

The rule for the derivative of the quotient of two functions, the sixth rule, follows immediately from the two previous rules once we note that $\frac{f(x)}{g(x)} = f(x)\frac{1}{g(x)}$.

---

[4]We have to emphasize that since $g(c) \neq 0$, and g is continuous at c, there is a neighborhood of c such that $g(x) \neq 0$.

It is time to discuss some examples. Given $h(x) = x^3 + x^2 + 4$ (Figure 1.27) we wish to calculate $h'(5)$. For any c, $h'(c) = 3c^2 + 2c$; thus, $h'(5) = 75 + 10 = 85$. Next suppose that $h(x) = \frac{x^3 + x + 1}{x^4 + x + 1}$ and we are asked to calculate $h'(0)$. Applying the Quotient Rule we get

$$h'(c) = \frac{(3c^2+1)(c^4+c+1) - (4c^3+1)(c^3+c+1)}{(c^4+c+1)^2}.$$

Therefore $h'(0) = \frac{1-1}{1} = 0$. This fact is telling us that the tangent to the function at $c = 0$ is parallel to the horizontal axis.

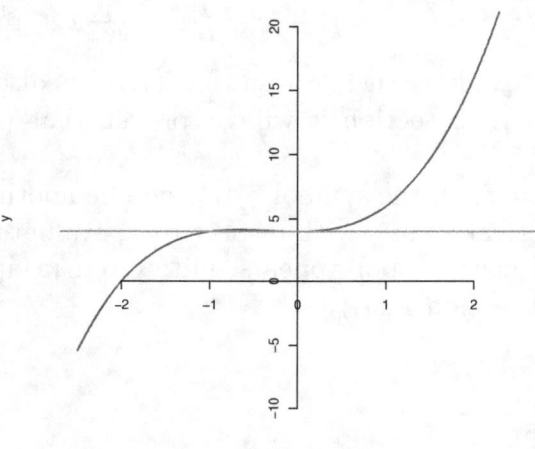

Figure 1.27: The function $x^3 + x^2 + 4$ and its tangent at $x = 0$

The derivative function

Given a function f, defined on an open interval I, it may well happen that it is differentiable at every x in I with the possible exception of one or more isolated points. Then we can define a new function, to be denoted f' and called the 'derivative function,' which makes each x correspond to the number $f'(x)$. For instance, if n is a natural number and $f(x) = x^n$, then, thanks to (1.11) it follows that $f'(x) = nx^{n-1}$ for every real number x. Using **Leibniz's notation** we would write

$$\frac{d}{dx} x^n = nx^{n-1}. \tag{1.13}$$

Therefore

$\frac{d}{dx}(x^5 + x + 1)(x^4 + 3x^2 + 7) = (5x^4 + 1)(x^4 + 3x^2 + 7) + (x^5 + x + 1)(4x^3 + 6x)$

and

$$\frac{d}{dx}\frac{x}{x^2+1} = \frac{(x^2+1)-2x^2}{(x^2+1)^2} = \frac{1-x^2}{(x^2+1)^2}.$$

For any natural number n we have

$$\frac{d}{dx}(x^{-n}) = \frac{d}{dx}\frac{1}{x^n} = \frac{-nx^{n-1}}{x^{2n}} = -nx^{-n-1}, \, x \neq 0.$$

Thus, (1.13) is valid for any integer n provided that, when $n < 0$, zero is not in the domain of differentiability. Moreover, since $\frac{d}{dx}\sqrt{x} = \frac{1}{2\sqrt{x}}$, $x > 0$, for every positive x we will have

$$\frac{d}{dx}x^{3/2} = \frac{d}{dx}(x^{1/2}x) = (\frac{d}{dx}x^{1/2})x + x^{1/2}\frac{d}{dx}x = \frac{1}{2\sqrt{x}}x + x^{1/2} = \frac{3}{2}\sqrt{x}.$$

The reader may note that we have just proven that (1.13) is also true when $n = 3/2$. As a matter of fact, in a later section it will be proven that (1.13) is true for any *rational* number n.

Let us agree that if $y = f(x)$, the symbol $\frac{dy}{dx}|_{x=c}$ will be another way to denote the number $f'(c)$: it is the derivative of y with respect to x, evaluated at c. For instance, $\frac{d}{dx}x^{3/2}|_{x=5} = \frac{3}{2}\sqrt{5}$. This new symbol appears quite often in applied problems (see exercise 10, at the end of section 1.11.)

Lateral derivatives

Let us consider the function $f(x) = \sqrt{x}$. As we know, it is differentiable at every $c > 0$. What happens at $x = 0$? We cannot expect the function to have a derivative there because it is not even defined on a neighborhood of 0. Since

$$\lim_{x \to 0+} \frac{\sqrt{x}-\sqrt{0}}{x-0} = \lim_{x \to 0+} \frac{1}{\sqrt{x}} = \infty,$$

the square root function does not even have a 'right derivative' at $x = 0$. However, the function $g(x) = x^2$ for $x \geq 0$ and $g(x) = -x$ for $x < 0$, does have a right derivative at $x = 0$. Indeed

$$\lim_{x \to 0+} \frac{x^2-0}{x-0} = \lim_{x \to 0+} x = 0.$$

We then write $g'_+(0) = 0$, where the plus sign reminds us that we are dealing with a right derivative. On the other hand, $g'_-(0)$, the 'left derivative' of g at 0, equals -1 because

$$\lim_{x \to 0-} \frac{g(x)-g(0)}{x-0} = \lim_{x \to 0-} \frac{-x-0}{x-0} = -1.$$

Of course, if $h(x) = x^2$, $0 \leq x \leq 1$, then $h'(x) = 2x$ on $[0, 1]$, where one has to keep in mind that $h'(0)$ and $h'(1)$ are really $h'_+(0)$ and $h'_-(1)$.

The headlights of an automobile

So far we have interpreted the derivative at a point as the slope of the tangent at that particular point. This is an important interpretation, especially in the applications of calculus to optics. Let us consider a two-dimensional simplification of a headlight of an automobile, whose shape is a mirror described by the parabola $y = x^2$ (Figure 1.28). Assume that the light bulb is at the focus, namely $(0, 1/4)$. A ray of light emerges from the light bulb and hits the mirror at a point C and is reflected upwards. We wish to show that the reflected ray is parallel to the axis of the parabola. The coordinates of C are (c, c^2), thus the tangent to the parabola at C is given by the equation $y - c^2 = 2c(x - c)$. We note that the tangent intersects the y-axis at $(0, -c^2)$. On the other hand, the normal line at C is the perpendicular to the tangent at C. According to the law of reflection, a basic law of optics, the measure of the angle of incidence is equal to the measure of the angle of reflection. That is to say, $m(\angle BCN) = m(\angle MCN) = \rho$. We observe that $BC = AB$ (both segments have length $c^2 + 1/4$). Therefore, the base angles of $\triangle ABC$ are congruent. Let β be their measure and let $\alpha = m(\angle ABC)$. Since $\alpha + 2\beta = 180^o$ and $\rho + \beta = 90^o$ we can conclude that $\alpha = 2\rho$. Having shown that $\angle ABC$ is congruent to $\angle BCM$, it is possible to conclude that the reflected ray is parallel to the axis of the mirror. So, all the rays of light that start from the bulb and hit the parabolic mirror will emerge in such a way that they are parallel to each other and parallel to the mirror's axis.

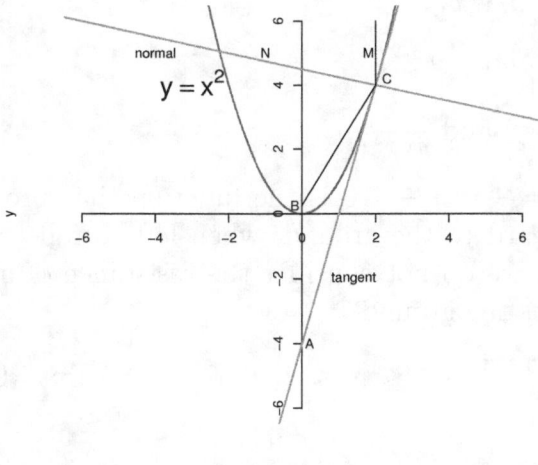

Figure 1.28: The headlights of an automobile

The derivative as an instantaneous rate of change

We should keep in mind that often the derivative is to be interpreted as an instantaneous rate of change where, as expected, the independent variable is t (time). For instance, if $s(t)$ denotes the distance traveled by a falling object, $s'(t)$ is the instantaneous velocity. If $T(t)$ is the temperature of a solution being cooled, $T'(t)$ is the instantaneous rate at which the column of mercury of the thermometer is descending; that is to say,

$$T'(t) = \lim_{h \to 0} \frac{T(t+h) - T(t)}{h}.$$

Similarly, if $y(t)$ denotes the weight of a radioactive substance, $y'(t)$ represents the instantaneous rate at which the substance is decaying. And if $W(t)$ denotes the mass of a plant, $W'(t)$ is the instantaneous rate at which the plant is growing.

As we will discuss in later sections, many models in the natural sciences involve instantaneous rates of change. For instance, $s'(t) = v_o - gt$ where v_o is the initial velocity of a descending body and g is the acceleration due to gravity, $T'(t) = -k(T(t) - T_M)$ where T_M is the constant temperature of the cooling liquid, $y'(t) = -ky(t)$ (the model for radioactive decay) and $W'(t) = k_1 W(t) - k_2 W^2(t)$ (a model for plant growth). These are all examples of **differential equations**, where the unknown is hidden inside a derivative. Eventually, we will learn the steps needed to solve these equations in order to validate the model, estimate the respective parameter k (or k_1, k_2), and analyze concrete problems.

Exercises for Section 1.7

1. Calculate $\frac{d}{dx}\left(\frac{x+3}{x^2+x}\right)$ and $\frac{d}{dx}x^{5/2}$.

2. Calculate $\frac{d}{dx}\left(\sqrt{x} + \frac{1}{x}\right)$ and $\frac{d}{dx}\frac{1}{x^2+1}$.

3. Suppose that $s(t) = -16t^2 + 420$ is the function that provides the position of an object, with regard to the ground, when left to fall from the top of a 420 ft tall building and we do not consider the resistance of air. What will be its velocity when it hits the ground?

4. Given $f(x) = \sqrt{1 - x^2}$, $-1 \le x \le 1$, find $f'(x)$ (Hint: Use the definition of derivative as a limit.)

5. Suppose that (c, c^2) is a point on the curve $y = x^2$, such that the tangent at it passes through $(2, 1)$. Find the two values of c that have this property.

6. Let $f(x) = x^{1/3}$. Show that $f'(c) = \frac{1}{3}c^{-2/3}$ for any $c \ne 0$ (Hint: Remember that $(a - b)(a^2 + ab + b^2) = a^3 - b^3$.)

7. Is the function $f(x) = x|x|$ differentiable at $x = 0$?

8. Assume that $f(t) = \frac{150(t+1)}{t+25}$ provides the weight (in pounds) of an individual, as a function of time (in years). Calculate the instantaneous rate of growth when $t = 1$ and $t = 15$.

9. The function $f(x) = \frac{x}{1+0.015x^2}$, $x > 0$, provides the yield of a crop as a function of the amount of nitrogen in the soil. Calculate $f'(x)$.

10. The concentration of a reactant, which undergoes a second order chemical reaction, is given by $c(t) = \frac{c_o}{1+kc_o t}$, where c_o is the initial amount of reactant and k is the kinetic constant of the reaction. Calculate $c'(t)$ and show that $c'(t) = -k(c(t))^2$.

11. Poiseuille's law of laminar flow for a cylinder of radius R, namely $v(r) = k(R^2 - r^2)$, $0 \le r \le R$, gives the velocity at which a fluid flows, as a function of the distance r to the axis of the cylinder (we are assuming that the difference of pressure, between the ends of the cylinder, is constant). An artery fits this description. Calculate $v'(r)$, which is known as the *velocity gradient*, and provides the rate of change of the velocity as a function of the distance to the axis.

12. Assume that two strains of bacteria, x and y, grow in the same environment. The rate at which the population of each of them grows is proportional to the amount present, that is, $x'(t) = \alpha x(t)$, $y'(t) = \beta y(t)$. Let $z(t)$ denote the proportion of x at any instant t, i.e.,

$$z(t) = \frac{x(t)}{x(t)+y(t)}$$

and let $w(t)$ denote the proportion of y at any instant t, that is,

$$w(t) = \frac{y(t)}{x(t)+y(t)}.$$

Calculate $z'(t)$ in terms of $z(t)$, $w(t)$, and the parameters α, β.

13. Although a first course on calculus deals primarily with functions of one independent variable, it is worthwhile to say something about **partial derivatives**, a concept linked to functions of more than one independent variable. Let us consider the law of ideal gases, namely $PV = nRT$, where P denotes pressure, V denotes volume, and T denotes temperature (n is the number of moles of gas; R is a universal constant). We have $V = nR\frac{T}{P}$. Then V is a function of two variables, T and P. The symbol $\frac{\partial V}{\partial T}$ denotes the derivative of V with respect to T when we keep P constant, thus $\frac{\partial V}{\partial T} = \frac{nR}{P}$. Calculate $\frac{\partial V}{\partial P}$, the derivative of V as a function of P when T is kept constant.

14. Suppose that we have a parabolic telescope whose shape follows the curve $y = x^2$ (this is a two-dimensional simplification of an actual parabolic telescope). Assuming that a ray of light reaches the telescope on a path parallel to the axis of it, an assumption that makes sense when the ray comes from a far distant star and the axis of the telescope has been pointed toward the star, show that the ray will pass through the focus of the parabola after bouncing off the mirror (Hint: This problem is quite similar to the problem of the headlights of an automobile).

15. Accepting the reciprocal rule, provide a proof of the quotient rule.

16. Using the definition of derivative as a limit, prove that (1.10) and (1.12) are two equivalent ways of defining the derivative at a point.

17. Find an example of a function that is differentiable everywhere, but whose derivative function is not differentiable at $x = 0$.

18. Define two functions that are differentiable everywhere except at the origin, but whose sum is differentiable everywhere (including the origin).

19. Define two functions such that one is differentiable everywhere but the other is differentiable everywhere with the exception of the origin, and their product is differentiable everywhere (including the origin).

20. Define two functions that are differentiable everywhere with the exception of the origin, but their product is differentiable everywhere.

1.8   Derivatives of Trigonometric Functions

Radians

In High School geometry we deal with degrees when measuring an angle. But there is another way of measuring angles, namely in radians. What is a radian? Given any angle, draw a circle with center at its vertex and an arbitrary radius r. A well-known geometrical result asserts that $\frac{p}{2\pi r} = \frac{\alpha^\circ}{360^\circ}$, where p is the length of the arc subtended by the central angle with measure α°. Thus,

$$\frac{p}{r} = \frac{\pi}{180^\circ}\alpha^\circ.$$

We note that the quotient p/r is a dimensionless quantity that depends only on the angle. It is called 'the measure of the angle in radians,' which is often denoted by the symbol θ. Hence $p = r\theta$, a relationship that has considerable importance in mathematics. In particular, it allows us to conclude that the measure of the corresponding central angle is $\theta = 1$ radian if the subtended arc has length r. Moreover, since the area of a sector, of a circle of radius r and subtended arc of length p, is $\frac{1}{2}pr$, we can conclude that the area becomes $\frac{1}{2}r^2\theta$ whenever the central angle is measured in radians. This formula adopts the simple expression $\frac{1}{2}\theta$ if the circle under consideration has unit radius and the central angle is measured in radians. It is also important to keep in mind that if the circle has unit radius then $p = \theta$; thus, under these circumstances the length of the subtended arc is the same as the measure of the angle measured in radians.

Sometimes we have to make conversions between degrees and radians. For this purpose we have to keep in mind that

$$\theta = \tfrac{\pi}{180}\alpha$$

where θ is the measure in radians and α is the measure, of the same angle, in degrees. For instance, if $\alpha = 60^{\circ}$ then $\theta = \pi/3$ radians, while if $\theta = \pi$ radians then $\alpha = 180^{\circ}$.

What is the advantage of radians vis a vis degrees? It is a matter of comparing the expressions $p = \frac{\pi}{180}r\alpha$ and $p = r\theta$. The latter is simpler than the former, so working with radians is the natural choice when we have to deal with trigonometric functions in Calculus. *From now on it is to be understood that we will work with radians.*

A brief review of some trigonometric facts

Let us consider a unit circle on the Cartesian plane. A point B on the circle moves in a counterclockwise direction describing an angle θ. We know that $\sin\theta = BD$ and $\cos\theta = OD$ (see Figure 1.29) and realize that $\sin 0 = 0$, $\sin\frac{\pi}{2} = 1$, $\sin\pi = 0$, $\sin\frac{3\pi}{2} = -1$, $\sin 2\pi = 0$, while $\cos 0 = 1$, $\cos\frac{\pi}{2} = 0$, $\cos\pi = -1$, $\cos\frac{3\pi}{2} = 0$, and $\cos 2\pi = 1$. Moreover, we note that $\sin\theta$ increases from 0 to 1 in the first quadrant, then decreases from 1 to 0 in the second quadrant, and further decreases from 0 to -1 in the third quadrant and finally increases from -1 to 0 in the fourth quadrant. In a similar fashion, $\cos\theta$ decreases from 1 to 0 in the first quadrant, decreases from 0 to -1 in the second quadrant, increases from -1 to 0 in the third quadrant and finally increases from 0 to 1 in the fourth quadrant. This pattern repeats itself as the angle goes beyond 2π. It should be clear, from the definition, that $\sin(\theta + 2\pi) = \sin\theta$ and $\cos(\theta + 2\pi) = \cos\theta$. Furthermore, one can show that $\sin(\theta + \pi) = -\sin\theta$, $\cos(\theta + \pi) = -\cos\theta$, $\cos\theta = \sin(\theta + \frac{\pi}{2})$, $\sin(\alpha + \beta) = \sin\alpha\cos\beta + \cos\alpha\sin\beta$, and $\cos(\alpha + \beta) = \cos\alpha\cos\beta - \sin\alpha\sin\beta$, and the well-known identity $\sin^2\theta + \cos^2\theta = 1$. On the other hand, we are dealing with negative angles when the point B on the unit circle moves in a clockwise direction. As we did for positive angles, the sine and cosine of negative angles are defined through the unit circle. Evidently, $\sin(-\theta) = -\sin\theta$ and $\cos(-\theta) = \cos(\theta)$. We should always keep in mind the unit circle and the moving point B.

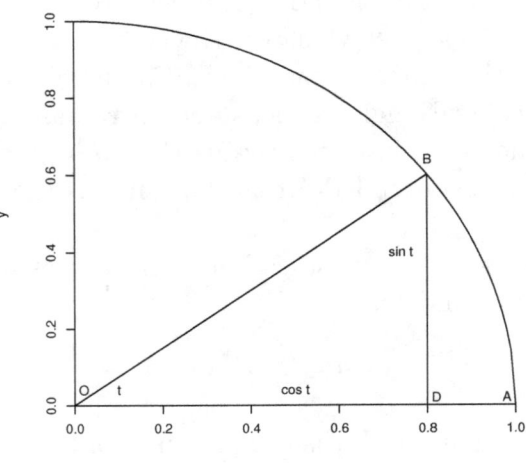

Figure 1.29: The sine and cosine of an angle

Next we define a function $t \to \sin t$ and a function $t \to \cos t$ for all real numbers t (often we will use the variable t instead of θ to emphasize that the independent variable is *time*). Both functions, with the domain being the set of all real numbers and the range being the interval $[-1, 1]$, are denoted simply sin and cos or $\sin(t)$ and $\cos(t)$ (Figure 1.30). One may observe that cos has the same shape as sin, but shifted $\pi/2$ to the left. Both functions are periodic, with period 2π, because every 2π units of time they take the same value, that is to say, it takes 2π units of time to complete a cycle. Thus, one unit of time contains $\frac{1}{2\pi}$ of the cycle; this is the inverse of the period and is called the 'frequency.' The sine and cosine function oscillate between 1 and -1, so we may say that their 'amplitude' is 1. If we wish to work with periodic functions like sine and cosine but with an amplitude different from 1, we can multiply them by a constant as in $2\cos t$ or $3\sin t$.

If we wish to obtain periodic functions that have periods longer or shorter than 2π, we can do so by multiplying the independent variable t by a value w (called **angular frequency**). As a matter of fact, the functions $f(t) = A\cos(wt)$ and $g(t) = A\sin(wt)$ are periodic functions with period $2\pi/w$ because

$$A\cos w(t + \tfrac{2\pi}{w}) = A\cos(wt + 2\pi) = A\cos(wt).$$

In a similar fashion we can prove that $A\sin(wt)$ has period $\frac{2\pi}{w}$. In summary: given $A\cos(wt)$ or $A\sin(wt)$, A is the amplitude, w is the angular frequency, the period is $\frac{2\pi}{w}$, and the frequency is $\frac{w}{2\pi}$. For instance, given $f(t) = 3\cos(\frac{\pi}{6}t)$ we can assert that its amplitude is 3 (the function oscillates between 3 and -3), the angular frequency is

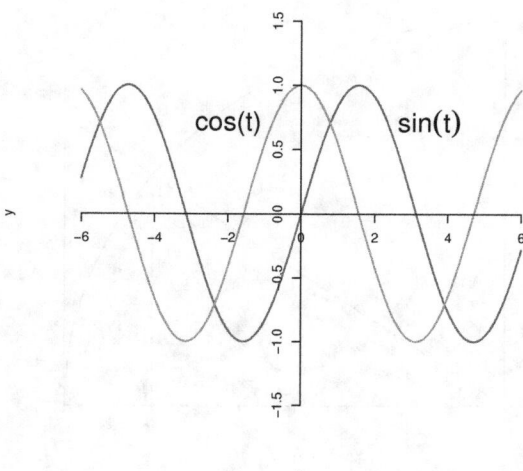

Figure 1.30: The sine and cosine functions

$\frac{\pi}{6}$, the period is $\frac{2\pi}{\pi/6} = 12$ (that is, it takes 12 units of time to complete a cycle), and the frequency is $1/12$ (that is, there are $1/12 = 0.08$ cycles per unit of time or, what is the same, 0.08 of a cycle is completed in one unit of time). Having exhausted our review of the sine and cosine function, it is time to pay attention to the tangent of an angle.

Let us consider a unit circle on the Cartesian plane, wherein a point B moves in a counterclockwise direction (Figure 1.31). We define $\tan\theta = \frac{\sin\theta}{\cos\theta}$ (the 'tangent of θ'), a definition that makes sense whenever $\cos\theta \neq 0$. Since triangles OBD and OCA are similar, it follows that $\frac{AC}{BD} = \frac{1}{OD}$, that is to say,

$$AC = \frac{BD}{OD} = \frac{\sin\theta}{\cos\theta} = \tan\theta.$$

As θ increases so will AC, in fact AC will become bigger and bigger as we approach $\pi/2$. If the point B moves in a clockwise direction, we will have that $\tan(-\theta) = AC'$ where C' is the reflection, about the horizontal axis, of the point C. This graphical interpretation leads to the conclusion that if the central angle varies from 0 to $-\pi/2$, the corresponding tangent increases dramatically in absolute value but preceded by the negative sign.

We can define a function $t \to \tan t$ on the whole real line except where the cosine adopts the value zero, that is, except values of the form $(2n+1)\frac{\pi}{2}$, n an arbitrary integer. Note that we changed the variable θ, and wrote t instead, to emphasize the fact that the independent variable quite often will refer to time.

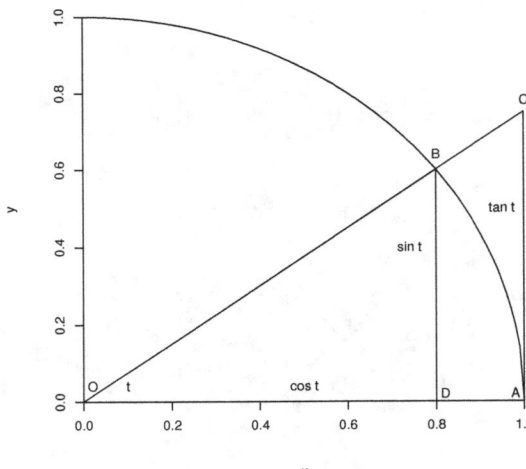

Figure 1.31: Geometrical interpretation of the tangent of an angle

From the unit circle interpretation of the tangent it follows that tan has period π. Indeed,

$$\tan(t + \pi) = \tfrac{\sin(t+\pi)}{\cos(t+\pi)} = \tfrac{-\sin t}{-\cos t} = \tan t.$$

A graph of the tangent function, between $-3\pi/2$ and $3\pi/2$, can be seen in Figure 1.32. Given that t belongs to $(-\frac{\pi}{2}, \frac{\pi}{2}) \cup (\frac{\pi}{2}, \frac{3\pi}{2})$ (the union of two intervals), how do we solve the equation $\tan t = b$ for a given real number b? For instance, let us consider the equation $\tan t = 1$. If we use the $\tan^{-1}$ option on a graphing calculator, we will get the answer $\pi/4$ provided that we are in 'exact' mode, which makes perfect sense since $\tan(\pi/4) = 1$. But tan has period π, hence $\frac{\pi}{4} + \pi$ is also a good answer. Which one among the two possibilities should we choose? We need information on whether t lies on the first quadrant or the third quadrant! In section 1.12 we will develop the $\tan^{-1}$ function in detail. Right now we only have to keep in mind that the $\tan^{-1}$ option of a graphing calculator provides the solution of the equation $\tan t = b$ on the interval $(-\frac{\pi}{2}, \frac{\pi}{2})$.

A closer look at the main trigonometric functions

Let us recall that the sine function has as its domain the whole real line, is periodic with period 2π, and oscillates between -1 and 1. Moreover, $\sin x$ approaches the value zero as x tends to zero due to the way that $\sin x$ is defined through the unit circle. In a similar fashion, we can assert that $\cos x$ approaches the value 1 as x tends to zero.

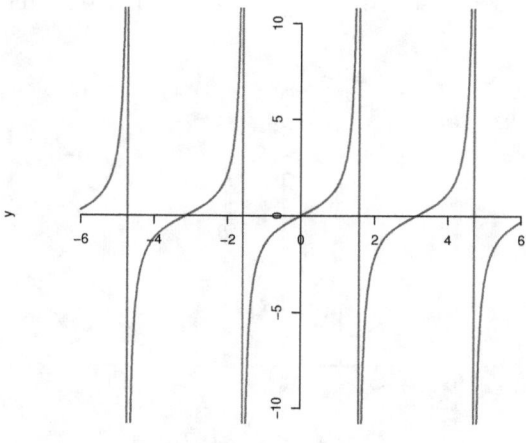

Figure 1.32: The tangent function

For any fixed x, $\sin(x + h) = \sin x \cos h - \cos x \sin h \to \sin x \times 1 - \cos x \times 0 = \sin x$ as $h \to 0$. Therefore $\sin x$ is continuous at every point on the real line.

The cosine function is defined over the whole real line, is periodic with period 2π, and oscillates between -1 and 1. Since $\cos x = \sin(x + \pi/2)$ we can conclude that it has the same shape as the sine function but it is shifted to the left by $\pi/2$, no surprise then that it is continuous everywhere. Strictly speaking, cos is continuous because it is the composition of the continuous functions sin and $x \to x + \frac{\pi}{2}$. Or we could just note that $\cos(x + h) = \cos x \cos h - \sin x \sin h \to \cos x \times 1 - \sin x \times 0 = \cos x$ as $h \to 0$.

The tangent function is defined as the quotient of the sine and cosine functions; this means that $\tan x = \sin x / \cos x$. Being the quotient of two continuous functions, $\tan x$ is continuous on its domain. The fact that $\lim_{x \to \pi/2-} \sin x = 1$ and $\lim_{x \to \pi/2-} \cos x = 0$ implies that $\lim_{x \to \pi/2-} \tan x = \infty$. In a similar fashion we can observe that $\lim_{x \to -\pi/2+} \sin x = -1$ while $\lim_{x \to -\pi/2+} \cos x = 0$, therefore $\lim_{x \to -\pi/2+} \tan x = -\infty$. This behavior of the tangent function repeats itself at each point of discontinuity:

$$\ldots - 5\pi/2, -3\pi/2, -\pi/2, \pi/2, 3\pi/2, 5\pi/2, \ldots$$

The well-known secant function is defined as $\sec x = 1/\cos x$. Thus, its domain is the same as the domain of the tangent function; namely, every point on the real line except those points of the form $(2n + 1)\pi/2$ (n an arbitrary integer). Obviously sec is continuous on its domain and has period 2π since

$$\sec(x + 2\pi) = \frac{1}{\cos(x+2\pi)} = \frac{1}{\cos x} = \sec x.$$

A graph of the secant function can be seen in Figure 1.33. We observe that $\sec(0) = 1$ while $\lim_{x \to \pi/2^-} \sec x = \infty$ because $\lim_{x \to \pi/2^-} \cos x = 0$. Similarly, $\lim_{x \to -\pi/2^+} \sec x = \infty$ because $\lim_{x \to -\pi/2^+} \cos x = 0$.

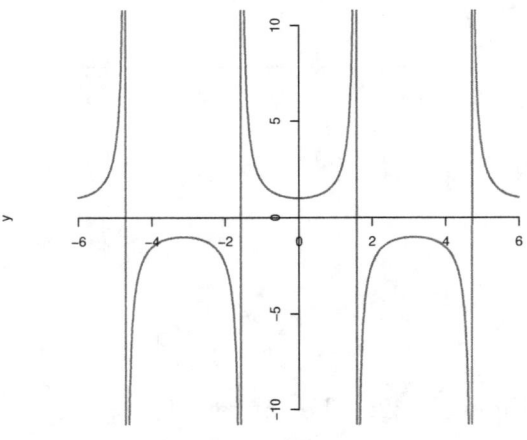

Figure 1.33: The secant function

On $(\pi/2, 3\pi/2)$ the graph of the secant function looks like the graph on $(-\pi/2, \pi/2)$ but opening downwards due to the fact that $\sec \pi = -1$, $\lim_{x \to \pi/2^+} \sec x = -\infty$ and $\lim_{x \to 3\pi/2^-} \sec x = -\infty$. Thereafter the graph of the secant function repeats itself by periodicity.

Calculating derivatives

Now it is time to find the derivative of the sine and cosine function at the origin. We know that $\lim_{x \to 0} \sin x = 0$ and $\lim_{x \to 0} \cos x = 1$ because the two basic trigonometric functions are continuous everywhere; in particular at the origin. In order to find the derivative of the sine function at $x = 0$ we need to calculate $\lim_{x \to 0} \frac{\sin x - \sin 0}{x - 0}$, that is to say $\lim_{x \to 0} \frac{\sin x}{x}$, a task that we accomplished in section 1.4 using a graphing calculator. Indeed, we found that $\lim_{x \to 0} \frac{\sin x}{x} = 1$.

Let us try to provide a geometrical argument to justify our previous assertion that $\lim_{x \to 0} \frac{\sin x}{x} = 1$. From Figure 1.34, a quarter of a circle of unit radius and central angle of x radians ($0 < x < \pi/2$), we can observe that area $\triangle OBD$ <area sector OAB < area $\triangle OAC$ [5]. The area of each triangle is $\frac{1}{2} \sin x$ and $\frac{1}{2} \tan x$ respectively.

---

[5]Figure 1.34 is identical to Figure 1.31, except that the variable x is used instead of t.

What is the area a of the sector? The area of the whole circle is π, which corresponds to a central angle of 2π radians, therefore we can set up the proportion

$$\frac{a}{x} = \frac{\pi}{2\pi}.$$

Hence $\frac{1}{2}\sin x < \frac{1}{2}x < \frac{1}{2}\tan x$, which in turn leads to $1 < \frac{x}{\sin x} < \frac{1}{\cos x}$ and consequently $\cos x < \frac{\sin x}{x} < 1$ (recall that $\cos x$ and $\sin x$ are positive whenever $0 < x < \frac{\pi}{2}$). Since $\lim_{x\to 0^+}\cos x = 1$, the squeeze property of limits implies that $\lim_{x\to 0^+}\frac{\sin x}{x} = 1$. On the other hand,

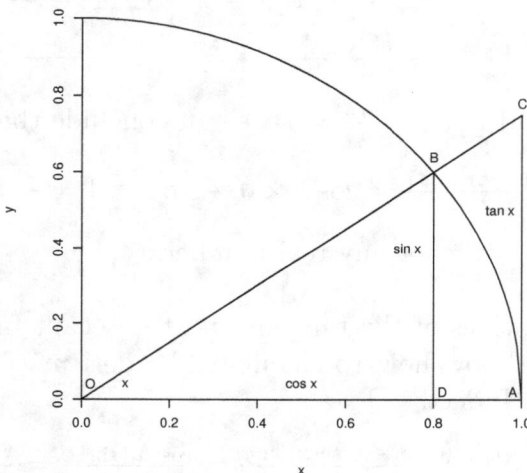

Figure 1.34: Diagram related to the justification of $\lim_{x\to 0}\frac{\sin x}{x}$

$$\lim_{x\to 0^-}\frac{\sin x}{x} = \lim_{x\to 0^+}\frac{\sin(-x)}{-x} = \lim_{x\to 0^+}\frac{-\sin x}{-x} = \lim_{x\to 0^+}\frac{\sin x}{x} = 1.$$

Since both the limit from the right and the left are equal to 1 we can conclude that, as expected, $\lim_{x\to 0}\frac{\sin x}{x} = 1$.

Having the derivative of the sine function at the origin, namely $\sin'(0) = 1$, finding the derivative of the cosine function at $x = 0$ should not be an onerous task. Indeed,

$$\frac{\cos x - \cos 0}{x - 0} = \frac{\cos x - 1}{x} = \frac{(\cos x - 1)(\cos x + 1)}{x(\cos x + 1)} = \frac{-\sin^2 x}{x(\cos x + 1)} = \frac{-\sin x}{x}\frac{\sin x}{\cos x + 1}.$$

But $\lim_{x\to 0}\frac{-\sin x}{x} = -\lim_{x\to 0}\frac{\sin x}{x} = -1$ and $\lim_{x\to 0}\frac{\sin x}{\cos x + 1} = \frac{0}{2} = 0$. Therefore $\lim_{x\to 0}\frac{\cos x - \cos 0}{x - 0} = -1 \times 0 = 0$. We have succeeded in showing that $\cos'(0) = 0$.

The next natural step is to calculate the derivatives of sin and cos at any point x on the real line. First we will concentrate our attention on trying to calculate $\sin'(x)$;

that is to say, $\lim_{z \to x} \frac{\sin z - \sin x}{z - x}$, which is equivalent to $\lim_{h \to 0} \frac{\sin(x+h) - \sin x}{h}$. A little bit of trigonometry will do the job. Indeed, for $h \neq 0$ we have

$$\frac{\sin(x+h) - \sin x}{h} = \frac{\sin x \cos h + \cos x \sin h - \sin x}{h} = \sin x \frac{\cos h - 1}{h} + \cos x \frac{\sin h}{h}.$$

Let us recall that $\lim_{h \to 0} \frac{\cos h - 1}{h} = 0$ while $\lim_{h \to 0} \frac{\sin h}{h} = 1$, thus

$$\lim_{h \to 0} \frac{\sin(x+h) - \sin x}{h} = \sin x \times 0 + \cos x \times 1 = \cos x.$$

We have just shown that $\sin'(x) = \cos x$ for any real number x. Next let us try to find the derivative of the cosine function at any point x. The path that we will follow resembles what we did before when dealing with the sine function: For $h \neq 0$

$$\frac{\cos(x+h) - \cos x}{h} = \frac{\cos x \cos h - \sin x \sin h - \cos x}{h} = \cos x \frac{\cos h - 1}{h} - \sin x \frac{\sin h}{h}.$$

Since $\lim_{h \to 0} \frac{\cos h - 1}{h} = 0$ and $\lim_{h \to 0} \frac{\sin h}{h} = 1$ we can conclude that

$$\lim_{h \to 0} \frac{\cos(x+h) - \cos x}{h} = \cos x \times 0 - \sin x \times 1 = -\sin x.$$

That is to say, $\cos'(x) = -\sin x$ for any real number x.

What about the derivatives of the tangent and the secant function? Our task is pretty simple because we know how to calculate the derivative of a product and a quotient of two functions. Indeed

$$\frac{d}{dx} \tan x = \frac{d}{dx} \frac{\sin x}{\cos x} = \frac{\cos x (\cos x) - (-\sin x) \sin x}{\cos^2 x} = \frac{\cos^2 x + \sin^2 x}{\cos^2 x} = \frac{1}{\cos^2 x} = \sec^2 x,$$

$$\frac{d}{dx} \sec x = \frac{d}{dx} \frac{1}{\cos x} = \frac{-(-\sin x)}{\cos^2 x} = \frac{\sin x}{\cos x} \frac{1}{\cos x} = \tan x \sec x.$$

Of course, these derivatives exist whenever $\cos x \neq 0$. That is to say, they will exist for any x different from $(2n + 1)\frac{\pi}{2}$, n being an arbitrary integer.

Exercises for Section 1.8

Calculate the following derivatives:

1. $\frac{d}{dx} x^2 \sin x$

2. $\frac{d}{dx} \frac{\cos x}{\sin x}$

3. $\frac{d}{dx} (\sin x)^2$

4. $\frac{d}{dx} \left(x \cos x - 7 \frac{\cos x}{x} \right)$

5. $\frac{d}{dx} (\sec x \tan x)$

6. $\frac{d}{dx}(\cos x)^2$

7. $\frac{d}{dx}\sin^2 x \times \cos x$

8. $\frac{d}{dx}\sin^3 x$

9. $\frac{d}{dx}\sec^2 x \tan x$

10. Given $f(t) = \sin(\pi t)$, find $f'(t)$ (Hint: Follow the steps that led to the derivative of $g(t) = \sin t$).

Enrichment Note: Harmonic Functions

Let us recall, from a pre-calculus course, that if $f(t)$ is a function and $c > 0$, $g(t) = f(t - c)$ has the same graph as $f(t)$ except that it has been shifted to the right c units, while $h(t) = f(t + c)$ has the same graph as $f(t)$ except that it has been shifted to the left c units. In particular, $A\sin(wt - c)$ has the same graph as $A\sin(wt)$ but shifted to the right c/w units. Why is this so? It follows from the fact that

$$A\sin(wt - c) = A\sin w(t - \tfrac{c}{w}).$$

In the same fashion, $A\cos(wt - c)$ has the same graph as $A\cos(wt)$ but shifted to the left c/w units (their frequency, namely $\frac{w}{2\pi}$, is the same). When working with either $A\sin(wt - c)$ or $A\cos(wt - c)$, the number c is called the **phase shift**, which should not be confused with the shift to the right of $A\sin(wt)$ or $A\cos(wt)$ by c/w.

Functions of the type $A\cos wt + B\sin wt$ are called **harmonics**. They play an important role in the natural sciences as well as in time series analysis. If we were to graph a harmonic we would notice that it looks like a shifted cosine with a certain amplitude and the same frequency $\frac{w}{2\pi}$. Thus, we might assume that

$$A\cos wt + B\sin wt = C\cos(wt - \delta)$$

for some constants C and δ. Then

$$A\cos wt + B\sin wt = (C\cos\delta)\cos wt + (C\sin\delta)\sin wt.$$

Consequently $A = C\cos\delta$ and $B = C\sin\delta$, which in turn leads to $A^2 + B^2 = C^2$; hence $C = \sqrt{A^2 + B^2}$, and $\tan\delta = B/A$. Indeed, let us check that $\sqrt{A^2 + B^2}\cos(wt - \delta)$, where $\cos\delta = \frac{A}{\sqrt{A^2+B^2}}$ and $\sin\delta = \frac{B}{\sqrt{A^2+B^2}}$, is equal to $A\cos wt + B\sin wt$:

$$(\sqrt{A^2 + B^2}\cos\delta)\cos wt + (\sqrt{A^2 + B^2}\sin\delta)\sin wt =$$

$$= \sqrt{A^2 + B^2}\,\frac{A}{\sqrt{A^2+B^2}}\cos wt + \sqrt{A^2 + B^2}\,\frac{A}{\sqrt{A^2+B^2}}\sin wt =$$

$$= A\cos wt + B\sin wt.$$

A couple of examples will help us understand how to find the amplitude and phase shift of a harmonic. Assume that $f(t) = \cos(3t) + \sin(3t)$. Then $C = \sqrt{2}$ and $\tan \delta = 1$, thus $\delta = \tan^{-1} 1 = \pi/4$. There is the possibility that $\delta = \tan^{-1} + \pi$, but since $\cos \delta = \frac{1}{\sqrt{2}} > 0$ and $\sin \delta = \frac{1}{\sqrt{2}} > 0$ we can assert that δ is in the first quadrant. Thus, δ has to be $\frac{\pi}{4}$ (not $\frac{\pi}{4} + \pi$). That is to say,

$$\cos 3t + \sin 3t = \sqrt{2}\cos(3t - \tfrac{\pi}{4}).$$

Therefore, the amplitude of the harmonic is $\sqrt{2}$ and its phase shift is $\pi/4$. Notice that the harmonic being discussed happens to be a cosine function with period $2\pi/3$, frequency $\frac{3}{2\pi}$, and phase shift $\pi/4$, whose graph has the same shape as the graph of $\sqrt{2}\cos 3t$ but shifted $\frac{1}{3}\frac{\pi}{4}$ units to the right.

Next, suppose that we have to deal with the harmonic function $f(t) = -\sqrt{3}\cos t - \sin t$. Then $C = \sqrt{1+3} = 2$, $\cos \delta = -\frac{\sqrt{3}}{2} < 0$, and $\sin \delta = -1/2$. Thus, δ is in the third quadrant, consequently

$$\delta = \tan^{-1}\tfrac{1}{\sqrt{3}} + \pi = \tfrac{\pi}{6} + \pi = \tfrac{7\pi}{6}.$$

That is to say, $f(t) = 2\cos(t - \tfrac{7\pi}{6})$. We can observe that the amplitude of the harmonic is 2, while the phase shift is $7\pi/6$. It happens to be a cosine function with period 2π, frequency $\frac{1}{2\pi}$, and phase shift $7\pi/6$. In this example the shift to the right coincides with the phase shift.

In summary, any harmonic $A\cos wt + B\sin wt$ has angular frequency w, period $2\pi/w$, frequency $w/(2\pi)$, amplitude $\sqrt{A^2 + B^2}$, and the phase shift is obtained solving $\tan \delta = B/A$.

1.9   The Derivative of the Exponential Function

In section 1.3 we mentioned the function $x \to a^x$, where $a > 1$. It is an increasing function, defined on the whole real line and with range $(0, \infty)$. To calculate its derivative at an arbitrary point x we would have to calculate

$$\lim_{h\to 0} \frac{a^{x+h} - a^x}{h}.$$

We note that $\frac{a^{x+h} - a^x}{h} = a^x\frac{a^h - 1}{h}$. Thus, all we need to do is calculate

$$\lim_{h\to 0} \frac{a^h - 1}{h}.$$

Using a graphing calculator we get $\lim_{h\to 0} \frac{2^h - 1}{h} \approx 0.69314718$, $\lim_{h\to 0} \frac{3^h - 1}{h} \approx 1.0986123$ while, interestingly (Figure 1.35)[6],

---

[6]This property of the number e is extremely important, so much so that some authors use it to define e.

$$\lim_{h \to 0} \tfrac{e^h - 1}{h} = 1.$$

Hence

$$\lim_{h \to 0} \tfrac{e^{x+h} - e^x}{h} = e^x.$$

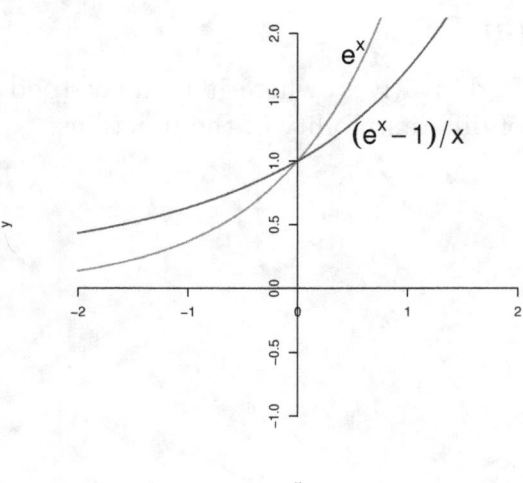

Figure 1.35: The exponential function and a closely related function.

Therefore

$$\frac{d}{dx} e^x = e^x. \tag{1.14}$$

That is to say, the derivative function of exp, where $\exp(x) = e^x$, is itself! Written in Newton's notation, $\exp' = \exp$. This fact has great significance in mathematics and its applications. In section 2.9 we will revisit the exponential function and will provide a thorough and complete mathematical argument, which does not depend on technology. We observe that the derivative of exp is never zero because e^x is always positive. We might also observe that

$$\frac{d}{dx} e^{-x} = \frac{d}{dx} \frac{1}{e^x} = \frac{-e^x}{(e^x)^2} = -\frac{1}{e^x} = -e^{-x}$$

while

$$\frac{d}{dx} e^{2x} = \frac{d}{dx} e^x e^x = e^x e^x + e^x e^x = 2e^{2x}.$$

In a similar fashion, we can show that $\frac{d}{dx} e^{-2x} = -2e^{-2x}$.

What can be said about $\frac{d}{dx} a^x$? Since $\ln 2 \approx 0.69314718$ and $\ln 3 \approx 1.0986123$ we can conclude that $\frac{d}{dx} 2^x = 2^x \ln 2$ and $\frac{d}{dx} 3^x = 3^x \ln 3$, and conjecture that for any $a > 1$ the equality

$$\frac{d}{dx}a^x = a^x \ln a$$

does hold, but we will have to wait until the next chapter in order to make sure that this conjecture is correct.

We might wonder, what is the derivative of $\log_a x$? An answer will be given in section 1.11 when $a = e$ and in section 2.10 for any $a > 1$.

Exercises for Section 1.9

Calculate the following ten derivatives, where it is understood that the derivations take place in the domain of differentiability of the functions:

1. $\frac{d}{dx}(e^x \sin x)$

2. $\frac{d}{dx}(x2^x + e^{-x}\sec x)$

3. $\frac{d}{dx}\frac{1}{e^x + x}$

4. $\frac{d}{dx}e^{-2x+5}$

5. $\frac{d}{dx}(x^2 + 1)e^{2x}$

6. $\frac{d}{dx}\frac{e^x}{x^2+6}$

7. $\frac{d}{dx}e^{3x}$

8. $\frac{d}{dx}\frac{\sin x}{e^x+1}$

9. $\frac{d}{dx}e^{3x}\cos x$

10. $\frac{d}{dx}\frac{e^{-x}\tan x}{x^2}$

11. Which points on the curve $y = e^x$ have the property that their corresponding tangent passes through $(-1, 0.1)$? (Hint: Once you have set the right equation, a graphing calculator will help you finish the problem.)

1.10   The Chain Rule

We are already acquainted with the formulas related to the derivative of the sum, difference, product, and quotient of functions. How about the derivative of the composition of two functions? Let us try to find the derivative of $\sqrt{\cos x}$, $-\pi/2 < x < \pi/2$, the composition of cos and the square root function; that is to say, $\sqrt{\cos x} = g(f(x)) = (g \circ f)(x)$ where $f(x) = \cos x$ and $g(x) = \sqrt{x}$. Assume that x_o is an arbitrary fixed number in the open interval $(-\pi/, \pi/2)$ and let x be any number in the above-mentioned interval. Then

$$\frac{\sqrt{\cos x}-\sqrt{\cos x_o}}{x-x_o} = \frac{\sqrt{\cos x}-\sqrt{\cos x_o}}{x-x_o}\frac{\sqrt{\cos x}+\sqrt{\cos x_o}}{\sqrt{\cos x}+\sqrt{\cos x_o}} = \frac{\cos x-\cos x_o}{x-x_o}\frac{1}{\sqrt{\cos x}+\sqrt{\cos x_o}}.$$

In the limit, as $x \to x_o$, the last expression converges to $\cos'(x_o)\frac{1}{2\sqrt{\cos x_o}}$. Keeping in mind that $\frac{d}{dx}\sqrt{x} = \frac{1}{2\sqrt{x}}$, it has been shown, in the particular example under consideration, that

$$(g \circ f)'(x) = g'(f(x))f'(x). \tag{1.15}$$

Let us go a little further and consider $(f(x))^2$, where f is any differentiable function. We are dealing with the composition of f and the function $g(x) = x^2$. As we did before, fix an arbitrary number x_o in the domain of differentiability of f. Then, for any x,

$$\frac{f(x)^2-f(x_o)^2}{x-x_o} = \frac{(f(x)+f(x_o))(f(x)-f(x_o))}{x-x_o} = \frac{f(x)-f(x_o)}{x-x_o}(f(x) + f(x_o)).$$

As $x \to x_o$, the last expression converges to $2f(x_o)f'(x_o)$. Once more, the validity of (1.15) has been checked. Remarkably, (1.15) is always true. We only have to make sure that g is differentiable at $f(x)$ and f is differentiable at x. For instance

$$\frac{d}{dx}\sin(x^3) = \cos(x^3) \cdot 3x^2,$$

$$\frac{d}{dx}(\sin x)^3 = 3(\sin x)^2 \cdot \cos x,$$

$$\frac{d}{dx}(x^2 + 1)^3 = 3(x^2 + 1)^2 \cdot 2x,$$

$$\frac{d}{dx}e^{\cos x} = e^{\cos x}(-\sin x),$$

$$\frac{d}{dx}e^{x^2+1} = 2xe^{x^2+1}.$$

The reader may observe that, when taking derivatives of the composition of functions, one has to proceed 'from the outside to the inside.' This practical way of applying (1.15) leads to consider it a rule, the 'chain rule' (a name of common usage). It should be noted that the rule can be applied to the composition of more than two functions. For instance,

$$\frac{d}{dx}\cos^2(\pi x) = 2\cos(\pi x)(-\sin(\pi x))\pi,$$

$$\frac{d}{dx}\sin(e^{x^2}) = \cos(e^{x^2})e^{x^2}2x.$$

In Leibniz's notation, the chain rule would be written

$$\frac{dy}{dx} = \frac{dy}{dt}\frac{dt}{dx} \tag{1.16}$$

where y is a function of t and t is a function of x. For instance, to calculate $\frac{d}{dx}\sin x^3$ let $y = \sin t$ and $t = x^3$. Then

$$\frac{dy}{dx} = \cos t(3x^2) = \cos(x^3) \cdot 3x^2.$$

In contrast to the formulas for the derivative of the sum, product, or quotient of two functions, the chain rule is not easy to prove in its most general context[7]. We will accept it and use it with great benefit, both in concrete examples and as a 'theoretical tool.' For instance, given any rational number r, let us try to justify the formula

$$\frac{d}{dx}x^r = rx^{r-1}. \tag{1.17}$$

This formula is a generalization of (1.13). Let $y(x) = x^{p/q}$, where p and q are integers. Then $y^q = x^p$ and by the chain rule we have $(y^q)' = qy^{q-1}y'$, which in turn leads to $qy^{q-1}y' = px^{p-1}$. But $y^{q-1} = (x^{p/q})^{q-1} = x^{p-p/q}$, consequently

$$y'(x) = \frac{p}{q}\frac{x^{p-1}}{x^{p-\frac{p}{q}}} = \frac{p}{q}x^{\frac{p}{q}-1}.$$

The preceding argument is not fully rigorous because we have accepted, without proof, the fact that the function $y(x) = x^{p/q}$ is indeed differentiable. Thus, we have only provided a plausible argument. In chapter 2 (section 11) we will analyze with care the behavior of x^r. **Let us accept that (1.17) holds true on $(0, \infty)$, r an arbitrary fixed real number**.

In forthcoming sections, especially in chapter 2, very often we will have to apply the chain rule in the process of solving problems. For the time being let us deal with the function $x(t) = 71 - 72.93e^{-0.23t}$, $1 \le t \le 10$, for the length of a specimen of fish (*Coregomus clupeaformis*), t in years and $x(t)$ in centimeters. In section 2.9 we will deduce the expression for $x(t)$ from data. Right now we are interested in finding the rate of growth at different times. Using the chain rule we obtain

$$x'(t) = -72.93(-0.23)e^{-0.23t} = 16.77e^{-0.23t}, \; 1 < t < 10.$$

For instance, $x'(2) = 10.59$ cm/year, while $x'(5) = 5.31$ cm/year. As expected, the rate of growth diminishes with time.

Another example linked to biology has to do with growth of plants (Figure 1.36). Suppose that $y(t) = ke^{-\cos(t/2)}$ describes the seasonal growth of a certain plant. What is its rate of growth? Using the chain rule we get

$$y'(t) = ke^{-\cos(t/2)}\sin(t/2)(1/2) = (k/2)e^{-\cos(t/2)}\sin(t/2).$$

---

[7]A proof can be found as an appendix at the end of the book.

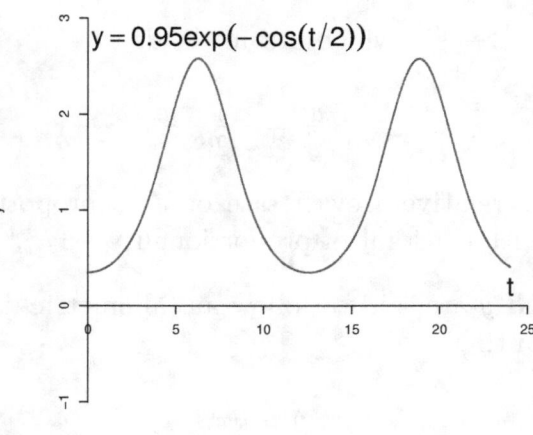

Figure 1.36: A seasonal growth function

Related rates

Often two or more variables are themselves functions of time, a fact that we have to take into consideration when doing the pertinent calculations involving rates. The problems that we will discuss in this subsection are generically called 'related rates problems.' All we have to do is proceed with care and apply the chain rule whenever needed.

1. Suppose we have a cylinder of fixed length L and variable radius $r(t)$, say an artery. Is there a simple relationship between the relative growth rates of r and V, $\frac{1}{V}\frac{dV}{dt}$ and $\frac{1}{r}\frac{dr}{dt}$? We have $V(t) = \pi L r^2(t)$, thus

$$\frac{dV}{dt} = L\pi 2r\frac{dr}{dt}.$$

Therefore

$$\frac{1}{\pi r^2 L}\frac{dV}{dt} = 2\frac{1}{r}\frac{dr}{dt},$$

which in turn leads to

$$\frac{1}{V}\frac{dV}{dt} = 2\frac{1}{r}\frac{dr}{dt}.$$

2. The adiabatic law for a gas, the law that governs the behavior of a gas that is expanding without gaining or losing heat, is given by the equation $P(t)V^\gamma(t) = k$, where γ and k are constants. Using the chain rule we obtain

$$\frac{dP}{dt}V^\gamma + P\gamma V^{\gamma-1}\frac{dV}{dt} = 0.$$

Therefore $\frac{dV}{dt} = -\frac{V^\gamma}{P\gamma V^{\gamma-1}}\frac{dP}{dt}$, which in turn leads to

$$\frac{1}{V}\frac{dV}{dt} = -\frac{1}{\gamma}\frac{1}{P}\frac{dP}{dt}.$$

That is to say, the relative growth rate of V is proportional to the relative growth rate of P, with constant of proportionality $-1/\gamma$.

3. Two variables x and y are said to follow an 'allometric' law if there exist constants k and α such that

$$y = kx^\alpha.$$

We will say more about allometry in section 1.22, where we will learn how to compare allometric models with data and thereafter estimate k and α. For the time being let us just stress the fact that, for instance, brain weight and body weight follow an allometric law, as well as limb length and body weight. Using the chain rule we get

$$\frac{dy}{dt} = k\alpha x^{\alpha-1}\frac{dx}{dt}.$$

Consequently $x\frac{dy}{dt} = k\alpha x^\alpha\frac{dx}{dt}$. Therefore

$$\frac{1}{y}\frac{dy}{dt} = \alpha\frac{1}{x}\frac{dx}{dt}.$$

We have just shown that the relative growth rates are proportional. It should be noted that the two previous examples, which dealt with arteries and gases, are particular cases of an allometric relationship because $V = \pi L r^2$ and $P = kV^{-\gamma}$.

4. Next let us suppose that a cylinder of radius a is empty, and at $t = 0$ water runs into it at a constant rate K through an opening at the top. If $V(t)$ denotes the volume and $h(t)$ the height of liquid inside the cylinder, we will have $V(t) = \pi a^2 h(t)$. According to the chain rule,

$$\frac{dV}{dt} = \frac{dV}{dh}\frac{dh}{dt}.$$

But $\frac{dV}{dt} = K$ and $\frac{dV}{dh} = \pi a^2$. Consequently $K = \pi a^2\frac{dh}{dt}$, which in turn leads to

$$\frac{dh}{dt} = \frac{K}{\pi a^2}.$$

As expected, the 'linear velocity,' $\frac{dh}{dt}$, is constant. That is to say, the rate at which the liquid is ascending is constant. Moreover, it does not depend on the original height of the cylinder.

5. There is a dramatic change of scenario when instead of a cylinder we have to deal with a cone. Assume that an inverted cone, of base radius a and height b, is empty and water starts to run into it, through an opening at the top, at the constant rate K. From the geometry of the cone we notice that

$$\frac{r(t)}{h(t)} = \frac{a}{b}$$

where $h(t)$ is the column of liquid and $r(t)$ is the variable radius of the cone formed by the liquid as it is ascending. Since $r(t) = \frac{a}{b}h(t)$ and $V(t) = \frac{1}{3}\pi r^2(t)h(t)$, we can conclude that

$$V(t) = \frac{\pi}{3}\frac{a^2}{b^2}h^3(t).$$

As in the problem with regard to a cylinder, the chain rule asserts that $\frac{dV}{dt} = \frac{dV}{dh}\frac{dh}{dt}$. Consequently

$$K = \frac{dV}{dt} = \frac{\pi a^2 h^2}{b^2}\frac{dh}{dt}.$$

Therefore

$$\frac{dh}{dt} = \frac{Kb^2}{\pi a^2}\frac{1}{h^2(t)}.$$

This time the linear velocity is not constant but inversely proportional to the square of the height of liquid inside the cone. In the early stages of the process, when $h(t)$ is relatively small, the linear velocity is large. However, as the process is coming to an end, $h(t)$ is becoming large and consequently the linear velocity diminishes.

Implicit differentiation

Let us consider the equation of a circle centered at the origin, namely $x^2 + y^2 = r^2$. The circle 'implicitly' defines two continuous functions:

$$y_1(x) = \sqrt{r^2 - x^2}, \ -r \le x \le r, \quad \text{and}$$

$$y_2(x) = -\sqrt{r^2 - x^2}, \ -r \le x \le r.$$

One is the upper part of the circle, the other is the lower part of the circle. We could take the derivative of each of them and obtain, through the chain rule, the following derivatives:

$$y_1' = \frac{-2x}{2\sqrt{r^2-x^2}} = \frac{-x}{\sqrt{r^2-x^2}}, \quad -r < x < r,$$

$$y_2'(x) = \frac{-2x}{-2\sqrt{r^2-x^2}} = \frac{x}{\sqrt{r^2-x^2}}, \quad -r < x < r.$$

There is an alternative approach that takes advantage of the fact that the relation $x^2 + y^2 = r^2$, although not a function, implicitly defines more than one function. We proceed to take the derivative, keeping in mind that x is the independent variable and y is the dependent variable. Indeed $2x + 2yy' = 0$, hence $y' = -\frac{x}{y}$ or, in Leibniz's notation, $\frac{dy}{dx} = -\frac{x}{y}$. For example, the slope m of the tangent to the circle $x^2 + y^2 = 17$ at $(1, -4)$ is $m = \frac{dy}{dx}\big|_{(1,-4)} = \frac{1}{4}$, where the symbol $\frac{dy}{dx}\big|_{(1,-4)}$ denotes the derivative, of y with respect to x, evaluated at the point $(-1, 4)$ on the corresponding curve. This is the technique of 'implicit differentiation.'

Let us use implicit differentiation to find the slope of the tangent at any point of an ellipse given by the equation

$$\frac{x^2}{a^2} + \frac{y^2}{b^2} = 1.$$

Right away we get

$$\frac{2x}{a^2} + \frac{2}{b^2}y\frac{dy}{dx} = 0.$$

Therefore

$$\frac{dy}{dx} = -\frac{b^2}{a^2}\frac{x}{y}.$$

Of course, when $y = 0$ we cannot use this formula. However, at both ends of the ellipse the tangent lines are $x = a$ and $x = -a$, respectively. The reader may realize that this approach takes considerably less work than trying to express y in terms of x at the beginning, and thereafter using the chain rule.

Implicit differentiation works well when dealing with any conic and its tangents, a topic of considerable interest in optics. Moreover, it can be applied to many other curves that are not conics. For instance, let us consider the curve (Figure 1.37) defined by the equation

$$y^2 + y = x^3.$$

We have $2y\frac{dy}{dx} + \frac{dy}{dx} = 3x^2$, consequently

$$\frac{dy}{dx} = \frac{3x^2}{2y+1}.$$

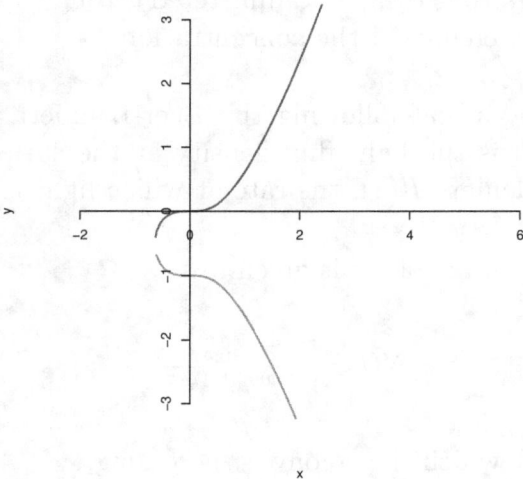

Figure 1.37: The curve $y^2 + y = x^3$

This formula allows us to calculate the slope of the tangent at any point of the curve, except when $y = -1/2$ (the curve has a vertical tangent when $y = -1/2$). It is to be noted that two functions are implicitly defined by $y^2 + y = x^3$, namely $y = -0.5 + 0.5\sqrt{1 + 4x^3}$ and $y = -0.5 - 0.5\sqrt{1 + 4x^3}$. They are obtained using the quadratic formula for equations.

Exercises for Section 1.10

In the following problems, use the chain rule whenever necessary.

1. $\frac{d}{dx}\cos(x^3 + 7)$, $\frac{d}{dx}\sqrt{x^2 + 1}$, $\frac{d}{dx}\frac{1}{\sqrt{4x^3 + 2}}$, $\frac{d}{dx}\cos(x^2 - 6)$.

2. $\frac{d}{dx}(\sin x)^4$, $\frac{d}{dx}\sin^3(\pi x + 6)$, $\frac{d}{dx}e^{5x}\cos x$, $\frac{d}{dx}(2x^2 + 3)^5$, $\frac{d}{dx}\tan(1/x)$.

3. $\frac{d}{dx}e^{x^2}$, $\frac{d}{dx}\sin(e^{x^2})$, $\frac{d}{dx}\sec x\tan(\sqrt{x})$, $\frac{d}{dx}\cos(\sin x)$, $\frac{d}{dx}x\cos(\frac{1}{x^2})$.

4. Suppose that $y(t) = ke^{-\cos(t/3)} + 1$, where k is a given constant. Calculate $y'(t)$.

5. Assume that the temperature (in the centigrade scale) in a certain day and place is given by the function $T(t) = 15 + 2\sin\frac{\pi}{15}t$ between 6 a.m. and 6 p.m. Calculate $T'(2)$, $T'(5)$, and $T'(8)$, that is, the rate of increase of temperature at 8 a.m., 11 a.m., and 2 p.m.

6. Assume that $f(t) = 5 + 10e^{-0.2t}\sin(1.2t)$, $0 \le t \le 5$, describes the concentration of a human hormone after a drug is administered through a vein. Calculate $f'(t)$, $0 < t < 5$, the rate of change of the concentration.

7. Light is absorbed by a leaf following the Beer-Lambert law, namely $I(z) = I(0)e^{-kz}$, where $I(0)$ is the light flux density at the surface of the leaf and z measures depth. Calculate $I'(z)$, the rate at which light is absorbed.

8. The growth of a bacterial colony is given by

$$N(t) = \frac{0.47}{0.009 + 1.47e^{-1.94t}}.$$

Calculate the rate at which the colony is increasing.

9. Suppose that two parts of the body of a mammal, say G and H, are related by the expression $G(t) = aH^{2.06}(t)$, at any time t. That is to say, $G(t)$ is proportional to $H^{2.06}(t)$. Calculate $G'(t)/G(t)$ in terms of $H'(t)/H(t)$.

10. The relationship $x^2 + y^3 = 2$ defines a function. Use the method of implicit differentiation to calculate $\frac{dy}{dx}$ at $(1, 1)$. Afterwards, express y in terms of x and find $y'(1)$. Which approach takes less work? Sketch a graph of the resulting curve.

11. Find the equation of the two tangent lines to the ellipse $\frac{x^2}{4} + y^2 = 1$, which pass through $(3, 0)$.

12. A curve is defined by the equation $y^2 = x^5$. Using implicit differentiation, calculate the equation of the tangent at $(1, 1)$ and sketch the curve.

13. Use implicit differentiation to calculate the slope of the tangent to the curve $\sin(xy) = \frac{1}{2}$ at $(1, \pi/6)$. Sketch a graph of the curve.

14. Suppose that a balloon, in the shape of a sphere, is being inflated in such a way that the rate at which its surface area changes with time is constant. Show that the rate at which the volume changes with time is proportional to the radius.

15. A gas is being compressed adiabatically ($\gamma = 1.2$). At a certain instant, $P = 50Kg/cm^2$, $V = 16cm^2$, and $\frac{dV}{dt} = -4cm^3/s$. Calculate $\frac{dP}{dt}$ at the above-mentioned instant.

1.11   The Derivative of the Natural Logarithm

First of all, let us accept the fact that if a function f is increasing or decreasing, and differentiable on an open interval I with $f'(x) \neq 0$ for every x in I, its inverse is also differentiable on the range of f. This is a reasonable assumption, called the theorem about the derivative of the inverse (**TDI**), whose proof can be found in an appendix at the end of the book[8].

For instance, let us consider the function $f(x) = x^2$, defined on $(0, \infty)$. It is increasing and differentiable on $(0, \infty)$ with $f'(x) \neq 0$ on $(0, \infty)$; indeed, $f'(x) = 2x$. Its inverse, $g(x) = \sqrt{x}$, is then differentiable on $(0, \infty)$. For any x in $(0, \infty)$ we have $f(g(x)) = x$. Using the chain rule, we get, for any $x > 0$,

$$f'(g(x))g'(x) = 1.$$

Thus $2\sqrt{x}g'(x) = 1$, which in turn leads to $g'(x) = \frac{1}{2\sqrt{x}}$. This way, we corroborate a fact known already from section 1.7, namely that $\frac{d}{dx}\sqrt{x} = \frac{1}{2\sqrt{x}}$. The technique that we just learned will be used next, and with great profit, to find the derivative of ln.

Let us recall that $\ln : (0, \infty) \to \Re$ is the inverse of $\exp : \Re \to (0, \infty)$, which is increasing and has a non-zero derivative. Then, TDI allows us to assert that ln is differentiable on $(0, \infty)$ (Figure 1.38). Since

$$\exp(\ln x) = x \text{ for every } x > 0$$

we can conclude that $\exp'(\ln x)\ln' x = 1$. But $\exp' = \exp$, hence $\exp(\ln x)\ln' x = 1$ for any $x > 0$. Therefore

$$\frac{d}{dx}\ln x = \frac{1}{x}, \; x > 0.$$

It should be noted that the function $x \to \ln(-x)$ is defined on $(-\infty, 0)$ (see Figure 1.39) and

$$\frac{d}{dx}\ln(-x) = \frac{1}{-x}(-1) = \frac{1}{x}, \; x < 0.$$

Putting together the preceding two derivatives we can write

$$\frac{d}{dx}\ln|x| = \frac{1}{x}, \tag{1.18}$$

where it is to be understood that the process of derivation takes place on an interval contained in either $(-\infty, 0)$ or $(0, \infty)$.

Using the chain rule we can obtain the derivative of many other functions that are compositions and involve the natural logarithm. For instance:

---

[8]Since every differentiable function is continuous, the range of f is an interval too. Moreover, since f is continuous and increasing (or decreasing), the range of f is also open; see enrichment note at the end of section 1.6.

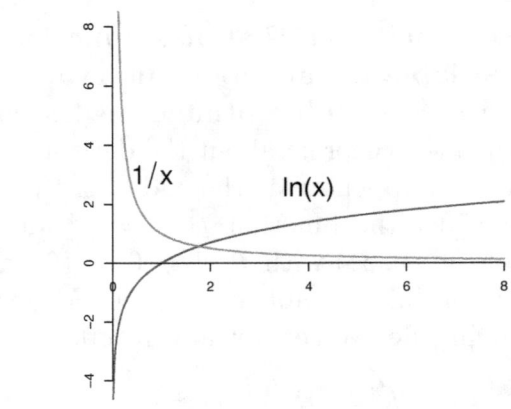

Figure 1.38: The natural logarithm and its derivative

$$\frac{d}{dx}\sin(\ln x) = \frac{\cos(\ln x)}{x}, \quad x > 0,$$

$$\frac{d}{dx}\ln(x^2) = \frac{1}{x^2}2x = \frac{2}{x}, \quad x > 0.$$

Exercises for Section 1.11

Calculate the following derivatives:

1. $\frac{d}{dx}\ln(\cos x)$, $-\pi/2 < x < \pi/2$.

2. $\frac{d}{dx}\cos(\ln x)$, $x > 0$.

3. $\frac{d}{dx}\ln(x^2 + 1)$.

4. $\frac{d}{dx}x(\ln x)^2$, $x > 0$.

5. $\frac{d}{dx}\frac{\ln x}{x}$, $x > 0$.

6. $\frac{d}{dx}\ln(\ln x)$, $x > 1$.

7. Find the point, on the curve $y = \ln x$, whose tangent passes through $(0,0)$.

8. Find the point, on the curve $y = \ln(-x)$, whose tangent passes through $(0,1)$.

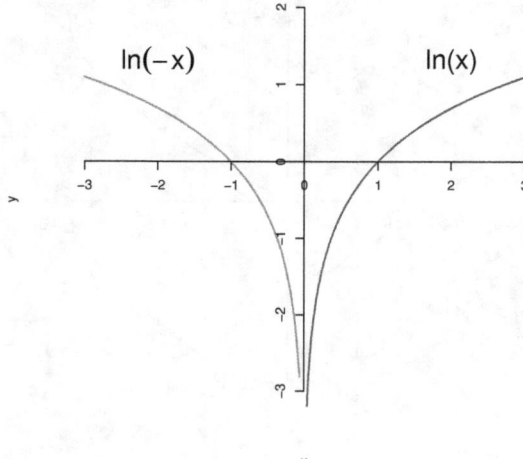

Figure 1.39: The natural logarithm and a closely related function

9. Suppose that a chemical reaction proceeds in accordance with the expression $\ln \frac{c(t)}{c(0)} = -kt$, where $c(t)$ denotes the concentration at any time t (k is a positive parameter, called the *kinetic constant* of the reaction). Show that $\frac{dc}{dt} = -kc(t)$; this means that the instantaneous rate of change of the concentration is proportional to the concentration itself.

10. The function $H(t) = 67.3 + 27.4 \ln(t)$ describes, approximately, the height of Japanese boys (1-13 years of age)[9]. Calculate $\frac{dH}{dt}$, the instantaneous rate at which height is changing across time ($1 < t < 13$). Furthermore, find $\frac{dH}{dt}|_{t=5}$ and $\frac{dH}{dt}|_{t=10}$. What do you notice?

1.12   Inverse Trigonometric Functions

The sine function is not one-to-one, thus it is not possible to define its inverse. However, we can restrict the sine function (sin) to $[-\pi/2, \pi/2]$. On this interval, sin is increasing; in particular it is one-to-one, its range is $[-1, 1]$ and $\sin'(x) = \cos x > 0$ on $(-\pi/2, \pi/2)$. Then its inverse $\sin^{-1} : [-1, 1] \to [-\pi/2, \pi/2]$ exists, is continuous and increasing on $[-1, 1]$, and differentiable on $(-1, 1)$ thanks to TDI (Figure 1.40). Quite often $\sin^{-1}$ is denoted arcsin.

---

[9]This function has been obtained from Table 1.2 (section 1.4), using the LnReg option of a graphing calculator.

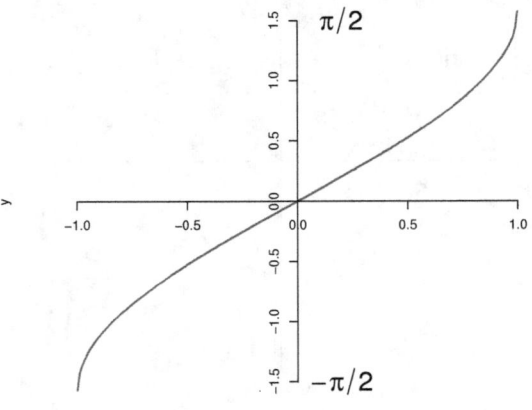

Figure 1.40: The inverse sine function

Since sin, restricted to $[-\pi/2, \pi/2]$, and arcsin are inverse functions, we can assert that $\sin(\arcsin x) = x$, $-1 \le x \le 1$. The chain rule then implies that $\sin'(\arcsin x) \arcsin'(x) = 1$ on $(-1, 1)$, that is, $\cos(\arcsin x) \arcsin'(x) = 1$. But

$$\cos(\arcsin x) = \sqrt{1 - \sin^2(\arcsin x)} = \sqrt{1 - x^2}.$$

Therefore

$$\arcsin'(x) = \frac{1}{\sqrt{1 - x^2}}, -1 < x < 1. \qquad (1.19)$$

This is a rather startling conclusion: the derivative of arcsin is not another trigonometric function!

The tangent function is not one-to-one either. However, if we restrict it to $(-\pi/2, \pi/2)$ the resultant function happens to be one-to-one, increasing, with range $(-\infty, \infty)$, and $\tan'(x) = \sec^2 x > 0$. Then its inverse $\tan^{-1} : (-\infty, \infty) \to (-\pi/2, \pi/2)$, also denoted arctan (Figure 1.41), is increasing and differentiable everywhere (we used TDI to reach the derivability of $\tan^{-1}$ on $(-\infty, \infty)$).

Since $\tan(\arctan x) = x$, $-\infty < x < \infty$, the chain rule leads to

$$\tan'(\arctan x) \arctan' x = 1 \text{ for every } x.$$

Thus $\sec^2(\arctan x) \arctan' x = 1$, which in turn implies $(1 + \tan^2(\arctan x)) \times \arctan' x = 1$. But $1 + \tan^2(\arctan x) = 1 + x^2$, consequently

$$\arctan' x = \frac{1}{x^2 + 1}, -\infty < x < \infty. \qquad (1.20)$$

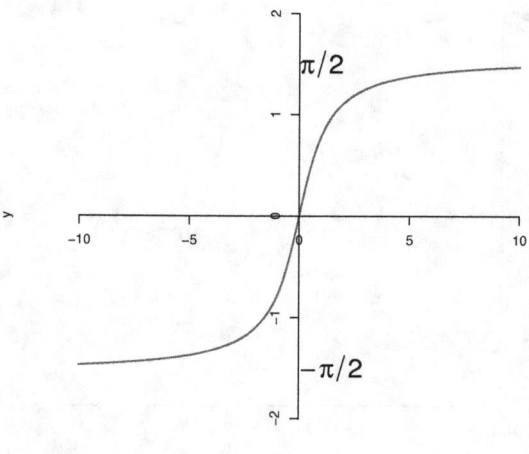

Figure 1.41: The inverse tangent function

The above-mentioned inverse trigonometric functions will play an important role in chapter 2.

We could also define arccos and $\sec^{-1}$. Indeed, restricting cos to the interval $[0, \pi]$ we get a one-to-one and onto function from $[0, \pi]$ to $[-1, 1]$. Its inverse arccos : $[-1, 1] \to [0, \pi]$ exists and is continuous on $[-1, 1]$, and differentiable on $(-1, 1)$ (see exercise 8 below). Similarly, if we restrict sec to $[0, \pi/2)$ we get a one-to-one and onto function from $[0, \pi/2)$ to $[1, \infty)$, consequently its inverse $\sec^{-1} : [1, \infty) \to [0, \pi/2)$ (also denoted arcsec) is well-defined and is continuous on $[1, \infty)$, and differentiable on $(1, \infty)$. Let us find an explicit formula for the derivative of $\sec^{-1}$:

Since $\sec(\sec^{-1} x) = x$, $x \geq 1$, the chain rule leads to

$$\sec'(\sec^{-1} x)(\sec^{-1})'(x) = 1, \quad x > 1.$$

Hence

$$\sec(\sec^{-1} x) \tan(\sec^{-1} x)(\sec^{-1})'(x) = 1.$$

Therefore

$$x\sqrt{\sec^2(sec^{-1}x) - 1}(\sec^{-1})'(x) = 1.$$

Consequently

$$x\sqrt{x^2 - 1}(\sec^{-1})'(x) = 1.$$

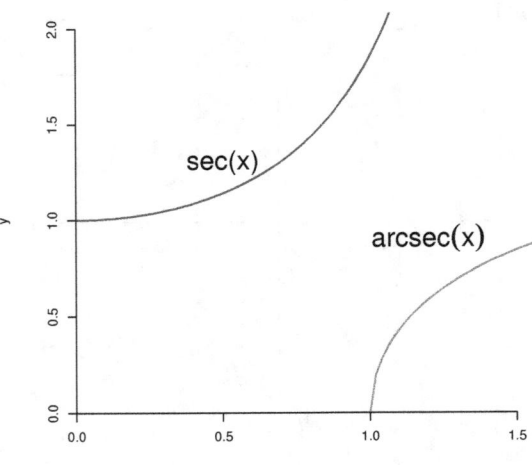

Figure 1.42: The inverse of the secant function

Finally we reach the interesting result that

$$\frac{d}{dx}\sec^{-1}(x) = \frac{1}{x\sqrt{x^2-1}}, \quad x > 1.$$

What happens when $x \le -1$? Under these circumstances it makes sense to write $\sec^{-1}(-x)$. In reality, we are dealing with the well-defined function $x \to \sec^{-1}(-x)$ defined on the interval $(-\infty, -1]$ (see Figure 1.43). Applying the chain rule we get

$$\frac{d}{dx}\sec^{-1}(-x) = \frac{1}{-x\sqrt{x^2-1}}(-1) = \frac{1}{x\sqrt{x^2-1}}, \quad x < -1.$$

Putting together the preceding two derivatives we can write

$$\frac{d}{dx}\sec^{-1}(|x|) = \frac{1}{x\sqrt{x^2-1}} \tag{1.21}$$

where it is to be understood that the process of derivation takes place on an interval contained in either $(1, \infty)$ or $(-\infty, -1)$; the reader may notice the striking similarity between (1.18) and (1.21). How could we compute $\sec^{-1}(x)$? It happens that $\sec^{-1}(x) = \cos^{-1}(1/x)$, $x \ge 1$, and $\sec^{-1}(-x) = \cos^{-1}(1/x)$, $x \le -1$ (exercise 10, section 1.15).

Now that we know the derivative of several inverse trigonometric functions, we can apply the chain rule in order to calculate many other derivatives. For instance:

$$\frac{d}{dx}\arcsin(\tfrac{1}{x}) = \frac{1}{\sqrt{1-\frac{1}{x^2}}} \cdot \frac{-1}{x^2}, \quad x < -1 \text{ or } x > 1,$$

$$\frac{d}{dx}\arctan(e^x) = \frac{e^x}{1+e^{2x}}.$$

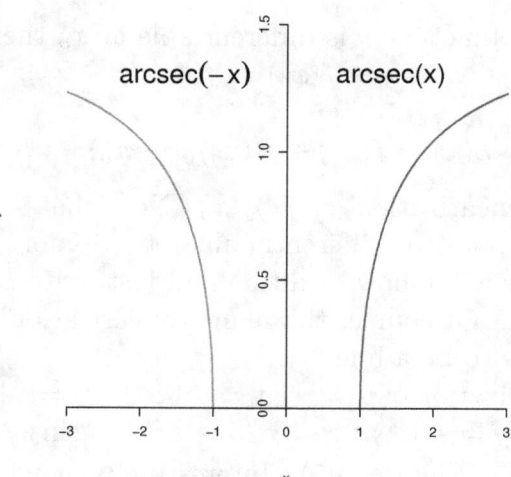

Figure 1.43: A closely related function of the inverse secant

Exercises for Section 1.12

1. Calculate $\frac{d}{dx}\arctan(x^2+1)$.

2. Calculate $\frac{d}{dx}\arcsin(e^x)$, $-\infty < x < 0$.

3. Calculate $\frac{d}{dx}\arccos(1/x)$, $x < -1$ or $x > 1$.

4. Calculate $\frac{d}{dx}\arctan(\ln(2x))$, $x > 0$.

5. Where does the curve $y = \arctan x$ have slope 1? In other words, find the point on the curve where the tangent to it has slope 1.

6. Calculate $\frac{d}{dx}x^2\arctan(x/3)$.

7. For what values of x does the tangent to the curve $y = \arctan x$ have slope α, $0 < \alpha \le 1$?

8. Show that $\arccos'(x) = \frac{-1}{\sqrt{1-x^2}}$, $-1 < x < 1$. Sketch a graph of $\arccos(x)$.

9. Show that $\frac{d}{dx}\arccos(\frac{1}{x}) = \frac{d}{dx}\sec^{-1}(x)$, $x > 1$.

1.13   Linearization

We have learned that if a function f is differentiable at x_o then the equation of the tangent line to $y = f(x)$ at x_o is $y - f(x_o) = f'(x_o)(x - x_o)$. We can now define a new function $L(x)$,

$$L(x) = f(x_o) + f'(x_o)(x - x_o). \tag{1.22}$$

It receives the name of 'linearization' of f at x_o. Note that L depends not only on f but also on the chosen point of differentiability x_0. A more appropriate notation would be L_{x_o}; however, we will follow tradition and just write L, keeping in mind the point x_o being considered. Of course, the name 'linearization' makes a lot of sense because $y = L(x)$ happens to be a line.

For instance, the linearization of $y = x^2$ at $x = 1$ is $L(x) = 1+2(x-1)$ (Figure 1.44) while the linearization of $y = \sin x$ at $x = \pi/3$ is $L(x) = \sin \pi/3 + \cos \pi/3(x - \pi/3)$, i.e., $L(x) = \frac{\sqrt{3}}{2} + \frac{1}{2}(x - \frac{\pi}{3})$ (Figure 1.45). Interestingly enough, the linearization of $y = \arctan x$ at the origin is $L(x) = x$ since $\arctan 0 = 0$ and $\arctan'(0) = 1$, exactly the same linearization of $y = \sin x$ at the origin.

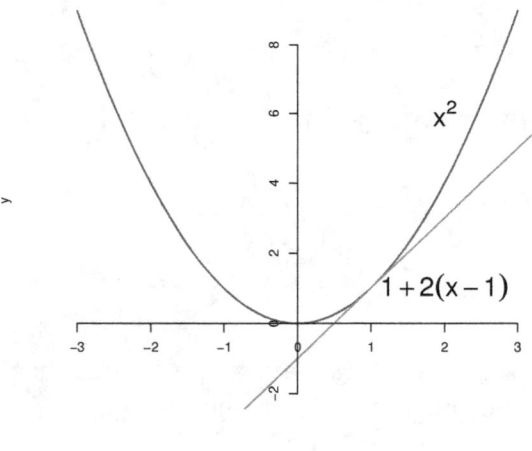

Figure 1.44: The linearization of the parabola at $x = 1$

Due to the fact that the tangent line to $y = f(x)$ at x_0 is very close to the curve when x is not far from x_0, we should expect that $L(x) \approx f(x)$ in a small neighborhood of x_o. Far away from x_o, $L(x)$ can be very different from $f(x)$.

Other important linearizations, at the origin, are $L(x) = \frac{1}{2}x + 1$ (corresponding to $\sqrt{1+x}$), $L(x) = x + 1$ (corresponding to e^x), and $L(x) = x$ (corresponding to

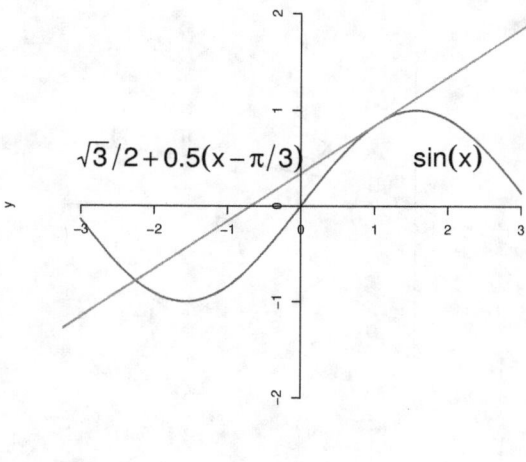

Figure 1.45: The linearization of the sine at $x = \pi/3$

$\ln(1+x)$). The last one (Figure 1.46) is particularly important in biochemistry (see section 4.5) because it allows us to replace $\ln(1 + \frac{S(t)-S_o}{S_o})$ by $\frac{S(t)-S_o}{S_o}$ at the beginning of the steady-state, when the amount of substrate $S(t)$ differs very little from the initial amount S_o.

In section 1.17 we will see that it is possible to obtain a 'better' approximation of a function if, instead of a line, one works with a parabola.

Exercises for Section 1.13

1. What is the linearization of $y = \cos x$ at $x = 0$ and at $x = \pi/4$?

2. Find the linearization of $y = x^n$ at $x = c$, where c is an arbitrary real number.

3. Find the linearization of $y = \ln x$ at $x = 1$.

4. Find the linearization of $y = \tan x$ at $x = \pi/4$.

5. What is the linearization of $y = e^{-ax}$ at $x = 0$? (a is an arbitrary non-zero constant). Sketch the curves corresponding to $y = e^{-ax}$ when $a > 0$ and $a < 0$, as well as their corresponding linearization.

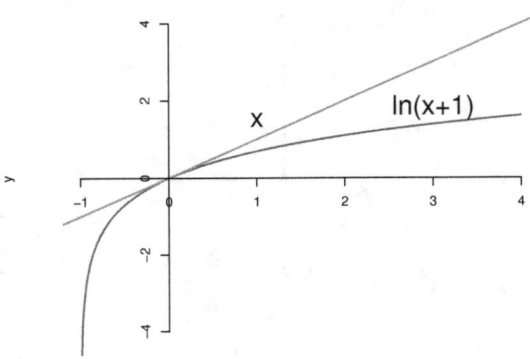

Figure 1.46: The linearization of $\ln(x+1)$ at $x = 0$

1.14   Newton's Method

The notion of derivative as the slope of the corresponding tangent was used by Newton to find an approximation to a root of a polynomial function $y = f(x)$. Newton's method is historically relevant and exemplifies one of the many applications of the fact that the derivative of a function at a point is the slope of the tangent at that point. We start by choosing a so that $f(a) > 0$ and b so that $f(b) < 0$. Since f is continuous, the curve $y = f(x)$ will cross the horizontal axis at some point between a and b. Let $x_o = a$. The tangent line to the curve $y = f(x)$ is

$$y - f(x_o) = f'(x_o)(x - x_o).$$

This line will cross the x-axis at a point x_1. Thus $0 - f(x_o) = f'(x_o)(x_1 - x_o)$, that is to say

$$x_1 = x_o - \frac{f(x_o)}{f'(x_o)}.$$

Next we consider the tangent line at $(x_1, f(x_1))$, namely $y - f(x_1) = f'(x_1)(x - x_1)$. Where does it cross the x-axis? It will do so at a point x_2 such that $0 - f(x_1) = f'(x_1)(x_2 - x_1)$. Then

$$x_2 = x_1 - \frac{f(x_1)}{f'(x_1)}.$$

A pattern is emerging. We repeat this process n times and reach the formula

$$x_n = x_{n-1} - \frac{f(x_{n-1})}{f'(x_{n-1})}.$$

Thus, we have a 'recursive formula.' The process starts with x_o, then we calculate x_1, thereafter x_2, and so on. Usually x_5 will be an excellent approximation to a root when dealing with a polynomial equation. Moreover, the method is 'robust' in the sense that it is not that important what initial x_o we choose, provided that the starting point is not too far away from the actual zero . Of course, there is the danger that, at some step, $f'(x_i) = 0$ and then the recursive process breaks down. In the unlikely event that such a thing happens, we just choose another x_o and the problem is usually fixed.

Let us try to find a positive solution of the polynomial equation $x^4 + x^2 + x - 1 = 0$ (obviously, -1 is a negative solution). With this purpose in mind we define the function

$$f(x) = x^4 + x^2 + x - 1.$$

Since $f(0) = -1 < 0$ and $f(1) = 2 > 0$ we can choose $x_o = 1$ as a starting point of the method. The recursive process becomes

$$x_n = x_{n-1} - \frac{x_{n-1}^4 + x_{n-1}^2 + x_{n-1} - 1}{4x_{n-1}^3 + 2x_{n-1} + 1}.$$

For instance

$$x_1 = x_o - \frac{x_o^4 + x_o^2 + x_o - 1}{4x_o^3 + 2x_o + 1} = 1 - \frac{2}{7} = \frac{5}{7} \approx 0.71428571,$$

$$x_2 = \frac{5}{7} - \frac{(\frac{5}{7})^4 + (\frac{5}{7})^2 + \frac{5}{7} - 1}{4(\frac{5}{7})^3 + 2(\frac{5}{7}) + 1} \approx 0.58954024,$$

and so on (Figure 1.47).

The calculations become lengthy, thus it is convenient to build a simple program with a graphing calculator (TI-89 or its equivalents):

```
: newt( )
: Prgm
: Input "x0", x
: Input "n", n
: For i,1,n
: x − y1(x)/y2(x) → x
: Disp x
: Endfor
: EndPrgm
```

For the specific function under consideration, namely $f(x) = x^4 + x^2 + x - 1$, we have to write $y1(x) = x^4 + x^2 + x - 1$ and $y2(x) = 4x^3 + 2x + 1$. Choosing $x_o = 1$ and

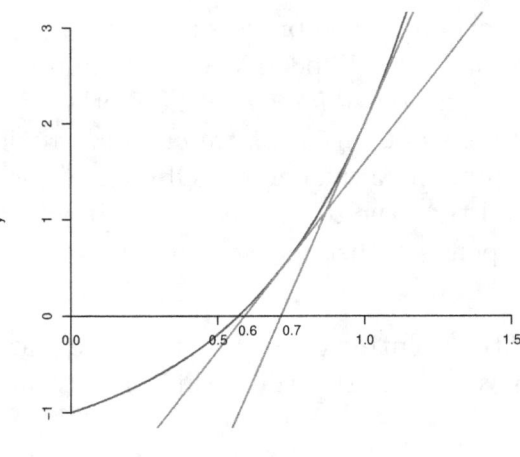

Figure 1.47: The first two steps of Newton's method

$n = 5$, and writing newt( ) under the Home option, once we press Enter we get five numbers on the screen: 0.71428571, 0.58954024, 0.57023363, 0.56984045, 0.56984029. The last one is precisely x_5. If we were to use the zero option of our calculator, we would get exactly the same number. If we were to choose $x_o = 2$ then we can get close to the root too, but we need to increase n, say to $n = 7$.

Is this too much work to approximate a root of a polynomial of fourth degree? A purely algebraic approach for quartics was developed by Lodovico Ferrari (1522-1565) and appeared in Gerolamo Cardano's *Ars Magna* (1545), but it is much longer than Newton's method (Helfgott and Helfgott, 2009). Newton's method can be applied to any differentiable function, not necessarily a polynomial. For instance, suppose we wish to solve the equation $\arctan x = x - 1$. With this purpose in mind we define

$$f(x) = \arctan x - (x - 1).$$

We observe that $f(1) = \arctan(1) = \pi/4 > 0$ while $f(2) = \arctan(3) - 2 \approx -0.75095423 < 0$. Thus, we can choose $x_o = 1$ and $n = 5$, and load our calculator with $y1(x) = \arctan x - (x - 1)$ and $y2(x) = 1/(x^2 + 1) - 1$. After writing newt( ) under the Home option and pressing Enter we will get five numbers on the screen: 2.5707963, 2.1436902, 2.13222786, 2.1322677, and 2.1322677. The last two numbers are identical, letting us know that we have achieved the 'true' first six decimals (always the last decimal, provided by a calculator, is an approximation to the irrational number being sought.) With confidence we can assert that the solution of the equation $\arctan x = x - 1$ is approximately 2.1322677. Of course, we can check this answer with the powerful Solve command under the F2 option of the calculator. Indeed, Solve $(\arctan x = x - 1, x) = 2.1322677$; no wonder because many calculators use a variation of Newton's method to solve equations of this kind.

Exercises for Section 1.14

In problems 1-5 you need to use the program *Newt*.

1. Calculate by hand a solution of $x^3 - x + 1 = 0$ using Newton's method (choose $n = 3$ and $x_o = -1$). Compare your answer with that obtained using the program *Newt*. Use the latter to calculate the first 5 'true' decimals of the solution.

2. Solve the equation $\cos x = x$. Your solution is expected to be correct to seven decimal places.

3. Find two negative solutions of $x^4 - 3x^2 + x + 1 = 0$ (the first six decimals should be 'true.')

4. Suppose that you wish to find a solution of $x^4 - 4x + 2 = 0$ and, with this purpose in mind, choose $x_o = 1$ and n=5. What happens? Change the selection of x_o in order to use Newton's method successfully.

5. Find the two solutions of $\sin(x^2) = x^2 - 3$. You are expected to find the first five 'true' decimals.

1.15 The Mean Value Theorem

Suppose that $f : [a, b] \to \Re$ is continuous on $[a, b]$ and differentiable on (a, b). Then there exists at least one point θ in (a, b) such that the slope of the tangent to the graph of f at $(\theta, f(\theta))$ is equal to the slope of the line that passes through $(a, f(a))$ and $(b, f(b))$ (Figure 1.48). That is to say, there exists θ in (a, b) such that

$$f'(\theta) = \frac{f(b) - f(a)}{b - a}.$$

This is the statement of the 'Mean Value Theorem' (MVT for short). There is compelling geometrical evidence with regard to the validity of MVT, so we will not provide a proof of it in this section. The interested reader can consult the appendix, at the end of the book, for a proof.

First of all, we should realize that MVT does not provide an explicit value θ. It merely says that it exists, although in some particular cases one might be able to explicitly find one or more values for θ. For instance, let $f(x) = x^2$, $-1 \le x \le 2$. Then $f'(x) = 2x$. Solving the equation

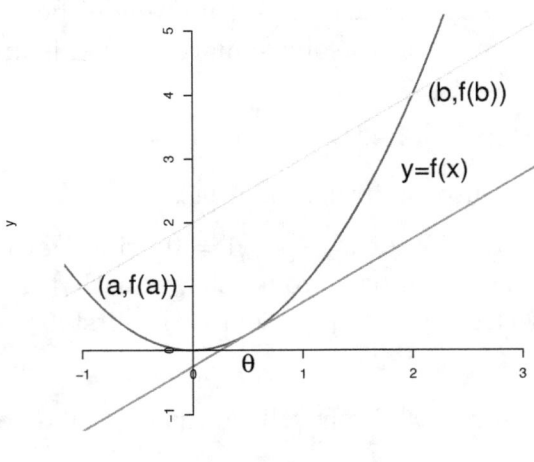

Figure 1.48: The Mean Value Theorem

$$2x = \frac{2^2 - (-1)^2}{2 - (-1)} = \frac{4-1}{3} = 1$$

we get $x = 1/2$. Thus, $\theta = 1/2$ does the job.

The point θ does not have to be unique. Indeed, let $f(x) = x^3$, $-1 \le x \le 1$. Then $f'(x) = 3x^2$. The equation

$$3x^2 = \frac{1^3 - (-1)^3}{1 - (-1)} = \frac{1+1}{1+1} = 2$$

has two solutions, namely $x = \pm\sqrt{2/3}$. Thus $\theta = \sqrt{2/3}$ or $\theta = -\sqrt{2/3}$ do the job.

We should stress that if we drop the hypothesis of continuity on $[a, b]$, or differentiability on (a, b), the thesis of MVT does not have to be true. For instance, let $f(x) = x$, $0 \le x < 1$, and $f(1) = 0$. This function is differentiable on $(0, 1)$ but not continuous on $[0, 1]$ (it loses continuity at $x = 1$). There is no point θ in $(0, 1)$ such that

$$f'(\theta) = \frac{f(1) - f(0)}{1 - 0} = 0.$$

Why? Because $f'(x) = 1$ for every x in $(0, 1)$. On the other hand, $g(x) = \sqrt{x}$, $0 \le x \le 1$, $g(x) = \sqrt{-x}$, $-1 \le x \le 0$ (Figure 1.49), is continuous on $[-1, 1]$ but not differentiable on $(-1, 1)$ (g loses differentiability at the origin). Clearly

$$g'(x) = \frac{1}{2\sqrt{x}}, 0 < x < 1$$

$$g'(x) = \tfrac{-1}{2\sqrt{-x}}, -1 < x < 0$$

while

$$\tfrac{g(1)-g(-1)}{1-(-1)} = \tfrac{1-1}{2} = 0.$$

We can see that there is no θ in $(-1, 1)$ such that $g'(\theta) = 0$.

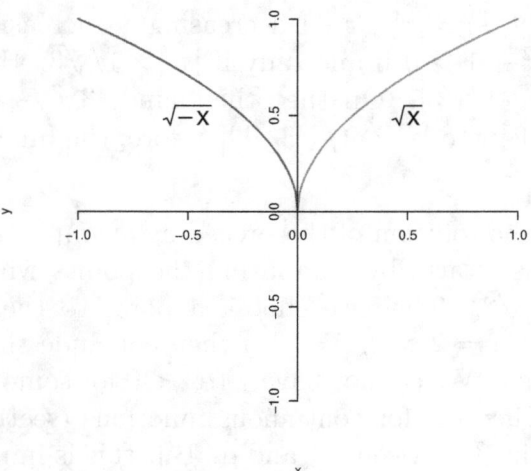

Figure 1.49: A special function

Some consequences of MVT

Interestingly enough, seldom we will use MVT as such but we will use four very important consequences that stem from it. Let f be a function defined on an interval I, in such a way that it is continuous on I and differentiable on $int(I)$ (the interior of I). Let us see the first three consequences, also called corollaries, of MVT.

(i) If $f'(x) > 0$ on $int(I)$ then f is increasing on I.
(ii) If $f'(x) < 0$ on $int(I)$ then f is decreasing on I.
(iii) If $f'(x) = 0$ on $int(I)$ then f is constant on I.

Why is this so? Suppose $x_1 < x_2$, where x_1 and x_2 lie on I. We apply MVT to f restricted to $[x_1, x_2]$ and are able to conclude that

$$f(x_2) - f(x_1) = f'(\theta)(x_2 - x_1) \tag{1.23}$$

for some θ in (x_1, x_2). A proof of (i), (ii), and (iii) follows almost immediately from (1.23). For instance, assume that $f'(x) > 0$ on $int(I)$. In particular $f'(\theta) > 0$, hence $f'(\theta)(x_2 - x_1) > 0$. Therefore $f(x_2) > f(x_1)$.

An example will help us understand the power of the above-mentioned corollaries of MVT. Let $f(x) = x^3 + 2x^2$ for every x in the real line. Where is it increasing? Where is it decreasing? We have $f'(x) = 3x^2 + 4x = x(3x + 4)$, hence $f'(x) > 0$ whenever $x < -4/3$ or $x > 0$, and $f'(x) < 0$ for $-4/3 < x < 0$. Therefore f is increasing on $(-\infty, -4/3]$ and $[0, \infty)$, while it is decreasing on $[-4/3, 0]$.

Where is the function $f(x) = x^3 - x + 1$ decreasing or increasing? We observe that $f'(x) = 3x^2 - 1$, thus $3x^2 - 1 > 0$ if and only if $|x| > 1/\sqrt{3}$. Hence, f is increasing on $[1/\sqrt{3}, \infty)$ and $(-\infty, -1/\sqrt{3}]$. On the other hand, $3x^2 - 1 < 0$ if and only if $x^2 < 1/3$, which is equivalent to $|x| < 1/\sqrt{3}$. Therefore, the function is decreasing on $[-1/\sqrt{3}, 1/\sqrt{3}]$.

Another approach to the solution of the preceding example, which can be applied whenever f' is continuous, starts by computing the points where the derivative is zero[10], namely $x = \pm 1/\sqrt{3}$. Then we choose a point to the left of $-1/\sqrt{3}$, say -1, and realize that $f'(-1) = 2 > 0$. We can then conclude that $f'(x) > 0$ for any $x < -1/\sqrt{3}$. Why is this so? We cannot have $f'(\alpha) < 0$ for some $\alpha < -1/\sqrt{3}$ because the Intermediate Value Theorem for continuous functions (section 1.6) would imply that $f'(\beta) = 0$ for a certain β between -1 and α. But this is impossible since $-1/\sqrt{3}$ and $1/\sqrt{3}$ are the only critical points. By the same token, since $f'(0) = -1 < 0$, we can conclude that $f'(x) < 0$ for every x in $(-1/\sqrt{3}, 1/\sqrt{3})$ and since $f'(1) = 2 > 0$ it follows that $f'(x) > 0$ for every $x > 1/\sqrt{3}$. The whole procedure consists in making a list of the critical points $c_1, ..., c_n$ and evaluating $f'(x)$ on a specific arbitrary number at each interval between the critical points, and to the left of c_1 and the right of c_n. Analyzing whether f' is either positive or negative will lead to the conclusion that f is either increasing or decreasing.

How about the function $g(x) = e^{ax}$, with $a > 0$? Since $g'(x) = ae^{ax}$ we can conclude that g is increasing due to the fact that $a > 0$ and $e^{ax} > 0$. Actually, $\lim_{x \to \infty} e^{ax} = \infty$ because, for any $M > 0$, $x > \frac{1}{a} \ln M$ implies $ax > \ln M$, which in turn leads to $e^{ax} > e^{\ln M} = M$. On the other hand, $h(x) = e^{-ax}$ is decreasing, if $a > 0$, since $h'(x) = -ae^{-ax}$ (recall that the exponential function always adopts positive values). Moreover, $\lim_{x \to \infty} e^{-ax} = \lim_{x \to \infty} \frac{1}{e^{ax}} = 0$.

One further important result stems from MVT:

(iv) If $f'(x) = g'(x)$ on $int(I)$ then there exists a constant K such that $f(x) =$

---

[10]Interior points where the derivative function adopts the value zero are called 'critical points,' a concept that will play an important role in sections 1.18 through 1.21.

$g(x) + K$ for every x in I.

That is to say, **if the derivatives of two functions are equal on** $int(I)$ **then the functions differ by a constant**. It should not be difficult to prove that (iv) is an immediate corollary of (iii). Indeed, we need only apply (iii) to the function $f - g$.

The third consequence of MVT is particularly important in the applications of mathematics to the natural sciences. Suppose that a strain of bacteria grows in such a way that its rate of growth is proportional to the amount present or, what is the same, the 'relative rate of growth' is constant:

$$x'(t) = \alpha x(t),$$

where $\alpha > 0$ is the constant of proportionality. What curve does $x(t)$ describe? We observe that $x'(t) - \alpha x(t) = 0$, consequently

$$e^{-\alpha t}(x'(t) - \alpha x(t)) = 0.$$

We could have multiplied both sides of $x'(t) - \alpha x(t) = 0$ by any function. However, we chose $e^{-\alpha t}$ because, thanks to the derivative of a product and the chain rule,

$$e^{-\alpha t}(x'(t) - \alpha x(t)) = (e^{-\alpha t}x(t))'.$$

Therefore $(e^{-\alpha t}x(t))' = 0$. By (iii) above, we can conclude that there exists a constant C such that $e^{-\alpha t}x(t) = C$ for any t. Consequently $x(t) = Ce^{\alpha t}$. In particular, at $t = 0$ we have $x(0) = C$, hence

$$x(t) = x(0)e^{\alpha t}.$$

We have reached the conclusion that the acceptance of $x'(t) = \alpha x(t)$ implies that $x(t)$ grows exponentially, namely $x(t) = x(0)e^{\alpha t}$. Of course, one can easily check that $x(t) = x(0)e^{\alpha t}$ has he property that $x'(t) = \alpha x(t)$. Due to its importance, we will return to the topic of exponential growth in section 2.15. Under the constraint of space and the competition for nutrients, bacteria do not usually grow exponentially but follow a 'sigmoidal' curve characteristic of logistic growth, a type of growth that will be analyzed in detail in section 2.16.

The fourth consequence of MVT also plays a very important role in the solution of some differential equations: Suppose that we wish to solve $y'(x) = \frac{1}{x^2+1}$, $y(0) = 2$. We observe that if there is a solution $y(x)$ then

$$y'(x) = \arctan'(x)$$

for any x. Then there exists a constant C such that $y(x) = \arctan(x) + C$ for all x. In particular, $2 = y(0) = C$. Consequently

$$y(x) = \arctan(x) + 2$$

for every x. The reader can check that this function does satisfy the differential equation as well as the given initial condition.

Let us analyze another example. Suppose that it is given that $y'(x) = 6x + 9$ and we wish to find $y(x)$. Since $(3x^2 + 9x)' = 6x + 9$ we can conclude that $y'(x) = (3x^2 + 9x)'$, then by corollary (iv) it follows that there exists a constant C such that $y(x) = 3x^2 + 9x + C$ for all x. The reader can verify that for *any* constant C the function $x \rightarrow 3x^2 + 9x + C$ satisfies the property that its derivative is the function $x \rightarrow 6x + 9$. If we were to add an initial condition, say $y(0) = 2$, then $C = 2$ and $y(x) = 3x^2 + 9x + 2$ happens to be the only function whose derivative is $x \rightarrow 6x + 9$ and adopts the value 2 at $x = 0$.

Remark

Interestingly, MVT can be generalized in the following sense: Assume that f and g are defined on $[a, b]$, where they are continuous. Furthermore, f and g are differentiable on (a, b) and $g'(x) \neq 0$ for every x in (a, b). Then there exists θ in (a, b) such that

$$\frac{f(b) - f(a)}{g(b) - g(a)} = \frac{f'(\theta)}{g'(\theta)}.$$

This theorem is called the **Generalized Mean Value Theorem** (GMVT for short), which will be used in section 3.14 and in the appendix B.6 on a proof of one of L'Hôpital's Rules , and can be proven without much effort. Indeed, let $h(x) = f(x) - \lambda g(x)$, where λ is chosen in such a way that $h(a) = h(b)$. Evidently, we must choose

$$\lambda = \frac{f(b) - f(a)}{g(b) - g(a)}.$$

There is no danger that $g(b) = g(a)$ because, if it were true, MVT would imply that there exists α in (a, b) such that $g'(\alpha) = 0$; but, by hypothesis, $g'(x) \neq 0$ for every x in (a, b). Next we apply MVT to the function $h(x)$, which is continuous on $[a, b]$ and differentiable on (a, b). Then there exists θ in (a, b) such that $0 = h(b) - h(a) = h'(\theta)(b - a)$, consequently $h'(\theta) = 0$. But $h'(\theta) = f'(\theta) - \lambda g'(\theta)$, therefore $f'(\theta) = \lambda g'(\theta)$; that is, $f'(\theta)/g'(\theta) = (f(b) - f(a))/(g(b) - g(a))$, as we wished to show.

Exercises for Section 1.15

1. Where is the function $f(x) = x^3 - x + 1$ increasing? Where is it decreasing?

2. Suppose that $f'(x) = \cos x$ for every x and there exists another function $g(x)$ such that $g'(x) = f'(x)$ for every x. Furthermore, $g(0) = \pi$. Can you find $g(x)$ explicitly?

3. Let $g(x) = x^3 + 1$ for every x. Where is $g(x)$ increasing?

4. Let $f(x) = -\arctan x$ for every x. Is it increasing? Is it decreasing?

5. Show that $h(x) = \ln x$ is increasing on its domain of definition. Where is $f(x) = \frac{\ln x}{x}$ increasing? Where is it decreasing? Sketch a graph of f.

6. Let $f(x) = x^4$, $-2 \le x \le 2$. Does there exist θ such that $f(2) - f(-2) = f'(\theta)(2 - (-2))$? Can you find θ explicitly?

7. In section 1.7 (exercise 12) we dealt with two strains of bacteria, x and y, that grow in the same environment and satisfy the differential equations $x'(t) = \alpha x(t)$, $y'(t) = \beta y(t)$. Let $z(t)$ denote the proportion of x at any instant t, that is,

$$z(t) = \frac{x(t)}{x(t) + y(t)}.$$

Calculate $\lim_{t \to \infty} z(t)$ under the assumption that $\alpha > \beta$. Does your answer coincide with your intuition about the problem at hand? (Hint: Recall that $\lim_{t \to \infty} e^{-\gamma t} = 0$ whenever $\gamma > 0$.)

8. The model $x'(t) = -\alpha x(t)$ is used to study radioactive decay, where $x(t)$ is the amount of radioactive substance at any instant t and α is a positive parameter. Calculate $x(t)$ in terms of the exponential function, $x(0)$, and α.

9. The half-life of a radioactive process is defined as $\tau = \ln 2/\alpha$, where α is the parameter of the process. Show that $x(\tau) = x(0)/2$.

10. Show that $\arccos \frac{1}{x} = arcsec(x)$, $x \ge 1$, and $\arccos \frac{1}{x} = arcsec(-x)$, $x \le -1$ (Hint: Verify that the derivative of both functions are the same).

11. Suppose that $f'(t) = \cos t + \sin t$ and $f(0) = 1$. Calculate $f(t)$.

12. Is it true that if $f : I \to \Re$ is increasing, and differentiable on $int(I)$, then $f'(x) > 0$ for every x in $int(I)$? Provide a proof or a counterexample (the converse of this statement is the first consequence of MVT, which was shown to be true).

13. Solve the differential equation $x'(t) + 2x(t) = 0$, with initial condition $x(0) = 5$.

14. Show that $\sec^{-1} x = \arctan \sqrt{x^2 - 1}$, $x \geq 1$ (Hint: Verify that $\frac{d}{dx}\sec^{-1}(x) = \frac{d}{dx}\arctan\sqrt{x^2 - 1}$, $x > 1$.)

15. Use what you have learned about parabolas in section 1.1 in order to find where the function $f(x) = x^3 + 3x^2 + x + 1$ is increasing or decreasing.

16. Where is the function $f(x) = \frac{1}{(x^2+1)\arctan x}$ increasing or decreasing?

1.16   Antiderivatives

Newton's second law, applied to a body of mass m under the force of gravity, disregarding the resistance of air, asserts that

$$mv'(t) = -mg$$

where g is the acceleration due to gravity at a certain fixed point on Earth and $v(t)$ is the velocity of the body[11]. Thus, $v'(t) = -g$, which in turn implies $v'(t) = (-gt)'$. Corollary (iv) from the previous section implies that there exists a constant L such that

$$v(t) = -gt + L \text{ for every } t \geq 0.$$

In particular $v_o = v(0) = L$, hence

$$v(t) = -gt + v_o. \tag{1.24}$$

That is to say, $s'(t) = -gt + v_o$ for any $t \geq 0$, where $s(t)$ denotes the position of the body. By simple inspection we realize that $(-\frac{1}{2}gt^2 + v_o t)' = -gt + v_o$, consequently

$$s'(t) = (-\tfrac{1}{2}gt^2 + v_o t)'.$$

Next, once again we apply corollary (iv) and conclude that, for some constant C,

$$s(t) = -\tfrac{1}{2}gt^2 + v_o t + C \text{ for every } t \geq 0.$$

In particular $s_o = s(0) = C$, therefore

$$s(t) = -\frac{1}{2}gt^2 + v_o t + s_o \tag{1.25}$$

at any time $t \geq 0$, where v_o is the initial velocity of the body and s_o is its initial position.

---

[11]The force of gravity is preceded by a negative sign because we are selecting the upwards direction as the positive one.

Many problems about kinematics, under the sole influence of gravity, can be solved using (1.24) and (1.25). For instance, let us suppose that an object is thrown upwards with an initial velocity v_o. Question: Is the time, t_1, it takes to reach its highest point the same as the time, t_2, it takes to reach the ground starting from the highest point? From (1.24) we get $0 = -gt_1 + v_o$, hence $t_1 = v_o/g$. Next we use (1.25) and obtain

$$s\left(\frac{v_o}{g}\right) = -\frac{1}{2}g\frac{v_o^2}{g^2} + v_o\frac{v_o}{g} = \frac{v_o^2}{2g}.$$

Using (1.25), once more, we get

$$0 = -\frac{1}{2}gt_2^2 + \frac{v_o^2}{2g}.$$

Consequently $t_2 = v_o/g$, which is the same value adopted by t_1.

The novelty about the problem that we just discussed resides on the fact that given a function f we mentally found another function F such that $F' = f$. Let us recall that given $f(t) = -g$ we realized that $F(t) = -gt$ has the property that $F' = f$, while given $f(t) = -gt + v_o$, the function $F(t) = -\frac{1}{2}gt^2 + v_o t$ has the property that $F' = f$. Finding an F such that $F' = f$ was very important because we had in mind applying corollary (iv) from the previous section.

Definition

Given a function f, defined on an interval I, an **antiderivative** of f is any differentiable function F defined on I and such that $F' = f$ on int(I).

For instance, the function $F(t) = \frac{t^5}{5}$ is an antiderivative of $f(t) = t^4$, while $G(t) = \frac{1}{3}e^{3t} + 7t^2$ is an antiderivative of $g(t) = e^{3t} + 14t$. Antiderivatives of a given f are certainly not unique. Indeed, if we can find one, say F, then $F(x) + C$ will also be an antiderivative of f. Moreover, any two antiderivatives differ by a constant. Why? Assuming $F' = f$ and $G' = f$ we get $F' = G'$, then by the fourth corollary of the Mean Value Theorem there exists a constant K such that $F(t) = G(t) + K$ for every t.

An example from chemical kinetics will help us to realize how important the notion of antiderivative is. A model of second order kinetics tells us that

$$c'(t) = -kc^2(t) \tag{1.26}$$

where $c(t)$ is the concentration of the substance under consideration and k is a parameter. Can we obtain a mathematical consequence in order to compare it with data on the variation of concentration over time? The first thing that comes to our mind is to write (1.26) as

$$c'(t)c^{-2}(t) = -k.$$

Is it possible to find an antiderivative of the function to the left and an antiderivative of the function to the right of the equal sign of the preceding equation? Obviously $(-kt)' = -k$, but it is more difficult to find an antiderivative of $c'(t)c^{-2}(t)$. After some 'guessing and checking,' we might conclude, thanks to the chain rule, that

$$(-c^{-1}(t))' = c'(t)c^{-2}(t).$$

Therefore,

$$(-c^{-1}(t))' = (-kt)'.$$

Hence, there exists a constant L such that

$$-\tfrac{1}{c(t)} = -kt + L \text{ for every } t \geq 0.$$

In particular, the preceding expression is true for $t = 0$. Consequently $-\tfrac{1}{c_o} = L$, which in turn leads to

$$-\tfrac{1}{c(t)} = -kt - \tfrac{1}{c_o}.$$

That is to say,

$$\frac{1}{c(t)} = kt + \frac{1}{c_o}. \tag{1.27}$$

Thus, a prediction of the model is that when we work with t on the horizontal axis and $1/c(t)$ on the vertical axis, the experimental points $(t, 1/c(t))$ will gather around a straight line. Using the 'line of best fit,' to be discussed in section 1.22, we will be able to estimate k. In chapter 2 we will analyze a systematic procedure to deal with problems of this and similar types. Indeed, we will revisit (1.26) in section 2.12.

First list of antiderivatives

Table 1.3 provides a first list of antiderivatives; it is to be understood that, in the first entry, n is a rational number different from -1 (the arbitrary constant C stems from the fourth corollary of MVT). Moreover, let us not forget the intervals where each of the functions are differentiable. The second entry is valid on any interval that does not contain the number zero, while in the first entry (when $n < 0$) the interval cannot contain the zero either. On the other hand, the third, fourth, seventh, and ninth entries are valid everywhere, but the eighth entry is valid only on $(-1, 1)$ and the tenth entry is valid only on $(1, \infty)$ or $(-\infty, -1)$. Finally, the fifth and sixth entries are valid on any interval where $\tan x$ and $\sec x$ are defined; that is to say, on any interval where the cosine function does not adopt the value zero.

We can check the validity of Table 1.3 by taking the derivative of each function from the second column. In section 2.18 we will present an enlarged list of antiderivatives.

Table 1.3: First list of antiderivatives

$f(x)$	$F(x)$
x^n	$x^{n+1}/(n+1) + C$
$\frac{1}{x}$	$\ln|x| + C$
$\cos x$	$\sin x + C$
$\sin x$	$-\cos x + C$
$\sec^2 x$	$\tan x + C$
$\sec x \tan x$	$\sec x + C$
$1/(x^2 + 1)$	$\arctan x + C$
$1/\sqrt{1 - x^2}$	$\arcsin x + C$
e^x	$e^x + C$
$1/(x\sqrt{x^2 - 1})$	$\sec^{-1}(|x|) + C$

Initial value problems

Equations like $v'(t) = -g$ or $c'(t) = -kc(t)^2$ are examples of first order differential equations. The unknown is a function, which is 'hidden' under a derivative. That is to say, we are given information about the derivative function and have to find the function itself. Often the differential equation is accompanied by an initial condition, say $v(0) = v_o$ or $c(0) = c_o$, in which case we are dealing with an **initial value problem** (IVP). Many models of the natural sciences appear as IVPs.

Another example of first order differential equation is $x'(t) + ax(t) = b$. From a strictly mathematical point of view, the technique of multiplying both sides of the above-mentioned differential equation by e^{at} is extremely useful, as we foreshadowed in the previous section. Actually, this technique sets the path to solve any equation of the type $x'(t) + ax(t) = b$, where a and b are constants. Such equations often appear in the study of electric circuits and mechanics. For instance, let us solve the differential equation $x'(t) + 3x(t) = 5$ assuming the initial condition $x(0) = 1$. Multiplying both sides of the equation by e^{3t}, we get $e^{3t}(x'(t) + 3x(t)) = 5e^{3t}$. But, thanks to the formula for the derivative of a product and the chain rule, $e^{3t}(x'(t) + 3x(t)) = (e^{3t}x(t))'$ and[12] $5e^{3t} = (\frac{5}{3}e^{3t})'$, therefore $(e^{3t}x(t))' = (\frac{5}{3}e^{3t})'$. The fourth corollary of MVT implies that there exists a constant C such that

$$e^{3t}x(t) = \tfrac{5}{3}e^{3t} + C \text{ for every } t.$$

Consequently $x(t) = \frac{5}{3} + Ce^{-3t}$ for every t. In particular, $1 = x(0) = \frac{5}{3} + C$; hence, $x(t) = \frac{5}{3} + (1 - \frac{5}{3})e^{-3t}$ for any t. The reader can verify that this function satisfies the given differential equation and the initial condition. In chapter 2, once we learn about integration techniques, an alternative approach will be discussed.

---

[12]In essence, we are doing a mental calculation that involves the idea of antiderivative.

Exercises for Section 1.16

1. Suppose that a stone is thrown up, with an initial velocity of 20 ft/s, from the top of a 100 ft building. How long will it take to reach its highest point? What will be its velocity when it reaches the ground? ($g = 32$ ft/s^2).

2. Assume that an individual is driving a car at a velocity v_o and applies the brakes, producing a constant deceleration K (we are neglecting forces of friction and damping); thus, $v'(t) = -K$. Let d be the distance traveled by the car before it comes to a stop. Find a formula relating d, v_o, and K (Hint: Take into consideration that the velocity of a car is zero when it comes to a stop).

3. Starting from rest, a light jet airplane can reach a take-off velocity of 195 mph in 30 seconds. What should be the minimum length of the runway? (Hint: Work with the differential equation $v'(t) = K$.)

4. A stone is dropped into a well and, after three seconds, one hears it hitting the bottom. Calculate d, the depth of the well, assuming that the only force acting on the stone is gravity ($g = 9.8 m/s^2$).

5. Solve the preceding problem if one also takes into consideration the speed of sound (340 m/s)(Hint: Note that $3 = t_1 + t_2$, where t_1 is the time it takes the stone to reach the bottom of the well and $t_2 = d/340$.)

6. Assume that a chemical reaction obeys second order kinetics, i.e., $c'(t) = -kc^2(t)$. Find τ such that $c(\tau) = \frac{c_0}{2}$ (τ is known as the 'half-life' of the reaction).

7. Find the family of all antiderivatives of $f(x) = 5\sin 3x$. Do the same for $g(x) = \cos \pi x + x^2 + 2$.

8. Is there a function $f(x)$ such that $f'(x) = 3\sin(\pi x)$ and $f(0) = 2$?

9. Find the solution to the IVP $y'(x) = \frac{1}{x^2+1} + 3$, $y(0) = \frac{\pi}{4}$.

10. Given that $y'(x) = xe^{x^2} + \frac{1}{x^2+1}$ and $y(0) = \pi$, find $y(x)$.

11. Given that $y'(x) = x^2 \sin(x^3) + \frac{x}{x^2+1}$, $y(0) = 1$, find $y(x)$.

12. At the end of section 1.15 we learned how to solve differential equations of the type $x'(t) = \alpha x(t)$. Follow the same procedure (not the end result) in the quest of the solution of the initial value problem $x'(t) = 3x(t)$, $x(0) = 7$.

13. Given $f(x) = \nu e^{-\nu x}$ (ν a constant), let $F(x)$ be an antiderivative of $f(x)$ such that $F(0) = 0$. Find $F(x)$ explicitly.

14. Solve the differential equation $x'(t) + 2x(t) = 3$ with initial condition $x(0) = 5$. Also solve $x'(t) + 3x(t) = 2$, $x(0) = 1$.

15. Let $f(x) = |x|$, $-1 < x < 1$. Find an antiderivative F of f on $(-1, 1)$. Then graph F and observe that it is 'smooth,' in the sense that it does not have a peak.

16. An object of mass m, dropped from a considerable height, is governed by the differential equation $mv'(t) = mg - cv(t)$ ($-cv(t)$ is the force due to the resistance of air). Assuming that the initial velocity is zero, calculate $v(t)$.

17. Find the family of all antiderivatives of $\frac{1}{\sin^2(x)}$.

1.17 Second Order Derivatives

If $f(x) = x^n$ then $f'(x) = nx^{n-1}$ for every x. $f'(x)$ is a new function, so we may write $g = f'$ and take the derivative of g. Obviously $g'(x) = n(n-1)x^{n-2}$. Usually g' is denoted f'' or, in Leibniz's notation, $\frac{d^2}{dx^2} f$. It is called the 'second derivative function.' For instance

$$\frac{d^2}{dx^2} \sin x = \frac{d}{dx} \cos x = -\sin x,$$

$$\frac{d^2}{dx^2} e^x = \frac{d}{dx} e^x = e^x,$$

$$\frac{d^2}{dx^2} \arctan x = \frac{d}{dx} \frac{1}{x^2+1} = -\frac{2x}{(x^2+1)^2}.$$

Although the second derivative of the sine function resembles the original function, and the second derivative of e^x is itself, the second derivative of the inverse tangent function is quite different from either the original function or its first derivative.

The number $f''(c)$, when it exists, is called 'the second derivative of f at c.' We should keep in mind that for $f''(c)$ to exist it is indispensable that f' be defined on a neighborhood of c and that

$$\lim_{x \to c} \frac{f'(x) - f'(c)}{x - c}$$

exists as a real number. Luckily, most common functions one encounters in a first calculus course are twice differentiable. As a matter of fact, they are often 'infinitely differentiable' in the sense that $f^{(n)}$ exists for any natural number n, where $f^{(n)} = (f^{(n-1)})'$.

The second derivative has a physical meaning when $s(t)$ denotes the position of an object: $s''(t)$ is the acceleration, the first derivative of the velocity. For instance, if

$s(t) = 3\cos \pi t$ then the velocity function is $s'(t) = -3\pi \sin \pi t$ while the acceleration function is $s''(t) = -3\pi^2 \cos \pi t$. It could happen that the acceleration is known and it is necessary to find the space function, a task that can be handled readily with the help of the notion of antiderivative learned in the previous section. For example, suppose that $s''(t) = \sin t$. Right away we note that $s''(t) = (-\cos t)'$, consequently $s'(t) = -\cos t + C$ for some constant C. But $v_o = s'(0) = -1 + C$, therefore $s'(t) = -\cos t + (v_o + 1)$. Consequently,

$$s'(t) = (-\sin t + (v_o + 1)t)'.$$

Hence, there exists a constant D such that $s(t) = -\sin t + (v_o + 1)t + D$ for every $t \geq 0$. Finally we make $t = 0$ and get $s_o = s(0) = D$. We have arrived to the answer, namely

$$s(t) = -\sin t + (v_o + 1)t + s_o \quad \text{for every } t \geq 0.$$

We are dealing with an initial value problem, this time associated with the second order differential equation $s''(t) = \sin t$ and two conditions (namely, $s'(0) = v_o$ and $s(0) = s_o$). IVPs of this type have just one solution, as their first order counterparts.

As we will see next, second order derivatives also play an important role in the quest of a parabola that may approximate a function close to a given point. In section 1.13 we learned that given a function f, differentiable at the origin, the tangent line to f at $x = 0$ is given by the equation $y - f(0) = f'(0)x$, which is a good approximation for values of x very close to the origin:

$$f(x) \approx f(0) + f'(0)x.$$

We observe that if $f(x) = a_o + a_1 x$ close to the origin, and $f'(0)$ exists, then $f(0) = a_o$ and $f'(0) = a_1$. Hence the first order polynomial becomes $f(0) + f'(0)x$, as expected. The next step, quite natural, is to assume that f can be approximated, close to the origin, by a second order polynomial; that is to say, $f(x) = a_o + a_1 x + a_2 x^2$ whenever $x \approx 0$. If $f'(0)$ and $f''(0)$ exist, then $a_o = f(0)$, $a_1 = f'(0)$, and $a_2 = f''(0)/2$ (it is enough to take the first and second derivative of f and make $x = 0$). Therefore

$$f(x) \approx f(0) + f'(0)x + \frac{f''(0)}{2}x^2 \text{ whenever } x \approx 0.$$

This second order polynomial happens to be a 'better' approximation *vis a vis* $f(0) + f'(0)x$, in the sense that the parabola is closer to $y = f(x)$ on a wider interval around the origin. In the framework of the theory of series, a topic to be studied in chapter 3, these qualitative ideas will be made more precise. In the meantime we settle for a heuristic approach.

For instance, let $f(x) = \ln(x+1)$, $x > -1$. Then $f'(x) = 1/(x+1)$ and $f''(x) = \frac{-1}{(x+1)^2}$. Thus $f'(0) = 1$ and $f''(0) = -1$. Consequently

$$\ln(x+1) \approx x \text{ and } \ln(x+1) \approx x - \tfrac{1}{2}x^2,$$

which are good approximations whenever x is very close to the origin. We may observe (Figure 1.50), that the parabola $y = x - 0.5x^2$ lies closer to $y = \ln(x+1)$ around the origin (when compared to the line $y = x$). Next let us analyze the function $f(x) = e^x$.

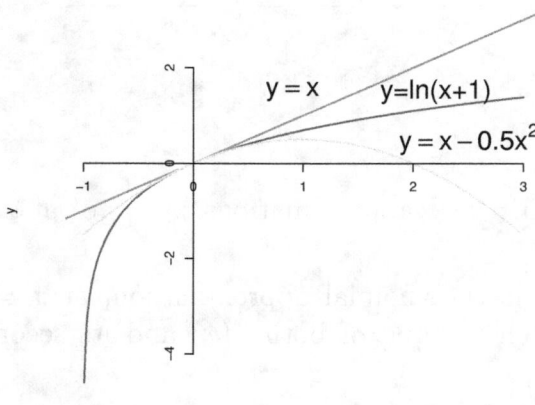

Figure 1.50: Two approximations to $y = \ln(x+1)$ at $x = 0$

Right away we can conclude that

$$e^x \approx 1 + x \text{ and } e^x \approx 1 + x + \tfrac{1}{2}x^2$$

whenever $x \approx 0$. We should keep in mind that the parabola $y = 1 + x + \frac{1}{2}x^2$ is a better approximation than the line $y = 1 + x$ (Figure 1.51).

Exercises for Section 1.17

1. Given $y(x) = \ln x$, calculate $y''(x)$. Do the same for $y(x) = e^{3x} + \cos(2x)$.

2. Suppose that $y''(t) = \sin(2t)$ and $y(0) = 1$, $y'(0) = -1$. Find $y(t)$.

3. Find all the functions $y(t)$ such that $y''(t) = e^{-t} + \cos(\pi t)$.

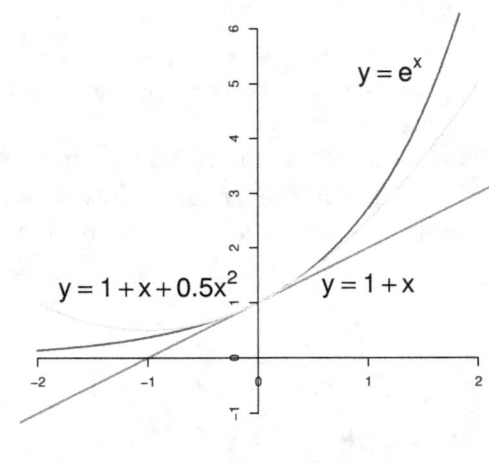

Figure 1.51: Two approximations to $y = e^x$ at $x = 0$

4. Find the second order polynomial approximation, at $x = 0$, of the function $f(x) = \cos x$. Sketch a graph of both $f(x)$ and its second order polynomial approximation.

5. What is the second order polynomial approximation, at $x = 0$, of the function $f(x) = \sec x$?

6. Find the function $y(x)$ such that $\frac{d^2y}{dx^2} = -x + \sin x$ and $y(0) = 1$, $\frac{dy}{dx}(0) = -2$.

7. Find all the functions $y(t)$ such that $\frac{d^2y}{dt^2} = 3t + \cos t$, and $y(0) = 0$.

8. Is there a function $y(t)$ such that $\frac{d^2y}{dt^2} = 3t + \cos t$ and $y(0) = 0$, $y(\pi) = 0$?

9. According to the law of universal gravitation, any object of mass m that starts from the surface of the earth with an initial velocity v_o, and travels vertically, is governed by the differential equation

$$my''(t) = \frac{-GMm}{(y(t)+R)^2}$$

where G is the constant of universal gravitation, M is the mass of the earth, R its radius, and $y(t)$ is the vertical position of the object with regard to the surface of the earth. Thus,

$$y''(t) = \frac{-K}{(y(t)+R)^2}$$

where $K = GM$. Find an expression that involves only $y'(t)$, $y(t)$ and the known constants R, K, v_o (Hint: Multiply both sides of the above-mentioned differential equation by $y'(t)$ and thereafter seek the corresponding antiderivatives, keeping in mind the chain rule).

10. If you are careful, the answer to the previous exercise should be

$$\tfrac{1}{2}v(t)^2 = \frac{K}{y(t)+R} + \tfrac{1}{2}v_o^2 - \frac{K}{R}.$$

After a certain time t the object will reach its maximum height y_{max}. Therefore $0 = \frac{K}{y_{max}(v_o)+R} + \tfrac{1}{2}v_o^2 - \frac{K}{R}$, where we have written $y_{max}(v_o)$ to emphasize the fact that the maximum height is a function of the initial velocity. Find a number c such that $\lim_{v_o \to c} y_{max}(v_o) = \infty$ (c is known as the 'escape velocity,' that is, c is the minimum velocity needed to escape earth's gravitational force).

11. Define a function that is twice differentiable everywhere, but its second derivative function is not differentiable at $x = 0$.

1.18 Maxima and Minima I

Let f be a function continuous on $[a, b]$ and differentiable on (a, b). As we saw in section 1.6, f will adopt its maximum and minimum on $[a, b]$. Let us assume that the maximum is adopted at an interior point c; this means that $a < c < b$. Under these circumstances, the tangent at c is parallel to the horizontal axis, which yields $f'(c) = 0$. In a similar fashion, if the minimum is attained at an interior point d we must have $f'(d) = 0$ (see Figure 1.22). This is an important mathematical proposition, known as **Fermat's Theorem** in honor of Pierre de Fermat, one of calculus' pioneers. Actually, mathematicians usually state Fermat's Theorem as follows: Suppose f is differentiable at an interior point c and f has a **local maximum or a local minimum** at c; that is to say, there exists $\delta > 0$ such that $f(c) \geq f(x)$ for every x in $(c - \delta, c + \delta)$ or $f(c) \leq f(x)$ for every x in $(c - \delta, c + \delta)$. Then $f'(c) = 0$ (a detailed proof can be found in an enrichment note at the end of the section). We hope that the reader will accept Fermat's Theorem on the basis of its pictorial interpretation: at an interior maximum or minimum, or at a local maximum or minimum, if the function is differentiable then the tangent line will be parallel to the x-axis. Obviously, if a maximum or minimum is attained at an interior point, this maximum or minimum is also local.

Points in (a, b) where the derivative is zero are called **critical points**. Thus, the maximum of f is adopted either at a, b or at a critical point. Similarly, the minimum is adopted either at a, b, or at a critical point. We have then a well-established procedure to calculate the maximum or the minimum.

For instance, let $f(x) = x^3 - x + 1$, $-2 \leq x \leq 1$. We note that $f'(x) = 3x^2 - 1$, $-2 < x < 1$. In order to find the critical points we need to solve the equation $3x^2 - 1 = 0$. Obviously, the two critical points are $1/\sqrt{3}$ and $-1/\sqrt{3}$. All that remains to do is to calculate f at -2, 1, $1/\sqrt{3}$, and $-1/\sqrt{3}$ and make a decision. Indeed, $f(-2) = -5$, $f(1) = 1$, $f(1/\sqrt{3}) \approx 0.62$, and $f(-1/\sqrt{3}) \approx 1.39$. Therefore, the maximum is adopted at $-1/\sqrt{3}$ and the minimum is adopted at -2 (Figure 1.52). The next three examples deal with real life problems.

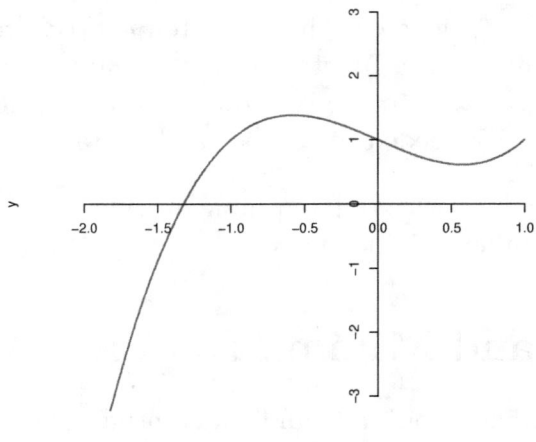

Figure 1.52: The function $f(x) = x^3 - x + 1$ on $[-2, 1]$

Applications

1. Assume that the temperature (in the centigrade scale) in a certain day and place is given by

$$T(t) = 20 + 3\sin \tfrac{\pi}{10} t$$

between 6 a.m. and 7 p.m. At what time does the temperature achieve its maximum? We note, right away, that $T(t)$ is defined on the interval $[0, 13]$. Applying the chain rule we obtain $T'(t) = \tfrac{3\pi}{10} \cos \tfrac{\pi}{10} t$. The critical points are found by solving the equation

$$0 = \tfrac{3\pi}{10} \cos \tfrac{\pi}{10} t.$$

Evidently, $\frac{\pi}{10}t = \frac{\pi}{2}$; thus, $t = 5$ (the possibility $\frac{\pi}{10}t = \frac{3\pi}{2}$ has to be discarded because it leads to $t = 15$, a number outside the domain of $T(t)$). We have $T(0) = 20$, $T(5) = 23$ and $T(13) = 20 + 3\sin\frac{13\pi}{10} \approx 17.57$, hence the maximum is attained at $t = 5$, that is, at 11 a.m. (Figure 1.53).

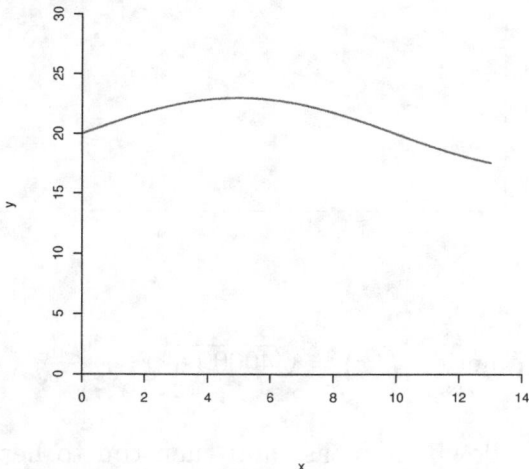

Figure 1.53: The function $T(t) = 20 + 3\sin\frac{\pi}{10}t$, $0 \le x \le 13$

2. A woman is swimming in a lake, at a location situated 200 meters from the shore, and wishes to reach her destination B, 500 meters down the coast, in the least amount of time. Suppose that she can swim at 1 m/sec and run at 5 m/sec. Where should she land? Let x be the unknown. Since time=distance/velocity we have to build the function

$$f(x) = \sqrt{40,000 + x^2} + \frac{500-x}{5} \quad 0 \le x \le 500.$$

The critical points are found by solving the equation

$$0 = f'(x) = \frac{2x}{2\sqrt{40,000+x^2}} - \frac{1}{5}.$$

Consequently $25x^2 = 40,000 + x^2$, which in turn leads to $x = \sqrt{\frac{40,000}{24}} \approx 40.82$ meters. This is the only critical point. Since $f(0) = \sqrt{40,000} + 100 = 300$ seconds, $f(500) = \sqrt{40,000 + 250,000} \approx 538.52$ seconds, $f(40.82) = \sqrt{40,000 + 40.82^2} + \frac{500-40.82}{5} \approx 295.96$ seconds, we can assert that the best option is to swim towards

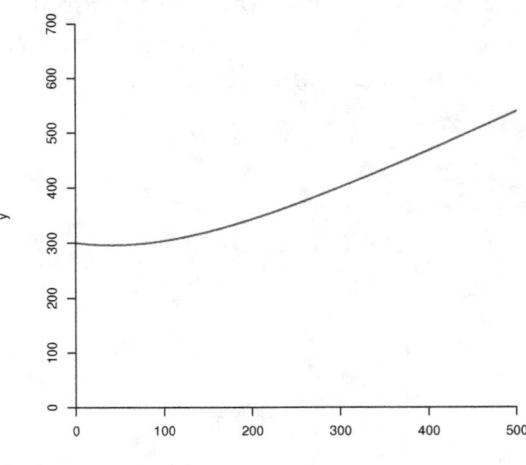

Figure 1.54: The function $f(x) = \sqrt{40000 + x^2} + \frac{500-x}{5}$, $0 \le x \le 500$

a point 40.82 meters down the coast and then run to her destination (Figure 1.54). Altogether it will take the swimmer $295.96/60 \approx 4.93$ minutes.

3. As the third example will illustrate, the number of critical points can be more than one. Let us assume that

$$f(t) = 10 + 15e^{-0.3t}\sin(1.7t) \qquad 0 \le t \le 6 \text{ (t in hours)}$$

describes the concentration of a human hormone after a certain drug is administered through a vein (Figure 1.55). When does the concentration reach its maximum or minimum? We observe that

$$f'(t) = 15[-0.3e^{-0.3t}\sin(1.7t) + 1.7e^{-0.3t}\cos(1.7t)].$$

The critical points are found by solving the equation $0 = -0.3\sin(1.7t) + 1.7\cos(1.7t)$, that is to say

$$\tan(1.7t) = \frac{1.7}{0.3}.$$

Therefore $t_1 = \frac{1}{1.7}\tan^{-1}\frac{1.7}{0.3} \approx 0.8212$ happens to be the first critical point. Since

$$\tan(1.7t) = \tan(1.7t + \pi) = \tan(1.7(t + \frac{\pi}{1.7}))$$

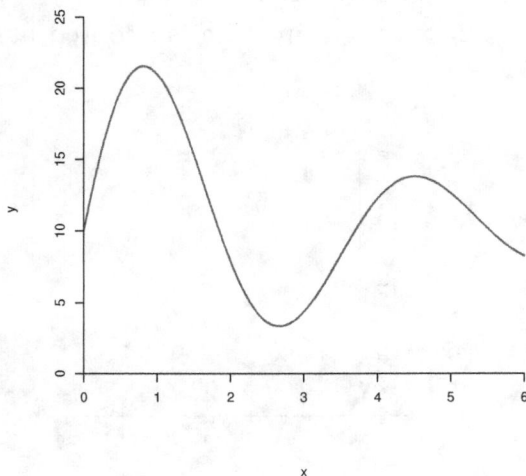

Figure 1.55: The function $10 + 15e^{-0.3t}\sin(1.7t)$, $0 \le t \le 6$

we can be sure that $t_2 = t_1 + \frac{\pi}{1.7} \approx 2.6692$ is also a critical point. Moreover

$$\tan(1.7t) = \tan(1.7t + 2\pi) = \tan(1.7(t + \tfrac{2\pi}{1.7})).$$

Hence $t_3 = t_1 + \frac{2\pi}{1.7} \approx 4.51715$ is the third critical point in the interval $[0,6]$. We have $f(0) = 10$, $f(6) \approx 8.265$, $f(0.8212) \approx 21.55$, $f(2.6692) \approx 3.368$, and $f(4.51715) \approx 13.81$. Therefore, the maximum is attained at 0.8212 and the minimum is attained at 2.6692.

Extending the notion of critical point

To bring this section to an end, let us mention that the methodology just learned, to find the maximum and the minimum of a continuous function defined on $[a, b]$ and differentiable in its interior, can be extended to continuous functions on $[a, b]$ that are not differentiable at a finite number of points in (a, b). For instance, let $f(x) = \sqrt{x}$, $0 \le x \le 2$, and $f(x) = \sqrt{-x}$, $-1 \le x < 0$ (Figure 1.56). This function is continuous on $[-1, 2]$ and differentiable on $(-1, 2)$ with the exception of $x = 0$. Being continuous on $[-1, 2]$, it adopts its maximum and its minimum at some points in $[-1, 2]$. Where? We observe that $f'(x) = \frac{1}{2\sqrt{x}}$, $0 < x < 2$, and $f'(x) = \frac{-1}{2\sqrt{x}}$, $-1 < x < 0$.

Therefore $f'(x) \ne 0$ at every point of the interval $(-1, 2)$ with the exception of $x = 0$ (where the derivative does not exist). Hence neither the maximum nor the minimum can be attained on $(-1, 0)$ or $(0, 2)$ because at either the maximum or the minimum the derivative would have to be zero since we are dealing with interior points. Thus, there are only three possibilities for the maximum or the minimum: the endpoints or $x = 0$. We have $f(0) = 0, f(-1) = 1, f(2) = \sqrt{2}$, therefore the maximum is adopted at $x = 2$ and the minimum is adopted at $x = 0$. An example

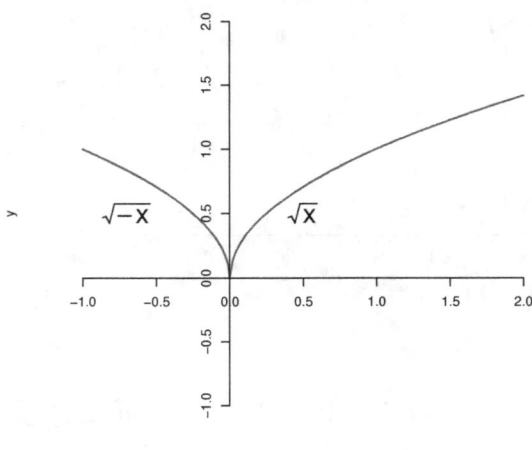

Figure 1.56: Extension of the idea of critical point

of this nature has led mathematicians to extend the notion of critical point, in the sense that an interior point c is called critical if either $f'(c) = 0$ or $f'(c)$ is not defined (that is, f is not differentiable at c). Thus, with all confidence we can assert that given a continuous function defined on $[a, b]$, with a finite number of critical points, that is to say interior points where either the derivative is zero or is not defined, there is a simple procedure to find the maximum or the minimum: evaluate f at a and b (the endpoints) and the critical points, and compare the resulting values. Sometimes the words **global maximum** or **global minimum** are used instead of maximum or minimum to emphasize the difference between them and a local maximum or a local minimum.

Exercises for Section 1.18

In exercises 1 through 6, determine where the function adopts its maximum or its minimum:

1. $f(x) = x^3 - 3x - 1, -2 \leq x \leq 1$.

2. $f(x) = -x^2 - x - 1, -2 \leq x \leq 3$.

3. $f(x) = \cos x + \sin(x), 0 \leq x \leq \pi$.

4. $f(x) = \sin(x^2) - x^2 - 3$, $-1 \leq x \leq \pi/2$.

5. $f(x) = e^{-x} \sin x$, $0 \leq x \leq 2\pi$.

6. $f(x) = \frac{\ln x}{x}$, $1 \leq x \leq 4$.

7. A person is swimming in the sea, at a location situated 300 meters from the shore. She wishes to reach her destination, 1,000 meters down the coast, in the least amount of time. Suppose that she can swim at 2m/s and run at 6 m/s. Where should she land?

8. Assume that $f(t) = 5 + 10e^{-0.2t} \sin(1.2t)$, $0 \leq t \leq 5$, describes the concentration of a human hormone after a drug is administered through a vein. When does the concentration reach its maximum or minimum?

9. A pump is to provide water to two cottages located at a distance of 300 ft and 450 ft, respectively, from the shore of a river. Given that the distance between the perpendicular lines, from the cottages to the river, is 1000 ft, where should the pump be located in order to minimize the amount of pipe needed?[13]

10. Let v_1 be the velocity of light in air and v_2 the velocity of light in water. Draw the coordinate axis in such a way that the interface between air and water lies on the x-axis. Show, in the first place, that

$$f(x) = \frac{\sqrt{x^2 + a^2}}{v_1} + \frac{\sqrt{(x-b)^2 + c^2}}{v_2}, \quad 0 \leq x \leq b,$$

   determines the time it takes a ray of light to travel from $(0, a)$, $a > 0$, to $(x, 0)$ and from there to (b, c), $b > 0$, $c < 0$ (the y-axis is built in such a way that it passes through $(0, a)$). According to **Fermat's Principle**, light will choose the path of least time, say from $(0, a)$ to a point $(x_o, 0)$ in the interface between air and water, and from there to (b, c). From experiments it is known that $0 < x_o < b$. Show that necessarily

$$\frac{\sin \alpha}{v_1} = \frac{\sin \beta}{v_2}$$

   where α=angle OAB and β=angle BCD ($O = (0, 0)$, $A = (0, a)$, $B = (x_o, 0)$, $C = (b, c)$, $D = (b, 0)$). This is the law of refraction of light.

11. Several species of birds fly faster over land than over water. Suppose that an adult bird flies at 10 km/h over land and 5 km/h over water. The bird wishes to fly from a location over the sea, 1 km offshore, to a point 15 km down the coast. What path should it use in order to minimize the total time?

---

[13]Readers knowledgeable about geometry will realize that this problem can also be solved using a non-calculus approach.

Enrichment Note: A Proof of Fermat's Theorem

Let us recall the statement of the theorem: Suppose f is a function, c is an interior point of its domain, and the function is differentiable at c. If f adopts a local extrema at c, then $f'(c) = 0$.

There is strong graphical evidence to the effect that the theorem is true. However, it is worthwhile to provide a detailed proof. Assume that f attains a local maximum at c; in other words, there exists $\delta > 0$ such that $f(x) \leq f(c)$ for all x in $(c - \delta, c + \delta)$. Then

$$\frac{f(x) - f(c)}{x - c} \geq 0, \quad \text{whenever } c - \delta < x < c$$

and

$$\frac{f(x) - f(c)}{x - c} \leq 0, \quad \text{whenever } c < x < c + \delta.$$

Therefore

$$f'_-(c) = \lim_{x \to c^-} \frac{f(x) - f(c)}{x - c} \geq 0$$

and

$$f'_+(c) = \lim_{x \to c^+} \frac{f(x) - f(c)}{x - c} \leq 0.$$

But $f'_-(c) = f'_+(c) = f'(c)$. Hence $f'(c) \geq 0$ and $f'(c) \leq 0$, which in turn implies that $f'(c) = 0$. A proof of Fermat's Theorem, when the function attains at c a local minimum instead of a local maximum, follows similar steps.

Implicitly, in the previous proof, we accepted the fact if $g(x) \geq 0$ for every x in $(c - \delta, c)$ and $\lim_{x \to c^-} g(x) = L$, it must be true that $L \geq 0$. Indeed, $L < 0$ would imply that there exists $\mu > 0$, $\mu < \delta$, such that $|g(x) - L| < -L$ for every x in $(c - \mu, c)$. In particular, we would have $g(x) < \frac{L}{2} < 0$ for every x in $(c - \mu, c)$, which is an impossibility because $g(x) \geq 0$ on $(c - \mu, c)$[14]. Similarly, if $g(x) \leq 0$ for every x in $(c, c + \delta)$ and $\lim_{x \to c^+} g(x) = L$, we can conclude that $L \leq 0$. Actually, these properties of limits are valid when instead of lateral limits we have to deal with two-sided limits.

1.19   Maxima and Minima II

If the interval of definition of f is not $[a, b]$, even if the function is continuous everywhere there is no certainty that it has a maximum or a minimum. For instance, $f(x) = 1/x$, $x > 0$, has neither a maximum nor a minimum, while $g(x) = x^2$, defined

---

[14]The squeeze property of limits (section 1.4) provides a straightforward alternative justification.

on $(-\infty, \infty)$, has a minimum at $x = 0$ but it has no maximum. In many applications, as we will soon see, a function is defined on $(0, \infty)$ and is differentiable on its domain. If the function is to have a maximum or a minimum at a point c, necessarily $f'(c) = 0$ because all the points of $(0, \infty)$ are interior points (recall Fermat's theorem, section 1.18). Thus, all the 'candidates' for a maximum or a minimum have to be critical points; there is no other possibility. Luckily, there is a simple test when one deals with a single critical point.

First Derivative Test

Let f de defined on an open interval I (often $I = (0, \infty)$ or $I = (0, b)$), differentiable everywhere in its domain. Suppose that c is the only critical point, that is to say, a point in I where the derivative is zero.

   (i) If $f'(x) < 0$ for $x < c$ and $f'(x) > 0$ for $x > c$, f adopts its minimum at c.
   (ii) If $f'(x) > 0$ for $x < c$ and $f'(x) < 0$ for $x > c$, f adopts its maximum at c.

Let us prove (i). Since $f'(x) < 0$ on $(0, c)$ and f is continuous on $(0, c]$, we can conclude that f is decreasing on $(0, c]$; thus, $f(x) > f(c)$ on $(0, c)$. On the other hand, since $f'(x) > 0$ on (c, ∞) and continuous on $[c, \infty)$, we can conclude that f is increasing on $[c, \infty)$, consequently $f(c) < f(x)$ on (c, ∞). We have just shown that $f(c) < f(x)$ for any $x \neq c$, which in turn implies that $f(c) \leq f(x)$ for any x in $(0, \infty)$. A proof of (ii) follows similar steps. Let us discuss four real-life problems.

1. Suppose we have L meters of wire and wish to build a rectangular pen, next to a river, of maximum area. Let x, y be the width and length of the pen. Then $2x + y = L$ and consequently the area function is given by

$$A(x) = x(L - 2x) = Lx - 2x^2, \quad 0 < x < \tfrac{L}{2}.$$

Therefore $A'(x) = L - 4x$. The only critical point happens to be $x = \frac{L}{4}$. Obviously, if $x < \frac{L}{4}$ then $L - 4x > 0$, that is, $A'(x) > 0$. Similarly, if $x > \frac{L}{4}$ we will have $A'(x) < 0$. The first derivative test (part (ii)) implies that the area function attains its maximum at $L/4$. Thus, the best we can do is build a rectangle of width $L/4$ and length $L/2$. The reader may want to compare this approach to the one we used at the beginning of chapter 1, wherein we used only algebra.

2. A second example has to do with the optimal time of harvesting (Katz, 1976). As one might surmise, models for harvesting are quite important. The yield of a certain cereal (kilograms per hectare) will be a function f of time (days). A model that could be used is

$$f(t) = t^2 - kt^3, t > 0$$

where k is a positive parameter. We observe that $f'(t) = 2t - 3kt^2$, hence the critical points stem from $0 = 2t - 3kt^2$. There is just one positive solution of this equation, namely $t = \frac{2}{3k}$. Assuming $t < \frac{2}{3k}$ we get $2t - 3kt^2 > 0$; that is to say, $f'(t) > 0$. Similarly, $f'(t) < 0$ provided that $t > \frac{2}{3k}$. The first derivative test allows us to conclude that $f(t)$ attains its maximum at $t_o = \frac{2}{3k}$.

Assuming that $k = 0.01$ we get

$$t_o = \frac{2}{3(0.01)} = \frac{2}{0.03} = \frac{200}{3} = 66.67 \text{ days.}$$

A slight change of k varies considerably the value of t_o (Figure 1.57). For instance, if $k = 0.011$ we get $t_o = \frac{2}{0.033} = 60.61$ days, while $k = 0.014$ leads to $t_o = \frac{2}{0.042} = 47.62$ days.

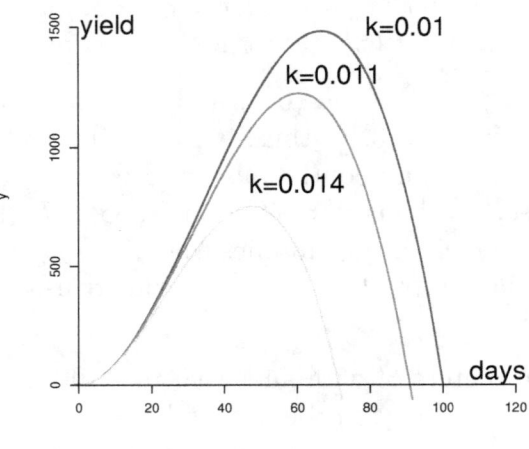

Figure 1.57: Optimal time of harvesting for different values of the parameter

3. The presence of nitrogen in the soil is an important factor in the yield of crops. Too little nitrogen is detrimental, but too much nitrogen can have an adverse effect. We could use the function

$$f(x) = \frac{x}{1+kx^2}, \ x > 0,$$

to study yield as a function of the amount of nitrogen in the soil, where k is a positive parameter[15] (Figure 1.58). Right away we get

$$f'(x) = \frac{1-kx^2}{(1+kx^2)^2}.$$

---

[15]Functions of this sort were analyzed in section 1.5; see Figure 1.18.

To obtain any critical point, we have to solve the equation $1 - kx^2 = 0$. It doesn't take much work to get the only critical point, namely $x_o = 1/\sqrt{k}$. If $x < 1/\sqrt{k}$ then $1 - kx^2 > 0$, thus $f'(x) > 0$. Similarly, if $x > 1/\sqrt{k}$ then $1 - kx^2 < 0$, which yields $f'(x) < 0$. The first derivative test implies that $f(x)$ attains its maximum at $x_o = 1/\sqrt{k}$.

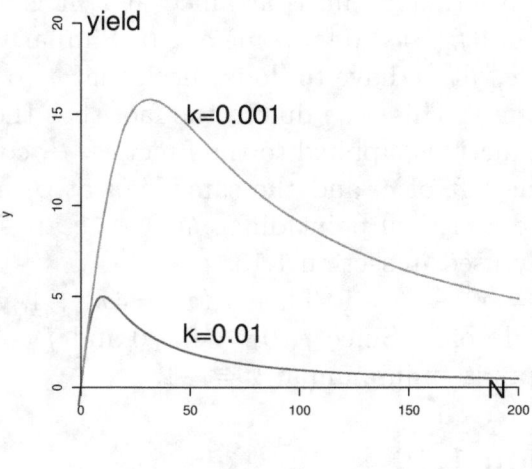

Figure 1.58: Yield as a function of the amount of nitrogen

4. Suppose we wish to build an open cylinder, of fixed volume V, with the least amount of material. Let x be the radius and y the height of an arbitrary cylinder of volume V. We have $V = \pi x^2 y$ and $S = \pi x^2 + 2\pi xy$, where S is the surface area of the cylinder. Therefore

$$S(x) = \pi x^2 + 2\pi x(\tfrac{V}{\pi x^2}).$$

That is to say, $S(x) = \pi x^2 + \frac{2V}{x}$, $x > 0$. Consequently

$$S'(x) = 2\pi x - \frac{2V}{x^2}, \ x > 0.$$

The only critical point is obtained by solving the equation $\frac{2V}{x^2} = 2\pi x$. Without much effort we get $x = (\frac{V}{\pi})^{1/3}$. Do we achieve a minimum at this critical point? Assuming $x < (\frac{V}{\pi})^{1/3}$ it follows that $x^3 < \frac{V}{\pi}$, therefore $\frac{V}{x^2} > \pi x$, which in turn leads to the inequality $2\pi x - \frac{2V}{x^2} < 0$; that is to say, $S'(x) < 0$. In a similar fashion, one can prove that if $x > (\frac{V}{\pi})^{1/3}$ then $S'(x) > 0$. The first derivative test implies that $S(x)$ achieves its minimum at $x = (\frac{V}{\pi})^{1/3}$. The reader can check that the corresponding value of y will be $y = (\frac{V}{\pi})^{1/3}$, exactly the same as that for x. In other words, to minimize the amount of material one has to build an open cylinder of given volume V in such a way that its radius is the same as its height.

Note

Assume that $f : I \to \Re$ is differentiable, I an open interval, and c is the only critical point. Furthermore, *suppose that f' is continuous on I.* To apply the first derivative test and find out whether the maximum is attained at c, it is enough to check that $f'(\alpha) > 0$ for some $\alpha < c$ and $f'(\beta) < 0$ for some $c < \beta$. Similarly, to find out whether a minimum is adopted at c, all we have to do is check that $f'(\alpha) < 0$ for some $\alpha < c$ and $f'(\beta) > 0$ for some $c < \beta$. This is so due to the fact that the Intermediate Value Theorem for continuous functions, applied to the function f', compels f' to have the same sign at α and to the left of c; and the same sign at β and to the right of c. Otherwise, we would have a critical point different from c. It should be noted that this technique was already used in section 1.15.

For instance, let $f(x) = x^4 + 4x + 1$. Then $f'(x) = 4x^3 + 4$, which in turn implies that -1 is the only critical point. Since $f'(0) = 4 > 0$ and $f'(-2) = 4(-8) + 4 < 0$, we can assert that f adopts its minimum at $x = -1$.

Exercises for Section 1.19

1. Suppose we wish to build, with the least amount of material, a cylinder of fixed volume V that has a top and a bottom (that is, a closed cylinder). What dimensions should it have?

2. Suppose we have S m^2 of material and wish to build a closed cylinder of maximum volume. What dimensions should it have?

3. Solve the same problem as in the preceding exercise, this time when the cylinder is open at the top.

4. Assume that $f(t) = t^2 - 0.01t^4$ is used as a model of harvesting. Where does f attain its maximum? Why would you prefer $g(t) = t^2 - 0.01t^3$ as a model of alfalfa harvesting rather than $f(t)$?

5. The function $f(x) = \frac{\ln x}{kx}$ could be used to model how the yield of a harvest varies with the amount of nitrogen in the soil. Calculate the point where $f(x)$ adopts its maximum and sketch a graph of the function when $k = 0.01$.

6. Biologists use the formula $E(v) = \frac{kdv^3}{v-u}$ to approximate the energy spent by a fish that swims a distance d at a velocity v against a river current that flows at a constant velocity u ($v > u$) (Batschelet, 1975). Find the value of v that minimizes E.

7. The amount of pollen collected by a bee from a flower can be described by many functions $f(t)$, all of which have in common one characteristic: their rate decreases with time because at the beginning pollen is easily accessible but it is harder to harvest as time passes (Adler, 2005). Assuming that 20 micro-grams is the total amount of pollen available per flower, we could put forward the function $f(t) = \frac{20t}{t+1}$. We can check that this function is increasing. concave down, and $\lim_{t\to\infty} f(t) = 20$. We wish to analyze the rate at which the bee is collecting pollen per each visit to a flower, that is to say $\frac{f(t)}{t+2}$, where we are assuming that she needs 2 seconds to travel from one flower to another. Let

$$r(t) = \frac{f(t)}{t+2} = \frac{20t}{(t+1)(t+2)}.$$

Where does this function attain its maximum? Sketch a graph of $r(t)$.

8. In relation to the preceding exercise, another possible function to describe the amount of pollen collected by a bee is $f(t) = 20 - 20e^{-0.5t}$. Then

$$r(t) = \frac{20 - 20e^{-0.5t}}{t+2}.$$

Where does this function attain its maximum? (Hint: Use technology to find the critical point c; thereafter choose a point α, $0 < \alpha < c$, and check that $r'(\alpha) > 0$. Similarly, choose a point β, $c < \beta$, and verify that $r'(\beta) < 0$. Since there is a unique positive critical point, this is all you need to do in order to apply the first derivative test.)

9. Among all right triangles with hypotenuse 3, which one encloses the largest area?

10. Try to find a non-calculus approach to the solution of the previous exercise. Compare both approaches.

1.20 Concavity

Let f be a differentiable function defined on an open interval I. We will say that f is 'concave up' on I if $f'(x)$ is increasing on I. This definition coincides with the intuitive idea of concave up that many of us have (Figure 1.59). Analogously, we will say that f is 'concave down' on I if $f'(x)$ is decreasing on I (Figure 1.60). We could show that f is concave up (down) if and only if, at each point c in I, the tangent lies below (above) the curve $y = f(x)$ in a neighborhood of c.

There is an easy criterion for concavity whenever f' is differentiable, that is, f'' exists. Namely, if $f''(x) > 0$ on I then f is concave up on I, while if $f''(x) < 0$ on I

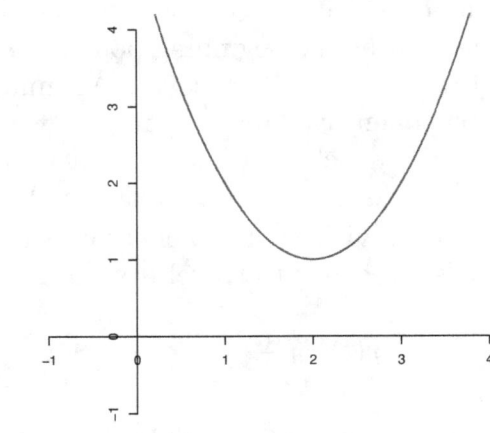

Figure 1.59: A concave up function

then f is concave down on I. The validity of the criterion for concavity stems from the fact that f'' positive on I implies that f' is increasing on I; similarly, f'' negative on I implies that f' is decreasing on I. For instance, let $f(x) = x^3$. Then $f'(x) = 3x^2$ and $f''(x) = 6x$ for any x, hence f is concave up on $(0, \infty)$ and concave down on $(-\infty, 0)$. The point where concavity changes is called a **point of inflection**, thus the function $f(x) = x^3$ has a point of inflection at 0 (Figure 1.61).

Let $f(x) = x^{1/3}$ and $g(x) = x^{2/3}$, functions that we studied in section 1.7 (Figure 1.25 and Figure 1.26, respectively). For $x \neq 0$ we have

$$f'(x) = \tfrac{1}{3x^{2/3}}, \qquad f''(x) = -\tfrac{2}{9x^{5/3}},$$

while for $x \neq 0$ we have

$$g'(x) = \tfrac{2}{3x^{1/3}}, \qquad g''(x) = \tfrac{-2}{9x^{4/3}}.$$

Thus, $f''(x) > 0$ for $x < 0$ and $f''(x) < 0$ whenever $x > 0$. Therefore, f is concave up on $(0, \infty)$ and concave down on $(0, \infty)$; the point $x = 0$ happens to be an inflection point. On the other hand, $g''(x) < 0$ if $x < 0$ or $x > 0$. Hence the function g is concave down both on $(-\infty, 0)$ and $(0, \infty)$. We can then assert that g does not have an inflection point at the origin (or anywhere else). Neither f nor g are differentiable at $x = 0$, a fact that was not an impediment in the analysis of concavity.

When a function is defined on an open interval I, is twice differentiable there and its second derivative is continuous on I, there is a nice criterion to find points of inflection as we shall see next.

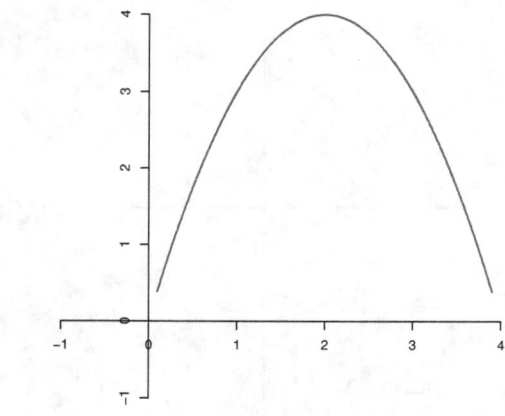

Figure 1.60: A concave down function

Candidates for points of inflection

Suppose f is defined and twice differentiable on an open interval I. Moreover, f'' is continuous on I. The following implication is true:

$$c \text{ is an inflection point of } f \implies f''(c) = 0.$$

We will prove the logical contrapositive, namely

$$f''(c) \neq 0 \implies c \text{ is not an inflection point of } f.$$

With this purpose in mind, let us assume that $f''(c) \neq 0$. Then there are two possibilities, either $f''(c) > 0$ or $f''(c) < 0$; suppose $f''(c) > 0$. Since f'' is continuous at c, thanks to the dragging property there exists $\delta > 0$ such that $f''(x) > 0$ for every x in $(c - \delta, c + \delta)$. Hence f is concave up on $(c - \delta, c + \delta)$. Thus, there is no change of concavity at c. Therefore c cannot be an inflection point of f. The case $f''(c) < 0$ can be treated in an analogous way.

Let $f(x) = x^5 - 3x^2 + 1$. Then $f'(x) = 5x^4 - 6x$ and $f''(x) = 20x^3 - 6$. Any change of concavity will take place at those points where f'' changes sign. Thus, any candidate for a point of inflection stems from the equation $20x^3 - 6 = 0$. This equation has only one solution, namely $(3/10)^{1/3}$. Without much effort one can show that if $x < (3/10)^{1/3}$ then $f''(x) < 0$, while if $x > (3/10)^{1/3}$ then $f''(x) > 0$. Therefore, f is concave down on $(-\infty, (3/10)^{1/3})$ and concave up on $((3/10)^{1/3}, \infty)$. The number $(3/10)^{1/3}$ happens to be a point of inflection.

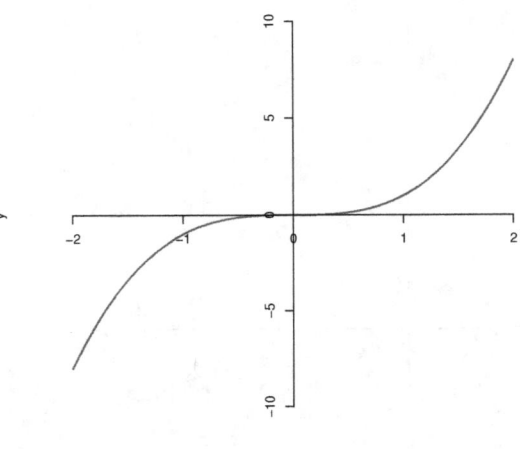

Figure 1.61: Change of concavity

We should stress that candidates for points of inflection stem from the equation $f''(x) = 0$, *but the equality* $f''(c) = 0$ *does not imply that* c *is a point of inflection.* For instance, let us consider the function $f(x) = x^4 + 3x + 1$. Since $f'(x) = 4x^3 + 3$ and $f''(x) = 12x^2$, for any x, we can conclude that f is concave up on the whole of the real line; it has no points of inflection, despite the fact that $f''(0) = 0$ (Figure 1.62).

Probably the most useful application of the idea of concavity is related to the determination of a maximum or a minimum of a function, and for this purpose we have a test that is an interesting alternative to the test we analyzed in the previous section.

The Second Derivative Test: a first version

Let f be defined on an open interval I and twice differentiable there. Suppose c is the only critical point, that is, $f'(c) = 0$. The following two implications are true:

(i) If $f''(x) < 0$ for all x in I then f adopts its maximum at c.
(ii) If $f''(x) > 0$ for all x in I then f adopts its minimum at c.

Is it possible to provide a plausible argument to justify this test? If $f''(x) < 0$ for all x in I then f is concave down on its domain. But $f'(c) = 0$, consequently f attains its maximum at c. A similar argument, albeit not a mathematical proof, can be used to justify part (ii) of the second derivative test.

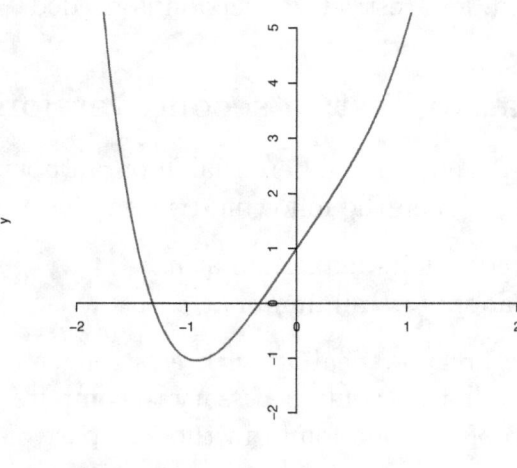

Figure 1.62: The function $f(x) = x^4 + 3x + 1$

Let us prove (i) in detail. According to the first derivative test (section 1.19), it would be enough to show that $f'(x) > 0$ for $x < c$ and $f'(x) < 0$ for $x > c$. Let us try to show that $f'(x) > 0$ for $x < c$. If this statement were not be true, there would exist d, $d < c$, such that $f'(d) \leq 0$. We cannot have $f'(d) = 0$ because c is the only critical point, hence $f'(d) < 0$. But f' is decreasing on I since $f''(x) < 0$ for every x in I. Consequently $f'(d) > f'(c) = 0$, contradicting the fact that $f'(d) < 0$. The proof of the inequality $f'(x) < 0$, for any $x > c$, follows the same pattern.

When should we use the second derivative test instead of the first derivative test? Evidently, the former is a better option than the latter when it is not difficult to calculate the second derivative. For instance, in example 4 of the previous section, the problem related to the construction of the cylinder of fixed volume V that uses the least amount of material, $c = \left(\frac{V}{\pi}\right)^{1/3}$ is the only critical point. We note that $S''(x) = 2\pi + \frac{4V}{x^3}$, thus S'' is positive everywhere. Consequently $S(x)$ adopts its minimum at the critical point. But in problem 3 of the previous section, related to the yield of a crop as a function of the amount of nitrogen in the soil, taking the second derivative of $f(x) = \frac{x}{1+kx^2}$ is rather lengthy; thus, the first derivative test is a better option.

Let us try to apply the second derivative test to the model of harvesting analyzed in the previous section. Let us recall that $f(t) = t^2 - kt^3$, $t > 0$. Then $f'(t) = 2t - 3kt^2$, and $f''(t) = 2 - 6kt$, $t > 0$, with $t_o = \frac{2}{3k}$ as the only critical point. We certainly do not have $f''(t) < 0$ for every $t > 0$, thus the first version of the second derivative test cannot be applied; a second, less restrictive, version is needed.

The Second Derivative Test: a second version

Let a function f be defined and twice differentiable on an open interval I, where it has only one critical point c. Then the following two implications are true:

(i) If $f''(c) < 0$ then f adopts its maximum at c.
(ii) If $f''(c) > 0$ then f adopts its minimum at c.

It can be seen that we are now dealing with a 'stronger' version of the second derivative test, in the sense that it is only necessary to compute the second derivative at the critical point. A proof is to be found in the last part of the enrichment note at the end of section 1.21.

Let us return to the model of harvesting. We observe that $f''(\frac{2}{3k}) = 2 - 6k\frac{2}{3k} = 2 - 4 = -2 < 0$; therefore, f attains its maximum at the only critical point. Next let us analyze an example of a geometrical nature.

Suppose we have a rectangular cardboard of length 3 ft and width 2 ft. and wish to build an open box, cutting squares of length x from its four corners and thereafter folding the structure. What value of x should we choose in order to maximize the volume of the box? The volume of the box will be

$$V(x) = (3 - 2x)(2 - 2x)x, \ 0 < x < 1.$$

We chose $(0, 1)$ as the domain because both $2 - 2x$ and $3 - 2x$ have to be positive, as well as x. Since $V(x) = 4x^3 - 10x^2 + 6x$, we can conclude that $V'(x) = 12x^2 - 20x + 6$. The equation $12x^2 - 20x + 6 = 0$ has only one solution in the interval $(0, 1)$, namely $x_1 \approx 0.39237478$. That is to say, x_1 is the only critical point. But $V''(x) = 24x - 20$, therefore $V''(x_1) = 24x_1 - 20 = 24 \times 0.39237478 - 20 = -10.583005 < 0$. The new version of the second derivative test leads to the conclusion that the box will have maximum volume if $x = x_1$. We could have used the first version too, indeed $24x - 20 < 0$ if $x < 5/6$. Thus, $V''(x) < 0$ if $0 < x < 1$, which implies that V adopts its maximum at x_1.

A third example has to do with the function $f(x) = e^{-x^2}$. We observe that $f'(x) = -2xe^{-x^2}$ and $f''(x) = -2e^{-x^2}(1 - 2x^2)$. The only critical point is 0 and $f''(0) = -2 < 0$, thus f adopts its maximum at 0 (Figure 1.63). We cannot apply the first version of the second test because f is neither concave up nor concave down on the whole real line: concavity changes at $\pm 1/\sqrt{2}$. However, we could have used the first derivative test since $f'(x) < 0$ for $x > 0$ and $f'(x) > 0$ for $x < 0$.

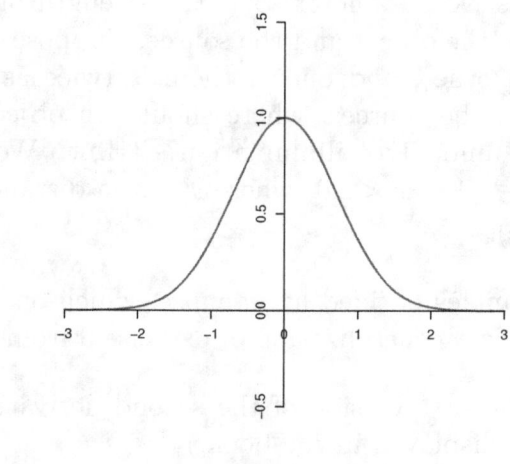

Figure 1.63: The function $f(x) = e^{-x^2}$

Exercises for Section 1.20

1. Determine where the function $f(x) = x^4 - x^2 + 1$ is concave up or concave down.

2. Find the inflection points of $g(x) = \frac{1}{1+x^2}$.

3. Does the function $f(x) = x^{4/3}$ have an inflection point?

4. What can be said about the concavity of $f(x) = x^{5/3}$?

5. A lamp hangs from the top of a circular table of radius r, directly above the center of the table. At what height should it be located in order to maximize the illumination I at any point on the border of the table? Keep in mind that $I = \frac{k}{l^2} \sin\theta$, where l is the distance between the lamp and any point on the border, and θ is the 'angle of elevation' from any point on the border and the lamp.

6. Which point on the parabola $y = x^2$ is closest to $(3, -2)$?

7. A farmer has L ft of fencing and wishes to build two identical adjacent rectangular pens (the pens share a fence). What dimensions of the pens provide the maximum area?

8. A farmer wishes to build a silo to hold corn. The silo has cylindrical shape, with a hemisphere at the top. Assuming that the farmer has S m^2 of material, how should he build the silo in order to maximize its volume?

9. The illumination that an object receives from a light source is given by $I(x) = \frac{kS}{x^2}$, where k is a positve parameter, S is the strength of the source, and x is the distance between the object and the source. Suppose that two light sources are 20 ft from each other, and one of them is twice as strong as the other. On the line between the sources, where should an object be placed in order to receive the least amount of illumination? (Hint: Work with the function $I(x) = \frac{2KS}{x^2} + \frac{kS}{(20-x)^2}$, $0 < x < 20$, where x is the distance from the object to the strongest source).

10. Among all right triangles of fixed hypotenuse, which one encloses the greatest area? This problem is a generalization of exercise 9 from the previous section.

11. Prove part (ii) of the first version of the second derivative test (Hint: Adapt the proof of part (i) displayed in the book.)

1.21   Local Extrema

Let us consider the function $f(x) = x^3 - 3x^2$, defined on the real line. It has no maximum or minimum (Figure 1.64). However, it has a 'local maximum' at $x = 0$ and a 'local minimum' at $x = 2$, in the sense that there exists a neighborhood of zero (say $(-1, 1)$) where $x^3 - 3x^2 \le 0 = f(0)$, and there exists a neighborhood of 2 (say $(1, 3)$) such that $f(2) = -4 \le x^3 - 3x^2$ for every x in $(1, 3)$. Recall that a neighborhood of a point, on the real line, is any open interval that contains the point.

Definition

Let f be a function defined on an open interval I. It has a 'local minimum' at c if there exists a neighborhood N_c of c such that $f(c) \le f(x)$ for every x in N_c. Similarly, f has a 'local maximum' at d if there exists a neighborhood N_d of d such that $f(x) \le f(d)$ for every x in N_d. *Implicit in the definition is the fact that a local extremum (that is, a local maximum or minimum) can take place only at an interior point of the domain of the function.*

In the first place, let us note that if f has a local maximum or local minimum at c, and f is differentiable at c, then $f'(c) = 0$. We found the same proposition when we studied maxima and minima in section 1.18, and gave it a name: **Fermat's Theorem**. Thus, if f is differentiable on an open interval I, all the candidates for local extrema stem from the equation $f'(x) = 0$. In consonance with the terminology we adopted in the past, any solution of the equation $f'(x) = 0$ will be called a *critical point* of the function f. For example, the critical points of $f(x) = x^3 - 3x^2$ are found

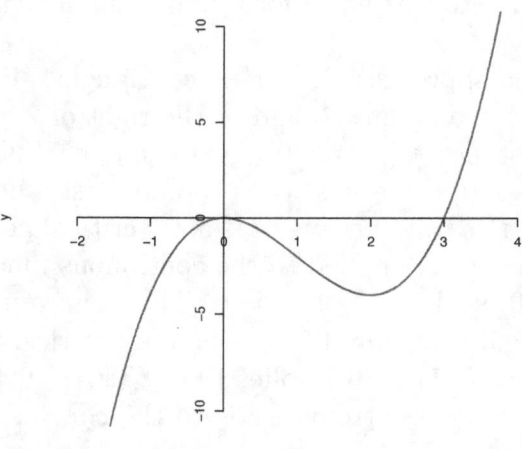

Figure 1.64: Local extrema of $f(x) = x^3 - 3x^2$

solving the equation $3x^2 - 6x = 0$. It has only two solutions, namely 0 and 2. These are the only candidates for local extrema.

First derivative test for local extrema

Let f be defined and differentiable on an open interval I. Suppose that c is a critical point, that is, $f'(c) = 0$.

(i) Assume that there exists $\delta > 0$ such that $f'(x) > 0$ for every x in $(c - \delta, c)$ and $f'(x) < 0$ for every x in $(c, c + \delta)$. Then f has a local maximum at c.

(ii) Assume that there exists $\mu > 0$ such that $f'(x) < 0$ for every x in $(c - \mu, \mu)$ and $f'(x) > 0$ for every x in $(c, c + \mu)$. Then f has a local minimum at c.

A proof follows the same pattern employed when we dealt with the first derivative test in section 1.19. Let us prove (i): $f'(x) > 0$ on $(c - \delta, c)$ implies that $f(x)$ is increasing on $(c - \delta, c]$; consequently $f(x) \leq f(c)$ for every x in $(c - \delta, c]$. On the other hand, $f'(x) < 0$ on $(c, c + \delta)$ implies that f is decreasing on $[c, c + \delta)$, thus $f(c) \geq f(x)$ for every x in $[c, c + \delta)$. Therefore $f(c) \geq f(x)$ for every x in $[c, c + \delta)$. In summary, we have shown that $f(x) \leq f(c)$ for every x in $(c - \delta, c + \delta)$, i.e., f attains a local maximum at c. A proof of (ii) follows a similar pattern.

Let us use the first derivative test when $f(x) = x^3 - 3x^2$, whose only two critical points are 0 and 2. We observe that if $-1 < x < 0$ then $3x - 6 < 0$, hence $f'(x) = 3x^2 - 6x = x(3x - 6) > 0$. In a similar fashion, one can prove that $f'(x) < 0$ whenever

$0 < x < 1$, $f'(x) < 0$ whenever $1 < x < 2$, and $f'(x) > 0$ whenever $2 < x < 3$. Thus, thanks to the first derivative test, f has a local maximum at 0 and a local minimum at 2.

But there is a different approach when f' is continuous. It is enough to find a number to the left of 0, between 0 and 2, and to the right of 2, and check the sign of their derivatives[16]. For instance $f'(-1) > 0$, $f'(1) < 0$ and $f'(3) > 0$. We must have $f'(x) > 0$ for every x in $(-1, 0)$, otherwise there would exist a point d in $(-1, 0)$ such that $f'(d) \leq 0$. Obviously $f'(d) \neq 0$ because d is not a critical point. If $f'(d) < 0$ then the Intermediate Value Theorem, applied to the continuous function f', implies that there exists d_1 $(d < d_1 < 0)$ such that $f'(d_1) = 0$. This is an impossibility because d_1 is not a critical point (recall that 0 and 2 are the only critical points). In a similar fashion one can prove that $f'(1) < 0$ implies that $f'(x) < 0$ for every x in $(0, 2)$. The first derivative test for local extrema leads to the conclusion that the function f attains a local maximum at 0. The same type of analysis can be applied to the intervals $(1, 2)$ and $(2, 3)$. Once understood, this approach to the first derivative test, for local extrema, takes less time.

Practical form of the first derivative test for local extrema

Suppose that f is defined on an open interval I and let us assume that f' exists and is continuous on I. Furthermore, assume that f has a finite number of critical points $c_1, ..., c_n$ in ascending order. Let c_i be an arbitrary critical point of this list, $1 < i < n$. If we can find a point r to the left of c_i and the right of c_{i-1} such that $f'(r) < 0$, and we can find a point s to the right of c_i and the left of c_{i+1} such that $f'(s) > 0$, then f attains a local minimum at c_i.

Why is this so? The fact that $f'(r) < 0$ implies that $f'(x) < 0$ for every x in (c_{i-1}, c_i) because otherwise the Intermediate Value Theorem (recall that f' is continuous) would imply that $f'(p) = 0$ for some p in (c_{i-1}, c_i). This is not possible because p cannot be a critical point of f. It is not part of the list $c_1, ..., c_n$. In a similar fashion, $f'(s) > 0$ implies that $f'(x) > 0$ for every x in (c_i, c_{i+1}). Then the first derivative test leads to the conclusion that f attains a **local minimum** at c_i. On the other hand, if $f'(r) > 0$ for some r in (c_{i-1}, c_i) and $f'(s) < 0$ for some s in (c_i, c_{i+1}), then f will attain a **local maximum** at c_i. The analysis at c_1 and c_n follows the same pattern.

In other words, if f' exists and is continuous, all we have to do is make a list of the critical points and check whether f' 'changes sign' at each of them. If the change is from negative to positive, when we go from the left to the right, a **local minimum** *is achieved at the corresponding critical point. If the change is from positive to negative, when we go from the left to the right, a* **local maximum** *is achieved at the*

---

[16]This technique was introduced, in a different context, in sections 1.15 and 1.19.

corresponding critical point. Evidently, if f' does not change sign at a critical point c, we can conclude that f either is increasing or decreasing; thus, it does not attain a local extremum at c.

An example will show how easy it is to apply the practical form of the first derivative test. Let $f(x) = x^4 - x^2 + 1$. Then $f'(x) = 4x^3 - 2x$, which leads to the critical points $0, 1/\sqrt{2}, -1/\sqrt{2}$. We have $f'(-1) < 0$, $f'(-1/2) > 0$, $f'(1/2) < 0$, and $f'(1) > 0$. Thus, f attains a local minimum both at $-1/\sqrt{2}$ and $1/\sqrt{2}$ while it attains a local maximum at $x = 0$ (Figure 1.65).

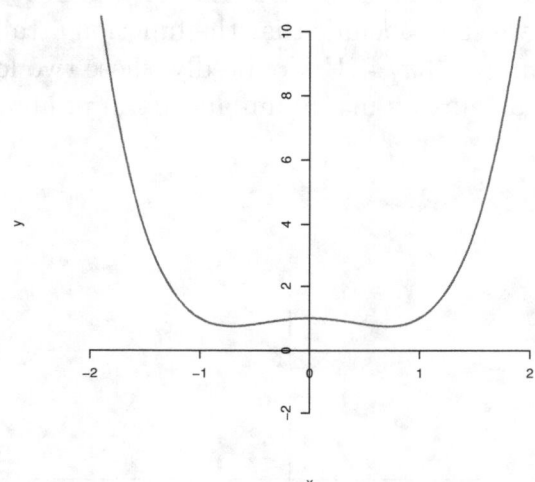

Figure 1.65: Local extrema of $f(x) = x^4 - x^2 + 1$

Second derivative test for local extrema

Let f be defined and twice differentiable on an open interval I and suppose that $f'(c) = 0$. The following two implications are true:

    (i) If $f''(c) < 0$ then f adopts a local maximum at c.

    (ii) If $f''(c) > 0$ then f adopts a local minimum at c.

This looks like the second version of the second derivative test, which was analyzed in the previous section in the context of discussing global maxima and global minima; all we have done is to adapt it to local extrema[17]. For a detailed proof, the reader can consult the enrichment note at the end of the section.

---

[17]Note that we are **not** assuming that c is the only critical point.

Undoubtedly, the second derivative test is very convenient if it is not difficult to calculate f''. Let us again consider the function $f(x) = x^4 - x^2 + 1$. Then $f'(x) = 4x^3 - 2x$ and $f''(x) = 12x^2 - 2$. The critical points are 0, $1/\sqrt{2}$, $-1/\sqrt{2}$. Since $f''(0) - 2 < 0$, $f''(1/\sqrt{2}) = 4 > 0$ and $f''(-1/\sqrt{2}) = 4 > 0$ we can conclude that f attains a local maximum at $x = 0$, and a local minimum both at $-1/\sqrt{2}$ and $1/\sqrt{2}$.

Next let $f(x) = \cos x + \sin x$, $-3 < x < 3$ (Figure 1.66). We observe that $f'(x) = -\sin x + \cos x$ and $f''(x) = -\cos x - \sin x$. The critical points stem from the equation $0 = -\sin x + \cos x$, that is to say $\tan x = 1$. Right away we get a solution, namely $x = \pi/4$. But $\tan$ has period π, therefore $\pi/4 - \pi = -3\pi/4$ is also a critical point ($\pi/4 + \pi$ lies outside the domain of the function). Since $f''(\pi/4) = -\sqrt{2} < 0$ and $f''(-3\pi/4) = \sqrt{2} > 0$ we can conclude that the function attains a local maximum at $\pi/4$ and a local minimum at $-3\pi/4$. Undoubtedly, these two local extrema are the points where the function attains its maximum and its minimum, respectively.

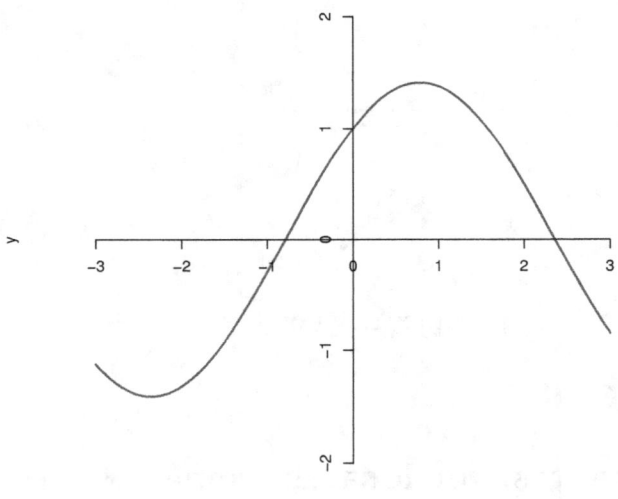

Figure 1.66: The function $f(x) = \cos(x) + \sin(x)$ on $(-3, 3)$

Exercises for Section 1.21

1. Find the local extrema of $f(x) = x^3 - 2x^2 + 0.5$ and sketch a graph of the given function.

2. Let $f(x) = x^4 - 2x^2 + 6$. Where does it attain a local maximum or a local minimum?

3. Given $f(x) = x^4 + 4x + 1$, find any local maximum or local minimum. Is any of the local extrema a minimum or maximum of the function? Provide a rough sketch of the function.

4. Find the local extrema of the function $f(x) = \sin(3x) + \cos(4x)$, $-1 < x < 1$. Use your graphing calculator to find the zeros of $f'(x)$ and thereafter use the practical form of the first derivative test.

5. In section 1.18 we considered the function $f(t) = 10 + 15e^{-0.3t}\sin(1.7t)$, $0 \leq t \leq 6$, which describes the concentration of a human hormone after a drug is administered through a vein. Find the local extrema of the function.

Enrichment Note: A Proof of the Second Derivative Test

Let us recall the statement of the second derivative test for local extrema: Suppose that f is defined and twice differentiable on an open interval I, and $f'(c) = 0$ [18].
   (i) If $f''(c) < 0$ then f adopts a local maximum at c.
   (ii) If $f''(c) > 0$ then f adopts a local minimum at c

Let us provide a proof of (i). By hypothesis $f''(c) < 0$, thus

$$\lim_{x \to c} \frac{f'(x) - f'(c)}{x - c} < 0.$$

Since $f'(c) = 0$ we will have

$$\lim_{x \to c} \frac{f'(x)}{x - c} < 0.$$

Due to the dragging property of limits (section 1.4), there exists $\delta > 0$ such that

$$\frac{f'(x)}{x - c} < 0 \tag{1.28}$$

provided $c - \delta < x < c + \delta$, $x \neq c$. Assuming $c - \delta < x < c$ we get $x - c < 0$, thus from (1.28) it follows that $f'(x) > 0$. If $c < x < c + \delta$ then $x - c > 0$. Again, from (1.28) we can conclude that $f'(x) < 0$. Then the first derivative test for local extrema implies that f adopts a local maximum at c. The proof of (ii) is similar. QED[19]

The preceding proof can be used to prove the second derivative test, second version, for global extrema. Under the hypotheses of this test, part (i) ($f'(c) = 0$, f twice

---

[18]The reader may notice that we are not assuming continuity of f'' at c; it is not necessary.

[19]QED stands for the Latin phrase quod erat demonstrandum, it has been demonstrated. It signals the completion of a proof.

differentiable on I and $f''(c) < 0$, with c the only critical point), we can conclude that f adopts a local maximum at c. But, besides c, no point of I is a critical point. Hence, it is intuitively clear that f adopts a global maximum at c.

If the reader wishes to see a formal proof, there is a path that can be followed. According to the first derivative test for global extrema, all we have to do is show that $f'(x) > 0$ for $x < c$ and $f'(x) < 0$ for $x > c$. Indeed, we already proved, at the beginning of this enrichment note, that there exists $\delta > 0$ such that $f'(x) > 0$ for $c - \delta < x < c$ and $f'(x) < 0$ for $c < x < c + \delta$. If there were to exist d, $d < c$, such that $f'(d) < 0$, then we can choose r in $(c - \delta, c)$ such that $f'(r) > 0$ and consequently $f'(d) < 0 < f'(r)$. The Intermediate Value Theorem would imply that for some s, between d and r, $f'(s) = 0$. This is impossible since c is the only critical point[20]. In a similar fashion, it can be proven that $f'(x) < 0$ for $x > c$. A proof of part (ii) can be patterned along the same lines. QED

1.22   The Least Squares Regression Line

Suppose we have n points (x_i, y_i), $1 \le i \le n$. Is there a line that passes through all these points? In general, this is not going to be the case. We lessen our goal and instead wish to find a line $y = a + bx$, to be called the 'least squares regression line,' in such a way that $\Sigma_{i=1}^n e_i^2$ has the smallest possible value, where $e_i = y_i - (a + bx_i)$, $1 \le i \le n$. That is to say, in such a way that the sum of the squares of the *residuals* has the smallest possible value. Letting

$$\hat{y}_i = a + bx_i,$$

we can write $e_i = y_i - \hat{y}_i$. The symbol $\hat{y}_i$ represents the value estimated for y_i by the regression line, given the value of x_i. We could try to minimize $\Sigma_{i=1}^n |e_i|$ instead of $\Sigma_{i=1}^n e_i^2$, but the presence of absolute values makes calculations difficult; numerical methods need to be used.

Let us first solve a simpler problem, namely trying to fit a least squares regression line that passes through the origin. This may well happen in practice because some models predict that a set of points will gather around a hypothetical line that passes through the origin.[21] Then we have to find the value of b that will minimize the function

$$g(b) = \Sigma_{i=1}^n (y_i - bx_i)^2.$$

As expected, we will apply the second derivative test for extrema. Indeed

$$g'(b) = \Sigma_{i=1}^n [-2(y_i - bx_i)x_i].$$

---

[20]For the same reason, the possibility $f'(d) = 0$ has to be discarded from the outset.
[21]The rationale behind such models is that y has to be 0 if x is 0.

The critical points are obtained solving the equation $0 = \Sigma_{i=1}^{n}[-2(y_i - bx_i)x_i]$ in the unknown b. That is to say, we have to solve the equation

$$0 = \Sigma_{i=1}^{n} y_i x_i - b\Sigma_{i=1}^{n} x_i^2.$$

Evidently,

$$b = \frac{\Sigma_{i=1}^{n} y_i x_i}{\Sigma_{i=1}^{n} x_i^2}. \tag{1.29}$$

This is the only critical point. Since $g''(b) = 2\Sigma_{i=1}^{n} x_i^2 > 0$ we can conclude that the minimum is attained precisely at the value of b given by (1.29).

The general case is more complicated, from a mathematical point of view, because we would have to deal with a function of two variables, namely

$$g(a, b) = \Sigma_{i=1}^{n}(y_i - a - bx_i)^2.$$

The minimum is achieved when

$$b = \frac{\Sigma_{i=1}^{n} x_i y_i - \bar{y}\Sigma_{i=1}^{n} x_i}{\Sigma_{i=1}^{n} x_i^2 - \bar{x}\Sigma_{i=1}^{n} x_i} \tag{1.30}$$

and

$$a = \bar{y} - b\bar{x}, \tag{1.31}$$

where $\bar{x}$ and $\bar{y}$ are the average of the x_i and y_i, respectively. That is to say, $\bar{x} = \frac{1}{n}\Sigma_{i=1}^{n} x_i$ and $\bar{y} = \frac{1}{n}\Sigma_{i=1}^{n} y_i$. A full justification of the steps that lead to (1.30) and (1.31) can be found in the literature (Gordon and Gordon, 2005). In summary, the least squares regression line is $\hat{y} = a + bx$, where a and b are given by (1.31) and (1.30), respectively, and the hat over y emphasizes that the values of y are not the observed values of y, but values estimated using the regression line. Sometimes the least squares regression line is called **'the line of best fit.'**

Luckily, nowadays nobody does these calculations by hand. Any graphing calculator has an option (LinReg for the TI's), which provides the least squares regression line and a number r (the 'correlation coefficient') that measures the strength and direction of the linear association of the two variables. A value of r close to 1 is desirable (when bigger values of the variable x correspond to bigger values of the variable y, that is, there is a 'positive association') or a value of r close to -1 is desirable (when bigger values of the variable x correspond to smaller values of the variable y, that is, there is a 'negative association'). The correlation coefficient, also known as Pearson's correlation coefficient, is defined by the following formula:

$$r = \frac{1}{n-1}\Sigma_{i=1}^{n}\left(\frac{x_i - \bar{x}}{s_x}\right)\left(\frac{y_i - \bar{y}}{s_y}\right) \tag{1.32}$$

where s_x and s_y are the standard deviations of the x_i and y_i, respectively. That is to say,

Table 1.4: Body Weight and Brain Weight

Body weight (x Kg)	Brain weight(y gr)	$\ln x$	$\ln y$
4.6	67	1.5260563	4.2046926
5.5	102	1.7047481	4.6249728
5.8	65.5	1.7578579	4.1820501
6	84.6	1.7917595	4.4379343
8.7	107	2.163323	4.6728288
21	183	3.0445224	5.2094862
32	179	3.4657359	5.1873858
37	343	3.6109179	5.8377304
45	360	3.8066625	5.886104
140	406	4.9416424	6.0063532

$$s_x = \sqrt{\frac{\Sigma_{i=1}^n (x_i - \bar{x})^2}{n-1}} \text{ and } s_y = \sqrt{\frac{\Sigma_{i=1}^n (y_i - \bar{y})^2}{n-1}}.$$

In the 'blink of an eye,' a graphing calculator finds (1.30), (1.31), and (1.32). The reader can learn more about the correlation coefficient, and linear regression, browsing through introductory statistics books (Seier and Joplin, 2011).

Let us illustrate the use of the least squares regression line with a particular data set relating body weight with brain weight for a family of 10 different mammals, either monkeys or apes (Table 1.4, Sacher and Staffeldt 1974).

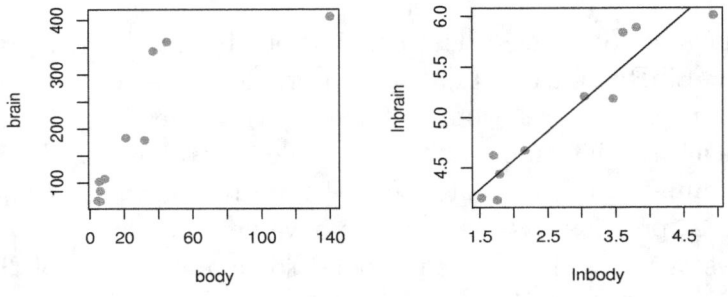

Figure 1.67: Brain weight versus body weight

We postulate a model of the type $y = Ax^\alpha$, called an 'allometric model,' where A and α are to be estimated. The model can be linearized by applying logarithms:

$$\ln y = \ln A + \alpha \ln x.$$

Thus, if $\ln x$ is on the horizontal axis and $\ln y$ is on the vertical axis, the corresponding points $(\ln x_i, \ln y_i)$ should gather around a straight line (Figure 1.67). Is this so? To answer this question we need to use the transformed values (third and fourth columns of Table 1.4) and the LinReg option of a graphing calculator or the appropriate command from a mathematical or statistical software.

On the screen we get the line of best fit $\hat{\ln} y = 0.575689 \ln x + 3.423787$ and a correlation coefficient $r = 0.950118$. Due to the fact that r is close to 1, we accept the model and proceed to estimate A and α. Indeed, $\alpha = 0.575689$ and $\ln A = 3.423787$. Hence $A = e^{3.423787} = 30.685401$. Thus, our allometric model becomes

$$y = 30.685401 x^{0.575689}.$$

For instance, we are able to predict that the brain weight of a 10 Kg. monkey will be $y = 30.685627(10)^{0.575682} \approx 115.51$ grams. The allometric model can be applied with confidence for weights ranging from 4.6 Kg. to 140 Kg., precisely the range of weights from the first column of table 1.4. Outside this range there is no assurance that the model will predict brain weight with any degree of certainty. In section 2.11, we will return to allometric models once we rigorously define x^p for any fixed real number p and $x > 0$.

We should stress that, whether or not a correlation coefficient is 'good enough,' is something that has to be determined by a scientist on the basis of the nature of the particular phenomenon under consideration. It is an important issue because it is linked to the acceptance or rejection of the model.

Exercises for Section 1.22

1. The body weight and heart weight, in kilograms, of five mammalian species have been collected (Table 1.5, Sacher and Staffeldt 1974). How well does an allometric model $y = Ax^\alpha$ agree with data? Estimate A and α.

2. The body weight and kidney weight, in kilograms, of five mammalian species have been collected (Table 1.5). How well does an allometric model $y = Ax^\alpha$ agree with data? Estimate A and α. Plot the original data and, in a separate graph, plot the natural logarithm of the data.

Table 1.5: Body Weight, Heart Weight, and Kidney Weight

Species	Body weight(x Kg)	Heart weight(y Kg)	Kidney weight(y Kg)
Grizzly Bear	140	1.106	0.532
African Buffalo	700	3.29	1.68
Caribou	98	0.882	0.1274
Holstein Cattle	600	2.22	1.44
Percheron Horse	635	5.588	1.7145

Chapter 2

Integral Calculus

2.1 Sequences

Suppose that the population of a species of insects grows 25% per week. Taking into account that the initial population is $1,000$, what will be the population size after n weeks? After one week we will have $1,000 + 0.25(1,000) = (1.25)1,000$. After two weeks we will have $(1.25)1,000 + 0.25(1.25)1,000 = (1 + 0.25)1.25(1,000) = 1.25^2(1,000)$ and so on. Thus, after n weeks we will have $1.25^n(1,000)$. For instance, if $n = 13$ we will have $1.25^{13}(1,000) = 18,190$ insects. In general, if we start with a population P_o and P_n denotes the population after n weeks, the same reasoning employed before leads to

$$P_n = (1+r)^n P_o \qquad (2.1)$$

where r is the fixed percentage of growth. Under these circumstances we say that the growth is Malthusian, in honor of the British scholar Thomas R. Malthus (1766-1834). Of course, n can denote weeks or days, or hours, depending on the nature of the problem. Bacteria in a culture with unlimited supply of nutrients and plenty of space often follow Malthusian growth. Mathematically speaking, we have a correspondence between n and P_n, which is usually denoted (P_n) and is called a 'sequence.'

Under ideal conditions, *Escherichia coli* bacteria subdivide after 20 minutes. How many will there be after 5 hours if we start with 100 bacteria? Applying (2.1), with $r = 1$ and $n = 15$ (the number of 'doubling times') we get $(1 + 1)^{15}100$; i.e., $2^{15}100$. The answer is $3,276,800$ bacteria. We can observe that the formula is $P_o 2^{t/\tau}$, where τ is the time it takes a bacterium to subdivide (in hours or fraction thereof) and t is the total time (in hours too).

Malthusian growth is closely related to compound interest. Indeed, suppose we deposit P_o dollars in a bank that offers a fixed annual rate r and compounds the interest n times per year. At the first compounding we will have $P_o + \frac{r}{n}P_o = (1+\frac{r}{n})P_o$. At the second compounding our account will amount to $(1 + \frac{r}{n})P_o + \frac{r}{n}(1 + \frac{r}{n})P_o = (1+\frac{r}{n})^2 P_o$. At the third compounding our account will have increased to $(1+\frac{r}{n})^2 P_o + \frac{r}{n}(1 + \frac{r}{n})^2 P_o$; that is to say, $(1 + \frac{r}{n})^3 P_o$ and so on. At the end of the first year, the final account will be worth $(1 + \frac{r}{n})^n P_o$.

A natural question to ask is whether compounding the interest on a monthly, or even daily basis, our account will be bigger at the end of the year. Let us analyze a particular case, namely when $r = 1$. The question is whether $(1 + \frac{1}{n})^n$ increases indefinitely. Using a graphing calculator, displaying five digits on its screen, we see that $(1 + \frac{1}{10})^{10} \approx 2.59374$, $(1 + \frac{1}{100})^{100} \approx 2.70481$, $(1 + \frac{1}{1,000})^{1,000} \approx 2.71692$, $(1 + \frac{1}{10^6})^{10^6} \approx 2.71828$.

The sequence $((1 + \frac{1}{n})^n)$ increases but 'tends to a limit when n goes to infinity,' in symbols we will write

$$\lim_{n \to \infty}(1 + \frac{1}{n})^n = e \qquad (2.2)$$

or $(1 + \frac{1}{n})^n \to e$ for short. In section 2.8 we will show that e is precisely the basis of the natural logarithm. But for now, all we know is that e is defined by (2.2) and its value is approximately 2.71828.

Convergent sequences

Given any sequence (a_n), we say that it converges to L or 'tends' to L, usually written $a_n \to L$, if a_n gets closer and closer to L when n increases. Well, not just closer but arbitrarily close! This intuitive idea of limit of a sequence is good enough for most purposes. For instance, $\frac{1}{n} \to 0$, $1 + \frac{3}{n} \to 1$, $\frac{1}{n^2+1} \to 0$, $\cos\frac{1}{n} \to 1$, $r^n \to 0$ whenever $|r| < 1$, etc. It is common to write $\lim_{n \to \infty} a_n = L$ instead of $a_n \to L$. Under these circumstances we will say that (a_n) is a convergent sequence with limit L.

Many years after Leibniz wrote the first ever published paper on the calculus, 19th century mathematicians put forward a precise definition of limit of a sequence: $a_n \to L$ if and only if for any given any $\epsilon > 0$ there exists a natural number N such that $|a_n - L| < \epsilon$ whenever $n \geq N$. Note that this definition is in agreement with our intuitive notion of limit of a sequence[1]. For example, let us try to show that $\frac{1}{n^2+1} \to 0$ using the rigorous definition of limit. Indeed, given any $\epsilon > 0$ choose a natural number N such that $N > \frac{1}{\sqrt{\epsilon}}$. We observe that if $n \geq N$ then $n > \frac{1}{\sqrt{\epsilon}}$, consequently $n^2 > \frac{1}{\epsilon}$, which in turn implies $\frac{1}{n^2+1} < \frac{1}{n^2} < \epsilon$. That is to say,

$$|\frac{1}{n^2+1} - 0| < \epsilon.$$

Convergent sequences have several important properties that will allow us to often avoid the use of the definition of limit, which is precise but not easy to employ in each

---

[1]This definition resembles the definition of limit of a function at infinity (section 1.5)

particular case. Suppose that $a_n \to L_1$ and $b_n \to L_2$. Then the following properties of limits of sequences are true: $a_n + b_n \to L_1 + L_2$, $a_n - b_n \to L_1 - L_2$, $a_n b_n \to L_1 L_2$, and $\frac{a_n}{b_n} \to \frac{L_1}{L_2}$ (provided that $L_2 \neq 0$). Moreover, if $a_n \leq b_n$, for all n, then $L_1 \leq L_2$. A particular case of the last property asserts that if $a_n \geq 0$, for all n, and $a_n \to L$, we can conclude that $L \geq 0$.

For instance,

$$\lim_{n \to \infty} \frac{1}{n} \frac{1}{3^n} = \lim_{n \to \infty} \frac{1}{n} \times \lim_{n \to \infty} \frac{1}{3^n} = 0 \times 0 = 0,$$

$$\lim_{n \to \infty} \left(\frac{1}{n^2} + \cos \frac{1}{n} \right) = \lim_{n \to \infty} \frac{1}{n^2} + \lim_{n \to \infty} \cos \frac{1}{n} = 0 + 1 = 1.$$

There is one more property, involving three sequences, that we have to keep in mind: Assume that $a_n \leq c_n \leq b_n$ for all n, and $a_n \to L$, $b_n \to L$. Then (c_n) has to converge to L too. This proposition is usually called the **squeeze property** of limits of sequences.

The squeeze property is very useful, as the next example will show. Suppose we wish to calculate $\lim_{n \to \infty} \frac{1}{n} \cos n$. We cannot use the fact that the limit of the product of two convergent sequences is equal to the product of the corresponding limits because $(\cos n)$ does not have a limit; the values of the sequence oscillate between positive and negative numbers, never clustering around a specific number. However, from pre-calculus mathematics we know that $-1 \leq \cos n \leq 1$, consequently

$$-\frac{1}{n} \leq \frac{1}{n} \cos n \leq \frac{1}{n}.$$

Since $\frac{1}{n} \to 0$ and $-\frac{1}{n} \to 0$, the squeeze property leads to the conclusion that $\frac{1}{n} \cos n \to 0$.

The above-mentioned six properties seem quite natural. Of course, a proof for each of them can be given. Let us prove the first and the last one, as an illustration of the techniques involved. Given any $\epsilon > 0$, since $a_n \to L_1$ there exists N_1 such that $|a_n - L_1| < \frac{\epsilon}{2}$ provided that $n \geq N_1$. Analogously, since $b_n \to L_2$ there exists N_2 such that $|b_n - L_2| < \frac{\epsilon}{2}$ provided that $n \geq N_2$. Let N be the maximum of N_1 and N_2. If $n \geq N$ then $|a_n - L_1| < \frac{\epsilon}{2}$ and $|b_n - L_2| < \frac{\epsilon}{2}$. Consequently,

$$|a_n + b_n - (L_1 + L_2)| = |(a_n - L_1) + (b_n - L_2)| \leq |a_n - L_1| + |b_n - L_2| < \frac{\epsilon}{2} + \frac{\epsilon}{2} = \epsilon.$$

We have succeeded in showing that $a_n + b_n \to L_1 + L_2$ if $a_n \to L_1$ and $b_n \to L_2$.

Next let us provide a proof of the squeeze property. Given any $\epsilon > 0$, since $a_n \to L$ there exists N_1 such that $L - \epsilon < a_n$ provided that $n \geq N_1$. Similarly, since $b_n \to L$ there exists N_2 such that $b_n < L + \epsilon$ provided that $n \geq N_2$. Let N be the maximum of the two natural numbers N_1, N_2. Then

$$L - \epsilon < a_n \leq c_n \leq b_n < L + \epsilon$$

provided that $n \geq N$, consequently $L - \epsilon < c_n < L + \epsilon$ whenever $n \geq N$. We have been able to show that given any $\epsilon > 0$ there exists a natural number N such that if $n \geq N$ then $|c_n - L| < \epsilon$, that is, $c_n \to L$. Luckily, for a first calculus course, we will seldom be compelled to use this level of rigor. For other proofs, the reader can browse through appendix B1[2].

Series

As the next example will show, sometimes a given sequence (a_n) leads to another sequence (b_n) where $b_n = a_1 + ... + a_n$. The sequence (b_n) is called a 'series' built from the initial sequence (a_n). Series do appear frequently in the analysis of real-life problems, as the next example will show.

A biologist has 30 cm^3 of a chemical reactant. She uses 2 cm^3 the first day and increases its use by 10% per day. How long will the reactant last? The second day she uses $2 + 0.1 \times 2 = 1.1 \times 2$ cm^3, and the third day she uses $1.1 \times 2 + 0.1(1.1 \times 2) = 1.1^2 \times 2$ cm^3. A pattern emerges and we are thus able to assert that on the nth day she will use $1.1^{n-1} \times 2$ cm^3. Then, from the first to the nth day she will use $2 + 1.1 \times 2 + 1.1^2 \times 2 + ... + 1.1^{n-1} \times 2$ cm^3, that is to say,

$$2(1 + 1.1 + 1.1^2 + ... + 1.1^{n-1}). \tag{2.3}$$

We observe that, for any $r \neq 1$,

$$1 + r + r^2 + ... + r^{n-1} = \frac{1 - r^n}{1 - r}. \tag{2.4}$$

Therefore (2.3) becomes $2(\frac{1.1^n - 1}{1.1 - 1})$. Since the total amount of the reactant is 30 cm^3 we can conclude that $20(1.1^n - 1) = 30$, that is, $1.1^n = 2.5$. We note that $1.1^{10} \approx 2.5937425$ while $1.1^{9.5} \approx 2.4730364$. Some guessing and checking leads to $1.1^{9.61} \approx 2.4991004$. Thus, we can choose $n = 9.61$ days as an approximate solution to the problem at hand. Readers knowledgeable about logarithms would get a better approximation quite fast: $n = \frac{\log 2.5}{\log 1.1} \approx 9.6138$ days.

Mathematicians have introduced a symbol for the sum $a_1 + ... + a_n$, namely $\Sigma_{i=1}^n a_i$. For instance, $\Sigma_{i=1}^n i = 1 + 2 + ... + n$ and $\Sigma_{i=1}^n i^2 = 1 + 2^2 + ... + n^2$. The reader can easily prove the following important properties, valid for any sequences (a_n), (b_n) and an arbitrary real number λ:

$$\Sigma_{i=1}^n (a_i + b_i) = \Sigma_{i=1}^n a_i + \Sigma_{i=1}^n b_i,$$

---

[2]There is a strong resemblance with the theory of limits for functions defined on an interval, as developed in chapter 1. As a matter of fact, both theories are intimately linked (see appendix B2).

$$\Sigma_{i=1}^n (\lambda a_i) = \lambda \Sigma_{i=1}^n a_i.$$

Having introduced the *Sigma* symbol, we can say that $(\Sigma_{i=1}^n a_i)$ is a series built from (a_n). The example about the biologist led to a special type of series, namely a geometric series. This is the topic of the next subsection.

Geometric series

Let us assume that a scientist uses 1 cc of a chemical reactant the first day and decreases its use by 50% per day. In the long run, how much reactant will she use? After two days she will have used $1 + \frac{1}{2}$ cc, after three days $1 + \frac{1}{2} + \frac{1}{2^2}$ cc, and after n days the total amount used will be

$$1 + \frac{1}{2} + \frac{1}{2^2} + ... + \frac{1}{2^{n-1}}.$$

We can then apply formula (2.4), with $r = 1/2$, and reach the equality

$$\Sigma_{i=0}^{n-1} \frac{1}{2^i} = \frac{1 - \frac{1}{2^n}}{1 - \frac{1}{2}} = 2 - \frac{1}{2^{n-1}}.$$

But the term $\frac{1}{2^{n-1}}$ becomes smaller and smaller as n becomes bigger and bigger. Strictly speaking, $\frac{1}{2^{n-1}} \to 0$. Consequently,

$$2 - \frac{1}{2^{n-1}} \to 2.$$

Thus, in the long run, the biologist will have used an amount very close to 2 cc.

Sequences of the form $1 + r + r^2 + ... + r^n$, that is, $\Sigma_{i=0}^n r^i$, appear quite often in mathematics and its applications (see exercise 6 at the end of this section, the example of a bouncing ball in section 3.2, and exercise 5 of section 2.9). So much so, that they have a special name: 'geometric series.' Since

$$\Sigma_{i=0}^n r^i = 1 + r + r^2 + ... + r^n = \frac{1 - r^{n+1}}{1 - r} = \frac{1}{1 - r} - \frac{r^{n+1}}{1 - r}$$

it follows that

$$\lim_{n \to \infty} \Sigma_{i=0}^n r^i = \frac{1}{1 - r}$$

whenever $|r| < 1$. This limit is usually denoted $\Sigma_{i=0}^\infty r^i$. Thus $\Sigma_{i=0}^\infty r^i = \frac{1}{1-r}$, $|r| < 1$. More generally, for any constant a we have

$$\Sigma_{i=0}^\infty a r^i = \frac{a}{1 - r}, \qquad |r| < 1. \tag{2.5}$$

Of course, we could equally write $\Sigma_{n=0}^\infty a r^n = \frac{a}{1-r}$, $|r| < 1$, because i and n are 'dummy variables.' For instance,

$$\Sigma_{n=0}^{\infty}\left(\tfrac{1}{4}\right)^n = \tfrac{1}{1-\tfrac{1}{4}} = \tfrac{4}{3},$$

$$\Sigma_{n=0}^{\infty}5\left(\tfrac{1}{3}\right)^n = \tfrac{5}{1-\tfrac{1}{3}} = \tfrac{15}{2}.$$

That is to say,

$$\lim_{n\to\infty}\left(1 + \tfrac{1}{4} + ... + \left(\tfrac{1}{4}\right)^n\right) = \tfrac{4}{3},$$

$$\lim_{n\to\infty}\left(5 + \tfrac{5}{3} + \tfrac{5}{3^2} + ... + \tfrac{5}{3^n}\right) = \tfrac{15}{2}.$$

A compact formula

Let us try to find a compact formula for $\Sigma_{i=1}^{n}i^2$, which will be very handy in section 2.2. Let $s_n = 1+2+...+(n-1)+n$, which can be written $s_n = n+(n-1)+...+2+1$. Adding both equalities we get $2s_n = n(n+1)$. Therefore

$$\Sigma_{i=1}^{n}i = \frac{n(n+1)}{2}. \tag{2.6}$$

Is there a simple expression for $\Sigma_{i=1}^{n}i^2$? Some algebraic manipulations are needed. Indeed,

$$\Sigma_{i=1}^{n}i^3 = n^3 + \Sigma_{i=1}^{n}(i-1)^3 = n^3 + \Sigma_{i=1}^{n}(i^3 - 3i^2 + 3i - 1) =$$

$$n^3 + \Sigma_{i=1}^{n}i^3 - 3\Sigma_{i=1}^{n}i^2 + 3\Sigma_{i=1}^{n}i - \Sigma_{i=1}^{n}1.$$

But $\Sigma_{i=1}^{n}1 = 1 + ...1 = n$ while $\Sigma_{i=1}^{n}i = \frac{n(n+1)}{2}$. Consequently

$$3\Sigma_{i=1}^{n}i^2 = n^3 + 3\frac{n(n+1)}{2} - n,$$

which in turn leads to

$$\Sigma_{i=1}^{n}i^2 = \frac{n}{6}(2n^2 + 3n + 1) = \frac{n}{6}(n+1)(2n+1). \tag{2.7}$$

An alternative derivation of (2.7) can be found in exercise 5 below.

Exercises for Section 2.1

1. Suppose that a population of bears in a park increases 5% per year. If there are 520 bears at the beginning, how many will there be after 10 years?

2. A biologist has 20 cm^3 of a chemical reactant. She uses 1 cm^3 the first day and increases its use by 4% per day. How long will the reactant last?

3. Using the definition of limit of a sequence, prove that $1 + \frac{3}{n} \to 1$.

4. Show that $\Sigma_{i=1}^{n}(a_i + b_i) = \Sigma_{i=1}^{n} a_i + \Sigma_{i=1}^{n} b_i$, $\Sigma_{i=1}^{n}(\lambda a_i) = \lambda \Sigma_{i=1}^{n} a_i$.

5. Compute $\Sigma_{i=1}^{n} i^2 / \Sigma_{i=1}^{n} i$ for $n = 1, 2, 3, 4, 5, \ldots$ Do you see a pattern? If you are careful, it is not hard to conjecture that

$$\Sigma_{i=1}^{n} i^2 / \Sigma_{i=1}^{n} i = \frac{2n+1}{3}.$$

Then show that $\Sigma_{i=1}^{n} i^2 / \Sigma_{i=1}^{n} i = \frac{n}{6}(n + 1)(2n + 1)$.

6. In a work on population genetics (Sved and Mayo, 1970), the series $\frac{v}{2} + \left(\frac{v}{2}\right)^2 + \left(\frac{v}{2}\right)^3 + \ldots$ is considered when analyzing the relative contributions of the heterozygote and homozygote to the gene pool at a distant time in the future; v, a proportion, is the heterozygote contribution to the next generation. Calculate $\Sigma_{n=1}^{\infty}\left(\frac{v}{2}\right)^n$.

7. Suppose that you wish to receive R dollars at the end of the first, second, ..., n^{th} year. How much do you have to deposit in a bank at the beginning, taking into consideration that the financial institution pays an interest of i percent per year? (Hint: According to the formula for compound interest we have $R = (1+i)A_1,\ldots$, $R = (1 + i)A_n$, where A_j is the amount to be deposited at the beginning if we wish to receive R dollars at the end of the jth year. Then $A_1 = R(1 + i)^{-1},\ldots$, $A_n = R(1 + i)^{-n}$ and the answer is $P = A_1 + \ldots + A_n$. Thereafter use (2.4) to reach the famous **annuities formula**, namely $P = \frac{R}{i}[1 - (1 + i)^{-n}]$.)

8. How much money do you need to deposit in a bank in order to receive 10,000 dollars indefinitely on a yearly basis? Assume that the bank agrees to use a yearly interest rate of 5 percent (Hint: Take $\lim_{n\to\infty}$ in the annuities formula).

9. Suppose that you buy a 5,000 dollars car, make a down payment of 1,000 dollars and hope to finance 4,000 dollars. The bank charges you an interest rate of 1 percent per month and expects to extend the loan to 36 months. How much would you have to pay at the end of each month? (Hint: The formula of annuities can be applied, interpreting i as 0.01 and n the number of months.)

10. A customer deposits 49,400 dollars in a bank, expecting to draw 5,000 dollars at the end of each year. Assuming that the bank offers an interest rate of 5 percent a year, for how many years will this annuity last? Round-off to the closest integer.

11. Suppose that (a_n) is a sequence of positive numbers such that $a_n \to c$, where $c > 0$. Show that $\sqrt{a_n} \to \sqrt{c}$ (Hint: Note that

$$|\sqrt{a_n} - \sqrt{c}| = \frac{|a_n - c|}{\sqrt{a_n} + \sqrt{c}} < \frac{1}{\sqrt{c}}|a_n - c|. \;)$$

.

12. Show that if $a_n \to L_1$ and $b_n \to L_2$ then $a_n - b_n \to L_1 - L_2$ (Hint: Follow the same steps used for the proof of the fact that $a_n + b_n \to L_1 + L_2$, with some needed minor modifications).

13. Show that if $a_n \geq 0$ for every n and $a_n \to L$, then $L \geq 0$.

14. Show that if $a_n \to L_1$ and $a_n \to L_2$, then $L_1 = L_2$; that is, the limit of a sequence, if it exists, has to be unique.

15. Show that if $a_n \to L$ then $a_{n-1} \to L$.

16. Assume that $a_n \to L$ and $L < M$. Show that $a_n < M$ 'eventually'. That is, there exists a natural number N such that $a_n < M$ for every $n \geq N$. In a similar fashion, if $a_n \to L$ and $M < L$ then $a_n > M$ eventually.

17. Show that if $a_n \to L$ then $|a_n| \to |L|$.

2.2   Calculating the Area under a Parabola

Suppose that we wish to find the area under the parabola $y = x^2$ between $x = 0$ and $x = b$, where $b > 0$. Letting n denote an arbitrary natural number, we divide the interval $[0, b]$ into n subintervals of length b/n and build n rectangles of equal width, namely b/n, and varying length. The ith rectangle will have width b/n and height $(ib/n)^2$, thus its area is $\frac{b}{n}\left(\frac{ib}{n}\right)^2$. Adding the area of the n rectangles we get $A_n = \Sigma_{i=1}^{n}\frac{b}{n}\left(\frac{ib}{n}\right)^2$, an approximation to the area we are looking for; the larger we choose n, the better the approximation will be. But $\Sigma_{i=1}^{n}i^2 = \frac{n}{6}(2n^2 + 3n + 1)$; therefore,

$$A_n = \frac{b^3}{n^3}\Sigma_{i=1}^{n}i^2 = \frac{b^3}{n^3}\frac{n}{6}(2n^2 + 3n + 1) = \frac{b^3}{6}\left(2 + \frac{3}{n} + \frac{1}{n^2}\right).$$

Since A_n tends to $\frac{2b^3}{6}$, when n becomes larger and larger, we can conclude that the area under the parabola, between $x = 0$ and $x = b$, is $\frac{b^3}{3}$. In the language of limits we would write $\lim_{n\to\infty} A_n = b^3/3$. Evidently, the area under the parabola, between $x = a$ and $x = b$ $(0 < a < b)$, will be $\frac{b^3}{3} - \frac{a^3}{3}$.

The success of the method to evaluate the area under the parabola depended heavily on the fact that we had a simple formula for $\Sigma_{i=1}^{n}i^2$, which was found in section 2.1. In principle, it is possible to find a closed formula for $\Sigma_{i=1}^{n}i^k$, thus the path to calculate the area under the curve $y = x^k$ between $x = 0$ and $x = b$ would be open.

However, extensive calculations are needed. One of the motivations behind the development of calculus was precisely the need to find a systematic procedure to calculate areas. Such a feat was accomplished toward the end of the 17th century through what we know nowadays as the fundamental theorem of calculus. Surprisingly, the problem of areas was shown to be intimately linked to the problem of tangents. Before we develop the theory any further, let us see next how far modern technology can help us in the task of computing areas under continuous functions.

2.3 The General Case

Let us again consider the function $f(x) = x^2$. Suppose this time we wish to calculate the area under the curve $y = x^2$ between $x = 1$ and $x = 2$. We subdivide the interval $[1, 2]$ into three subintervals of equal length, namely $1/3$, and then use the left endpoints to build three rectangles. The sum of the areas of these rectangles will be

$$L_3 = \tfrac{1}{3} + \tfrac{1}{3}(1 + \tfrac{1}{3})^2 + \tfrac{1}{3}(1 + \tfrac{2}{3})^2 = 1.85185.$$

If we rather choose the right endpoints, the sum of the areas of the rectangles will be

$$R_3 = \tfrac{1}{3}(1 + \tfrac{1}{3})^2 + \tfrac{1}{3}(1 + \tfrac{2}{3})^2 + \tfrac{1}{3}2^2 = 2.85185.$$

Approximating by trapezoids is equivalent to take the average of L_3 and R_3, that is, $T_3 = (L_3 + R_3)/2 = 2.35185$. It is to be expected that T_3 is a better approximation than either R_3 or L_3. What happens if we choose the midpoints of each subinterval and consider the corresponding rectangles? Since the midpoints are at $7/6$, $3/2$, and $11/6$ we can conclude that

$$M_3 = \tfrac{1}{3}(\tfrac{7}{6})^2 + \tfrac{1}{3}(\tfrac{3}{2})^2 + \tfrac{1}{3}(\tfrac{11}{6})^2 = 2.32407.$$

In the preceding section we learned that the area under the curve $y = x^2$, between $x = 1$ and $x = 2$, is $\tfrac{2^3}{3} - \tfrac{1^3}{3}$, that is, $7/3$, which is approximately equal to 2.33333. Thus, M_3 is a better approximation than T_3. If we define $A_3 = (2M_3 + T_3)/3$, a weighted average of M_3 and T_3, then $A_3 = 2.33333$. We note that, in the example under consideration, S_3 provides the 'exact' value of the area (up to five decimals). It should be clear that T_n and M_n become better approximations when n increases; for instance, $T_{10} = 2.335$ and $M_{10} = 2.3325$. On the other hand, L_{10} and R_{10} are also an improvement as compared to L_3 and R_3: $L_{10} = 2.185$, $R_{10} = 2.485$. Doing these calculations by hand is an onerous task, so we might well use the programming capabilities of a graphing calculator. This is what we will do next.

Let us consider a continuous function $f(x)$ defined on $[a, b]$. We divide $[a, b]$ in n subintervals of equal length; thus, the coordinates of each point of subdivision are $a, a + \tfrac{b-a}{n}, a + 2\tfrac{b-a}{n}, ..., a + n\tfrac{b-a}{n} = b$. We define:

$$L_n = \tfrac{b-a}{n} \Sigma_{i=1}^n f(a + (i-1)\tfrac{b-a}{n});$$

$$R_n = \tfrac{b-a}{n} \Sigma_{i=1}^n f(a + i\tfrac{b-a}{n});$$

$$T_n = \tfrac{L_n + R_n}{2}.$$

Thus, L_n is the approximation by left rectangles, R_n is the approximation by right rectangles, and T_n is the approximation by n trapezoids of equal height; namely $(b-a)/n$. How about M_n? The midpoint of the ith subinterval is

$$\tfrac{1}{2}(a + (i-1)\tfrac{b-a}{n} + a + i\tfrac{b-a}{n}).$$

That is to say, the midpoint is $a + (i - \tfrac{1}{2})\tfrac{b-a}{n}$. Consequently, we should define the midpoint approximation as

$$M_n = \tfrac{b-a}{n} \Sigma_{i=1}^n f(a + (i-\tfrac{1}{2})\tfrac{b-a}{n}).$$

Finally we define

$$A_n = \tfrac{2M_n + T_n}{3}.$$

This weighted average approximation is related to Simpson's method, a well-known approximation by parabolas to the area under a continuous function. In an enrichment note, at the end of this section, we will prove that $A_n = S_{2n}$, where S_{2n} is Simpson's method with $2n$ steps.

The challenge ahead of us is of a computational nature. Calculating L_n, R_n, T_n, M_n, and A_n by hand may require considerable effort. Luckily, we can build a program and use the capabilities of a graphing calculator (TI-89 or TI-92, but it can be adapted to any other similar calculator):

```
TMA()
Prgm
Input "a", a
Input "b", b
Input "n", n
a + (i − 1) ∗ (b − a)/n → p
a + i ∗ (b − a)/n → q
a + (i − 1/2) ∗ (b − a)/n → u
(b − a)/n ∗ Σ(y₁(p), i, 1, n) → l
(b − a)/n ∗ Σ(y₁(q), i, 1, n) → r
```

$$(b-a)/n * \Sigma(y_1(u), i, 1, n) \to m$$
Disp l
Disp r
Disp $(l+r)/2$
Disp m
Disp $(2m + (l+r)/2)/3$
Endprgm

It is to be noted that under the $y1$ option of the calculator we write the function that we are interested in. Once we write TMA() under the Home option and the program is run, five numbers will appear on the screen: the first corresponds to the left approximation, the next gives us the right approximation, the third is the trapezoidal approximation, while the fourth is the midpoint approximation. The last one, and the most accurate, is the weighted average between the midpoint and trapezoidal approximations.

For example, let us use the program to calculate the area under the function $f(x) = x^3$ between $x = 1$ and $x = 2$. If we choose $n = 10$ we will get on the screen the following numbers: 3.4075, 4.1075, 3.7575, 3.74625, 3.75. Choosing $n = 15$ we get 3.52, 3.98667, 3.75333, 3.74833, 3.75. We keep the last number, the weighted average of M_{10} and T_{10}, as our best approximation to the area. Indeed, under the F_3 option of the calculator we can access the command $\int$, which gives us $\int(x^3, x, 1, 2) = 3.75$. This number is the same as the weighted average obtained before. A value of $n = 15$ will be good in many cases, but a higher n might be needed. It is a sound policy to choose $n = 15$ and then $n = 20$. If the corresponding weighted averages, A_{15} and A_{20}, do not change, we can feel confident that a very good approximation has been reached.

Let us try to approximate the area under the function $f(x) = \sin x$ between 0 and π. If we choose $n = 15$ we will get on the screen 1.99268, 1.99268, 1.99268, 1.99268, 2.00366, while $n = 20$ leads to 1.99589, 1.99589, 1.99589, 1.99589, 2.00206 (make sure that your calculator is in radian mode). It seems that the answer might be 2. We try next $n = 200$ and get 1.99996, 1.99996, 1.99996, 2.00002, 2, and for $n = 500$ we get 1.99999, 1.99999, 1.99999, 2, 2. We are now confident that the answer is 2 because the weighted averages, for two different values of n, stay the same. As a matter of fact $\int(\sin x, x, 0, \pi) = 2$.

The program works well insofar as the function $f(x)$ is known explicitly. In real-life problems it may happen that we only have experimental values $y_0, y_1, ..., y_n$ obtained at intervals of time of equal length. Although the exact shape of f is unknown, we could still calculate by hand L_n, R_n, and T_n. Indeed, let us recall that if f were to be known and is defined on $[a, b]$, then

$$L_n = \frac{b-a}{n}(f(x_o) + ... + f(x_{n-1})), \quad R_n = \frac{b-a}{n}(f(x_1) + ... + f(x_n))$$

where $x_i = a + i\frac{b-a}{n}$. Therefore

$$T_n = \frac{R_n + L_n}{2} = \frac{b-a}{2n}(f(x_o) + 2f(x_1) + \dots + 2f(x_{n-1}) + f(x_n)).$$

This last expression suggests what to do when the function is not known explicitly. Suppose $n+1$ points $(x_o, y_o), \dots, (x_n, y_n)$ belong to the curve described by the function, such that all subintervals $[x_{i-1}, x_i]$ have equal length. If $x_o = a$ and $x_n = b$ then

$$T_n = \frac{b-a}{2n}(y_o + 2y_1 + \dots + 2y_{n-1} + y_n)$$

should be an approximation to the area under the the function that passes through the above-mentioned $n + 1$ points. For instance, suppose that we have 7 pairs of points $(0,28)$, $(1,32)$, $(2,40)$, $(3,45)$, $(4,49)$, $(5,62)$, $(6,70)$ that lie on the graph of a function whose explicit formula is unknown. Then

$$T_6 = \frac{6 - 0}{12}(28 + 64 + 80 + 90 + 98 + 124 + 70) = 277 \qquad (2.8)$$

provides an approximation to the area under the function between $x = 0$ and $x = 6$.

Exercises for Section 2.3

1. Use the TMA program to approximate the area between the curve $y = \cos x$, $0 \le x \le \pi/2$, and the x-axis. Write down the weighted average for $n = 5, 10, 15$ and make a conjecture about the above-mentioned area.

2. Use the TMA program to approximate the value of $\int_1^7 x^3 dx$. You are expected to find the first three 'true' decimals.

3. Suppose that we are given the points $(1,26)$, $(2,34)$, $(3,25)$, $(4,40)$, $(5,32)$. Use T_4 to approximate the area under the imaginary curve that passes through the above-mentioned five points.

4. Consider the function $f(x) = x$, $2 \le x \le 5$. Calculate R_n by hand and find $\lim_{n \to \infty} R_n$. Moreover, check your answer by finding the area of the appropriate trapezoid.

5. Use R_n to find an approximation to the area under $y = x^3$, $0 \le x \le 1$. Thereafter calculate $\lim_{n \to \infty} R_n$ (Hint: Use the fact that $\Sigma_{i=1}^n i^3 = [\frac{n(n+1)}{2}]^2$).

Enrichment Note: The Weighted Average and Simpson's Method

Let us now analyze A_n more closely. Assuming that f is defined on $[a,b]$, divide $[a,b]$ in n subintervals of equal length. Then

$$T_n = \frac{b-a}{2n}[f(a) + 2f(a + \frac{b-a}{n}) + ... + 2f(a + \frac{n-1}{n}(b-a)) + f(b)]$$

and

$$M_n = \frac{b-a}{n}[f(a + \frac{1}{2}\frac{b-a}{n}) + f(a + \frac{3}{2}\frac{b-a}{n}) + ... + f(a + (n - \frac{1}{2})\frac{b-a}{n})].$$

Thus,

$$2M_n = \frac{b-a}{2n}[4f(a + \frac{1}{2}\frac{b-a}{n}) + 4f(a + \frac{3}{2}\frac{b-a}{n}) + ... + 4f(a + (n - \frac{1}{2})\frac{b-a}{n})].$$

Defining $x_o = a$, $x_1 = a + \frac{1}{2}\frac{b-a}{n}$, $x_2 = a + \frac{b-a}{n}$, $x_3 = a + \frac{3}{2}\frac{b-a}{n}$, ... , $x_{2n-1} = a + (n - \frac{1}{2})\frac{b-a}{n}$, $x_{2n} = b$ we get

$$2M_n + T_n = \frac{b-a}{2n}[f(x_o) + 4f(x_1) + 2f(x_2) + ... + 2f(x_{2n-2}) + 4f(x_{2n-1}) + f(x_{2n})].$$

Therefore

$$A_n = \frac{2M_n + T_n}{3} =$$

$$\frac{\Delta x}{3}[f(x_o) + 4f(x_1) + 2f(x_2) + ... + 2f(x_{2n-2}) + 4f(x_{2n-1}) + f(x_{2n})]$$

where $\Delta x = \frac{b-a}{2n}$. We have succeeded in showing that $A_n = S_{2n}$, where S_{2n} is Simpson's approximation with $2n$ steps. It can be shown that Simpson's method consists in using parabolas instead of rectangles or trapezoids (Thomas, 2010). Thomas Simpson (1710-1761) was a British mathematician who wrote several noteworthy textbooks.

The last expression for A_n suggests a new approach to the problem that arises when f is not known explicitly. Suppose that $2n + 1$ points (x_o, y_o), (x_1, y_1), ..., (x_{2n-1}, y_{2n-1}), (x_{2n}, y_{2n}) lie on the curve defined by a function, where $x_i - x_{i-1} = \Delta x = \frac{b-a}{2n}$. Then, it is to be expected that

$$\frac{\Delta x}{3}[y_0 + 4y_1 + 2y_2 + 4y_3 + ... + 4y_{2n-3} + 2y_{2n-2} + 4y_{2n-1} + y_{2n}]$$

will be a good approximation to the area under the curve that passes through the above-mentioned $2n+1$ points. For instance, if we have to deal with the seven points $(0,28)$, $(1,32)$, $(2,40)$, $(3,45)$, $(4,49)$, $(5,62)$, $(6,70)$, which were used before, right away we get

$$A_3 = \frac{1}{3}[28 + 4 \times 32 + 2 \times 40 + 4 \times 45 + 2 \times 49 + 4 \times 62 + 70] \approx 273.333.$$

We might as well compare this answer with the one gotten in (2.8). It is to be expected that the weighted average will always be a better approximation than either the trapezoid or midpoint approximation.

2.4   The Definite Integral

An important proposition, which we will not prove, asserts that if f is a continuous function defined on $[a, b]$ then the three sequences (L_n), (R_n), and (M_n) converge to the same limit, which is denoted $\int_a^b f(x)dx$. Evidently, (T_n) and (A_n) will have to converge to the same limit too. Moreover, in advanced courses[3] it is proven that

$$\Sigma_{i=1}^n f(c_i)\frac{b-a}{n} \to \int_a^b f(x)dx$$

where c_i is **any** point in the ith subinterval of $[a, b]$. In other words, c_i could be the left endpoint, the right endpoint, the midpoint, or any point in the ith subinterval of common length $(b-a)/n$. Sums of the type $\Sigma_{i=1}^n f(c_i)\frac{b-a}{n}$ are called **Riemann sums**, honoring Bernhard Riemann (1826-1866), a notable 19th century mathematician.

A typical Riemann sum corresponding to $f(x) = x^2$, $0 \le x \le b$, is $\Sigma_{i=1}^n c_i^2 \frac{b}{n}$ where $c_i = i\frac{b}{n}$. In section 2.2 we found that $\Sigma_{i=1}^n c_i^2 \frac{b}{n} \to \frac{b^3}{3}$. Thus

$$\int_0^b x^2 dx = \frac{b^3}{3}.$$

The number $\int_a^b f(x)dx$ is known as the 'definite integral of f between a and b,' or **definite integral** for short. If $f(x) \ge 0$ on $[a, b]$, we define the area between f and the x-axis as $\int_a^b f(x)dx$. If $f(x) < 0$ on $[a, b]$ we can also develop a geometric interpretation of $\int_a^b f$: the number $\int_a^b f$ will be negative and it is $-1\times$area of the region between the x-axis and the curve $y = f(x)$, $a \le x \le b$. The symbol $\int$ reminds us that the definite integral stems from sums, while the symbol $f(x)dx$ says something about the nature of such sums. Sometimes we will simply write $\int_a^b f$ to denote the definite integral. It is advantageous to define $\int_a^a f = 0$ and $\int_b^a f = -\int_a^b f$.

Usually it is not easy to calculate $\int_a^b f(x)dx$ from its definition, except in simple cases such as $\int_3^8 7dx$ and $\int_2^5 xdx$. The value of the first integral is $7(8 - 3)$ while the value of the second integral is $\frac{3}{2}(5 + 2)$ because we are dealing with the area of a rectangle and a trapezoid respectively. By the same token, $\int_a^b \lambda dx = \lambda(b - a)$ (λ any real number) and $\int_a^b xdx = \frac{1}{2}(b - a)(b + a)$. In the next section we will study a systematic procedure for the calculation of $\int_a^b f(x)dx$. In the meantime, let us recall that we have the TMA program, which provides a way to reach a good approximation due to the fact that (A_n), the weighted average sequence, converges to $\int_a^b f$ 'pretty fast.'

We should point out that the definition of the definite integral as a limit is needed to show the following simple properties, valid for any number λ and any number c between a and b. Provided f, g are continuous on $[a, b]$,

---

[3]This is a topic that is covered in every course on real analysis for juniors.

(i) $\int_a^b (f \pm g) = \int_a^b f \pm \int_a^b g$;

(ii) $\int_a^b \lambda f = \lambda \int_a^b$;

(iii) $\int_a^b f = \int_a^c f + \int_c^b f$.

For instance, $\int_3^7 (x^2 - \cos x)dx = \int_3^7 x^2 dx - \int_3^7 \cos x dx$, $\int_0^4 6 \sin x dx = 6 \int_0^4 \sin x dx$, $\int_1^5 \arctan x dx = \int_1^2 \arctan x dx + \int_2^5 \arctan x dx$.

We might as well prove one of the properties to illustrate the techniques involved:

Divide the interval $[a, b]$ in n subintervals of length $(b - a)/n$ and let c_i be an arbitrary point in the ith subinterval. Then $\int_a^b (f + g) = \lim_{n \to \infty} \Sigma_{i=1}^n (f + g)(c_i)\frac{b-a}{n} = \lim_{n \to \infty}[\Sigma_{i=1}^n f(c_i)\frac{b-a}{n} + \Sigma_{i=1}^n g(c_i)\frac{b-a}{n}]$

$= \lim_{n \to \infty} \Sigma_{i=1}^n f(c_i)\frac{b-a}{n} + \lim_{n \to \infty} \Sigma_{i=1}^n g(c_i)\frac{b-a}{n} = \int_a^b f + \int_a^b g$.

One further property has to be remembered:

(iv) If $f(x) \leq g(x)$ for every x in $[a, b]$, then $\int_a^b f(x)dx \leq \int_a^b g(x)dx$.

Moreover, we can calculate the area between $y = g(x)$ and $y = f(x)$, where $f(x) \leq g(x)$ on $[a, b]$. It is precisely $\int_a^b g(x)dx - \int_a^b f(x)dx$, that is, $\int_a^b (g(x) - f(x))dx$. For instance, the area between the line $y = x$ and the curve $y = x^2$, $0 \leq x \leq 1$, is $\int_0^1 x dx - \int_0^1 x^2 dx = \frac{1}{2} - \frac{1}{3} = \frac{1}{6}$ (Figure 2.1).

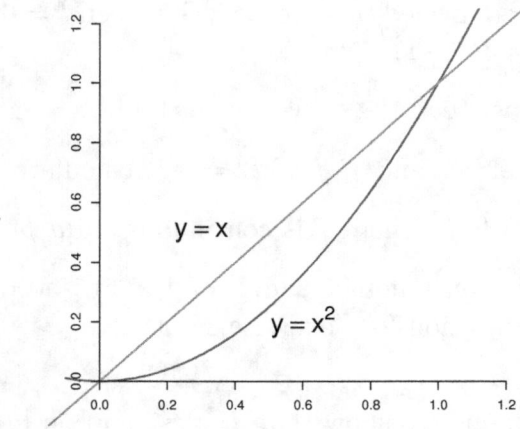

Figure 2.1: The area between two curves

Riemann sums appear quite often in the natural sciences. For instance, suppose that an object starts at $t = 0$, moves to the right, and its velocity is a continuous function $v(t)$. We would like to calculate the distance traveled between $t = 0$ and $t = T$, namely $s(T) - s(0)$, where $s(t)$ denotes the space function (as we know from differential calculus, $v(t) = s'(t)$). With this purpose in mind, we divide $[0, T]$ in n subintervals $[0, t_1]$, $[t_1, t_2]$,..., $[t_{n-1}, t_n]$ of equal length, namely T/n. Assuming that n is really large, we may think that $v(t)$ is practically constant on any arbitrary subinterval $[t_{i-1}, t_i]$; thus, $v(t_i) \times \frac{T}{n}$ should provide an approximation to the distance traveled by the object between $t = t_{i-1}$ and t_i. Consequently, $\Sigma_{i=1}^{n} v(t_i)\frac{T}{n}$ is an approximation to the distance traveled between $t = 0$ and $t = T$. This is a Riemann sum corresponding to the function $v(t)$, $0 \le t \le T$. Letting $n \to \infty$ we can conclude that

$$s(T) - s(0) = \int_0^T v(t)dt.$$

Exercises for Section 2.4

1. Suppose $v(t) = t$ is the velocity function of a body. Calculate the distance traveled by it between $t = 0$ and $t = 10$.

2. What is the area between the x-axis and the curve $y = -x^2$, $0 \le x \le 1$?

3. Based on the geometrical interpretation of the definite integral, calculate $\int_{-3}^{2} x dx$.

4. Calculate $\int_{-3}^{2} |x| dx$.

5. What is the distance traveled by an object, between $t = 0$ and $t = 5$, given that its velocity is $v(t) = 3t^2 + 1$?

6. Calculate the area between the curves $y = x$ and $y = -x^2$

7. Given that $\int_1^3 f(x)dx = 2$ and $\int_1^2 f(x)dx = -1$, calculate $\int_2^3 f(x)dx$.

8. Show that $\int_a^b \lambda f = \lambda \int_a^b f$, where f is continuous on $[a, b]$.

9. Show that if $h(x)$ is continuous on $[a, b]$ and $h(x) \ge 0$ on $[a, b]$ then $\int_a^b h \ge 0$ (Hint: Recall, from section 2.1, exercise 13, that $a_n \ge 0$ and $a_n \to L$ implies $L \ge 0$.)

10. Use the previous exercise to show that $f(x) \le g(x)$ on $[a, b]$ implies $\int_a^b f \le \int_a^b g$ (we are tacitly assuming that both functions are continuous on $[a, b]$).

11. Calculate $\int_0^3 (x-1)dx$ using areas for that purpose (Hint: Start by finding where the function $f(x) = x - 1$, $0 \le x \le 3$, crosses the $x-$ axis.)

2.5   The Evaluation Theorem

Although we know how to approximate $\int_a^b f$ for any continuous function defined on $[a, b]$, little has been said so far about the possibility of finding the 'exact' value of this integral. In the preceding section we found that $\int_0^t v(t)dt = s(T) - s(0)$, where $s'(t) = v(t)$. A startling theorem, called 'The Evaluation Theorem' (ET for short) tells us that this is true in general: If f is continuous on $[a, b]$ and F is an antiderivative of f on $[a, b]$, that is, F is continuous on $[a, b]$ and $F'(x) = f(x)$ on (a, b), we will have

$$\int_a^b f(x)dx = [F(x)]_a^b$$

where the symbol $[F(x)]_a^b$ denotes the number $F(b) - F(a)$.

For instance, $\int_1^2 x^3 dx = [\frac{x^4}{4}]_1^2 = \frac{1}{4}(2^4 - 1) = 3.75$ because $(\frac{x^4}{4})' = x^3$, while $\int_0^\pi \sin x dx = [-\cos x]_0^\pi = 2$ due to the fact that $(-\cos x)' = \sin x$. Obviously,

$$\int_a^b f'(x)dx = f(b) - f(a)$$

provided that f' is continuous, an immediate consequence of ET that we have to keep in mind.

The evaluation theorem will be useful insofar as it is possible to find an antiderivative of a given continuous function.

From table 1.3 (section 1.16) we can conclude, for example, that:

$$\int_a^b \frac{1}{\sqrt{1-x^2}}dx = [\arcsin x]_a^b,$$

$$\int_a^b \sec^2 x dx = [\tan x]_a^b,$$

$$\int_a^b \sec x \tan x dx = [\sec x]_a^b,$$

$$\int_a^b x^n dx = [\frac{x^{n+1}}{n+1}]_a^b, \; n \neq -1,$$

$$\int_a^b \frac{1}{x}dx = [\ln x]_a^b,$$

$$\int_a^b \frac{1}{x^2+1}dx = [\arctan(x)]_a^b$$

wherever the values of a and b 'make sense.'

Could we find the value of $\int_{-1}^1 \sqrt{1 - x^2}dx$? We have no idea on what could be an antiderivative of $f(x) = \sqrt{1 - x^2}$. Thus, the Evaluation Theorem cannot be applied; it remains silent. However, the geometrical interpretation of a definite integral, as an area, solves the problem: $y = \sqrt{1 - x^2}$, $-1 \le x \le 1$, is the equation of the upper semi-circle of radius 1, consequently $\int_{-1}^{1} \sqrt{1 - x^2} dx = \pi/2$. In section 2.22 we will learn that an antiderivative of $f(x) = \sqrt{1 - x^2}$ happens to be the rather complicated function

$$F(x) = \tfrac{1}{2} \arcsin x + \tfrac{1}{2} x \sqrt{1 - x^2}.$$

It is easy to check that, as expected, $[\tfrac{1}{2} \arcsin x + \tfrac{1}{2} x \sqrt{1 - x^2}]_{-1}^{1} = \pi/2$.

Is it possible to provide an accessible proof of ET? The answer is yes! Divide the interval $[a, b]$ in n subintervals of equal length, namely $\Delta x = (b - a)/n$. Let $a = x_o < x_1 < ... < x_{n-1} < x_n = b$ be the subdivision points. Applying the mean value theorem n times we can assert that there exist x_i^*, $1 \le i \le n$, with x_i^* satisfying the inequality $x_{i-1} < x_i^* < x_i$, such that

$$F(x_1) - F(x_o) = F'(x_1^*)\Delta x, ..., F(x_n) - F(x_{n-1}) = F'(x_n^*)\Delta x.$$

Adding these n equalities, and keeping in mind that $F'(x_i^*) = f(x_i^*)$, we get

$$F(b) - F(a) = \Sigma_{i=1}^{n} f(x_i^*)\Delta x.$$

But $\lim_{n \to \infty} \Sigma_{i=1}^{n} f(x_i^*)\Delta x = \int_a^b f$, consequently $F(b) - F(a) = \int_a^b f$. QED

Applications

As a way of applying the techniques of integral calculus learned so far, let us study five problems from the fields of physics, biology, and human demographics:

1. The first three problems have to do with the idea of **work**, a word that has a precise meaning in the realm of physics. Suppose that a *constant* force F acts on an object and moves it a distance d along the x-axis. Then we define $W = F \times d$, and call W the work done. What should we do if the force is not constant but a continuous function of position? Assume that the object is moved from a to b by a force $F(x)$ along the x-axis. We subdivide the interval $[a, b]$ into n subintervals of equal length, $[x_{i-1}, x_i]$, where $x_i = a + i\frac{b-a}{n}$. When n is large, $F(x_i)$ can be chosen as the constant value of F on $[x_{i-1}, x_i]$. Then $F(x_i)\frac{b-a}{n}$ is a 'good' approximation to the work done across the above-mentioned subinterval. We might all agree that

$$\Sigma_{i=1}^{n} F(x_i)\frac{b-a}{n}$$

is an approximation to the work done across $[a, b]$. But this is a Riemann sum, consequently

$$\lim_{n \to \infty} \Sigma_{i=1}^{n} F(x_i)\frac{b-a}{n} = \int_a^b F(x)dx.$$

It then makes a lot of sense to define the work W by the formula

$$W = \int_a^b F(x)dx. \tag{2.9}$$

Next let us consider an object that lies on a frictionless surface and is subjected to the restoring force $-kx$ of a spring, which is attached to a wall. To stretch the spring from $x = a$ to $x = b$ $(a < b)$ we need to apply a force to overcome the restoring force of the spring: $F(x) = kx$. Then the work done is

$$W = \int_a^b kxdx = k[\tfrac{x^2}{2}]_a^b = \tfrac{k}{2}(b^2 - a^2).$$

An analogous analysis can be made when we are compressing the spring.

2. Suppose that we have two point masses, m_1 and m_2, at a distance a from each other. We wish to put them at a distance b from each other, where $a < b$. We have to overcome the gravitational attraction, namely $k\frac{m_1 m_2}{x^2}$. Thus, the work done will be

$$W = \int_a^b \tfrac{km_1 m_2}{x^2}dx = km_1 m_2 \int_a^b x^{-2}dx = -km_1 m_2[x^{-1}]_a^b = -km_1 m_2(\tfrac{1}{b} - \tfrac{1}{a}) = km_1 m_2(\tfrac{1}{a} - \tfrac{1}{b}).$$

3. A gas is inside a cylinder of radius r, whose volume can be changed at will thanks to a movable piston on one end. Assuming that the temperature is kept constant, Boyle's law asserts that the pressure is inversely proportional to the volume, that is, $P = k\frac{1}{V}$. Assume that the piston is moved to the left compressing the gas, from a position x_1 to a position x_2. Evidently, V is a function of x (the position of the piston). We would like to know the work done in the process of compression.

Since the force on the piston, F, equals area $\times$ pressure and it opposes the direction of the movement, we will have

$$F = -k\tfrac{\pi r^2}{V} = -k\tfrac{\pi r^2}{\pi r^2 x} = -k\tfrac{1}{x}.$$

Consequently,

$$W = -k \int_{x_1}^{x_2} \tfrac{1}{x}dx = -k[\ln x]_{x_1}^{x_2} = -k \ln \tfrac{x_2}{x_1} = -k \ln \tfrac{\pi r^2 x_2}{\pi r^2 x_1} = k \ln \tfrac{V_1}{V_2}.$$

4. Let us consider a problem from human physiology, namely the flow of blood across an artery of radius R. Since the artery has a cylindrical shape, according to Poiseuille's law[4] the velocity at which blood flows is given by $v(r) = k(R^2 - r^2)$, $0 \leq r \leq R$, where r is the distance to the axis of the cylinder. The parameter k depends on three factors: the viscocity of blood, the length of the artery, and the difference of pressure between both ends. As expected, the velocity is zero at $r = R$ while it adopts a maximum value at $r = 0$. Our goal is to calculate the volume of blood that flows through any cross section in unit time.

We subdivide the circular cross section in n concentric circles with radii $r_i = i\frac{R}{n}$ ($i = 1, ..., n$). Let $\Delta r = R/n$ and let us look at the ring between the circles with radii r_i and $r_{i+1} = r_i + \Delta r$. The area of this ring is

$$\pi(r_i + \Delta r)^2 - \pi r_i^2 = 2\pi r_i \Delta r + \pi(\Delta r)^2 \approx 2\pi r_i \Delta r.$$

It is to be noted that in the last step we discarded $\pi(\Delta r)^2$ because it is a much smaller number than $2\pi r_i \Delta r$ when n, the number of subdivisions, is large. Since the ring is very narrow, the velocity of blood over the ring can be approximated by $v(r_i)$. Thus, the volume of blood flowing through the ring, per unit time, is approximately $2\pi r_i \Delta r v(r_i)$ (observe that area×velocity = area × distance/time = volume/time). Consequently $\Sigma_{i=1}^{n} 2\pi r_i \Delta r v(r_i)$ is an approximation to the volume of blood flowing through a cross section per unit time.

The approximation becomes better when n increases and, making $n \to \infty$, we get $\int_0^R 2\pi r v(r) dr$. All that remains to do is to calculate the integral $V = \int_0^R 2\pi r k(R^2 - r^2) dr$. Indeed

$$V = 2\pi k R^2 \int_0^R r\, dr - 2\pi k \int_0^R r^3 dr = 2\pi k R^2 \frac{R^2}{2} - 2\pi k \frac{R^4}{4} = \pi k \frac{R^4}{2}.$$

We have reached the conclusion that the volume of blood that flows through a cross section of an artery, per unit time, is proportional to the fourth power of its radius.

5. A closely related problem, from a mathematical point of view, has to do with human demographics. Suppose that $D(r)$ is the population density as a function of the distance to the center of a city (demographers would put forward, based on data, a decreasing function; say, $D(r) = 100,000 e^{-0.15r}$). What is the total population, Pop, within a radius R from the center?

As we did in the previous example, subdivide the circle of radius R in n concentric circles with radii $r_i = i\frac{R}{n}$ (i=1,...n) and let $\Delta r = \frac{R}{n}$, n a large natural

---

[4]This law was introduced in section 1.7 (exercise 11).

number. The area of the ring between the circles with radii r_i and $r_{i+1} = r_i + \Delta r$ is, as we found before,

$$\pi(r_i + \Delta r)^2 - \pi r_i^2 = 2\pi r_i \Delta r + \pi(\Delta r)^2 \approx 2\pi r_i \Delta r.$$

Since the ring is very narrow, we can assume that the population density, at any point of the ring, is approximately $D(r_i)$. Therefore, the population inside the ring will be $2\pi r_i \Delta r D(r_i)$. Thus, the population inside the circle of radius R will be, approximately,

$$\Sigma_{i=1}^n 2\pi r_i D(r_i) \Delta r.$$

This is a Riemann sum corresponding to the function $2\pi r D(r)$. Consequently, when $n \to \infty$ we get

$$\text{Pop} = \int_0^R 2\pi r D(r) dr.$$

Having found an expression for the total population within a distance R from the center, we can proceed to calculate Pop for a given density function $D(r)$ (see exercise 15 in section 2.19).

Exercises for Section 2.5

1. Calculate $\int_0^\pi \cos x \, dx$, $\int_{\pi/2}^\pi \cos x \, dx$. Interpret both definite integrals in terms of areas.

2. What is the exact value of $\int_{-1}^4 \frac{dx}{x^2+1}$?

3. Calculate the area of the region between $y = x^2$ and $y = x^3$, $0 \le x \le 1$.

4. We learned that the volume of blood that flows through a cross section of an artery, per unit time, is $\frac{\pi k R^4}{2}$. Suppose that a stent is introduced into the artery and the radius of it is increased by 10%. By how much is the volume of blood increased per unit time?

5. Calculate $\int_{3\pi/4}^\pi \sec^2 x \, dx$ and sketch the corresponding area.

6. Calculate $\int_0^{\pi/4} \tan^2 x \, dx$ (Hint: Recall that $1 + \tan^2 x = \sec^2 x$).

7. Assuming that the radius of an artery varies with time, calculate $\frac{1}{V(t)} \frac{dV}{dt}$ in terms of $\frac{1}{R(t)} \frac{dR}{dt}$, that is, calculate the relative growth rate of V in terms of the relative growth rate of R.

8. Calculate $\int_1^\pi 2xe^{x^2}$ (Hint: Observe that $2xe^{x^2} = (e^{x^2})'$.)

9. What is the work done by the force $F(x) = \frac{2}{x}$ in moving an object along the x-axis from $x = 2$ to $x = 4$?

10. Calculate $\int_0^{-1} \sin(2x)dx$.

11. Calculate the area of the region between the curves $y = x^2$ and $y = \frac{1}{x^2+1}$ (Hint: Use technology to find the points of intersection of both curves; if a graphing calculator is not available, solve the pertinent quartic through the transformation $z = x^2$).

12. Let $f(x) = x$, $0 \le x \le 1$ and $F(x) = \frac{x^2}{2}$, $0 \le x < 1$, $F(1) = 0$. We do have $F'(x) = f(x)$, $0 < x < 1$, $F(1) - F(0) = 0 - 0 = 0$, and $\int_0^1 xdx = \frac{1}{2}$. Thus, $\int_0^1 xdx \ne F(1) - F(0)$. Why the Evaluation Theorem does not hold true in this particular example?

2.6   Indefinite Integrals

Since the problem of evaluating $\int_a^b f$ becomes straightforward if an antiderivative of f can be found, it is worth having at hand a list of antiderivatives. The symbol

$$\int f(x)dx$$

is called the 'indefinite integral' of f, and will denote the family of all antiderivatives of the given function f. Let us recall from differential calculus that if F is an antiderivative of f then all antiderivatives of f are given by $F(x) + C$, where C is a constant. Thus

$$\int f(x)dx = F(x) + C.$$

Then Table 1.3 (section 1.16) leads to the following list of indefinite integrals:

1. $\int x^n dx = \frac{x^{n+1}}{n+1} + C$, $n \ne -1$ (n rational)

2. $\int \frac{1}{x}dx = \ln|x| + C$, x in I, I any interval that does not contain the number 0.

3. $\int \cos xdx = \sin x + C$

4. $\int \sin xdx = -\cos x + C$

5. $\int \sec^2 xdx = \tan x + C$, $-\frac{\pi}{2} < x < \frac{\pi}{2}$.

6. $\int \sec x \tan xdx = \sec x + C$, $-\frac{\pi}{2} < x < \frac{\pi}{2}$.

7. $\int \frac{1}{x^2+1}dx = \arctan x + C$

8. $\int \frac{1}{\sqrt{1-x^2}} dx = \arcsin x + C,\ -1 < x < 1.$

9. $\int e^x dx = e^x + C$

10. $\int \frac{1}{x\sqrt{x^2-1}} dx = \sec^{-1}|x| + C,\ \ x > 1 \text{ or } x < -1.$

Due to the Evaluation Theorem, the symbol $\int f(x)dx$ makes a lot of sense. For instance, $\int \frac{1}{x^2+4} dx = \frac{1}{2}\arctan \frac{x}{2} + C$. Why? The derivative of $\frac{1}{2}\arctan \frac{x}{2}$ is precisely $\frac{1}{x^2+4}$, so we will have $\int_{-2}^{1} \frac{1}{x^2+4} dx = [\frac{1}{2}\arctan \frac{x}{2}]_{-2}^{1}$.

We could have written $\frac{1}{2}\arctan \frac{x}{2} + 7$ inside the brackets, or for that matter $\frac{1}{2}\arctan \frac{x}{2} + C$ where C is any constant; after all, $[F(x) + C]_a^b = [F(x)]_a^b$. But evidently it is easier to choose $C = 0$ when evaluating $\int_a^b f$. We will make extensive use of indefinite integrals in forthcoming sections.

It is to be noted that if $u(x)$ is a function whose derivative is continuous, and the composition under consideration makes sense, the chain rule allows us to build a new list of indefinite integrals:

1. $\int u^n(x)u'(x)dx = \frac{u^{n+1}(x)}{n+1} + C,\ n \neq -1\ (n \text{ rational})$

2. $\int \frac{u'(x)}{u(x)} dx = \ln|u(x)| + C$

3. $\int \cos(u(x))u'(x)dx = \sin(u(x)) + C$

4. $\int \sin(u(x))u'(x)dx = -\cos(u(x)) + C$

5. $\int \sec^2(u(x))u'(x)dx = \tan(u(x)) + C$

6. $\int \sec(u(x))\tan(u(x))u'(x)dx = \sec(u(x)) + C$

7. $\int \frac{u'(x)}{u^2(x)+1} dx = \arctan(u(x)) + C$

8. $\int \frac{u'(x)}{\sqrt{1-u^2(x)}} dx = \arcsin(u(x)) + C$

9. $\int e^{u(x)}u'(x)dx = e^{u(x)} + C$

10. $\int \frac{u'(x)}{u(x)\sqrt{u^2(x)-1}} = \sec^{-1}(|u(x)|) + C$

Of course, there are some restrictions that have to be imposed on $u(x)$. For instance, in the second entry we need to have either $u(x) > 0$ or $u(x) < 0$ on the interval under consideration. If $u(x) > 0$ for every x in an interval I, then

$$\tfrac{d}{dx}\ln|u(x)| = \tfrac{d}{dx}\ln u(x) = \tfrac{u'(x)}{u(x)},$$

while if $u(x) < 0$ for every x in an interval I then

$$\tfrac{d}{dx}\ln|u(x)| = \tfrac{d}{dx}\ln(-u(x)) = \tfrac{-u'(x)}{-u(x)} = \tfrac{u'(x)}{u(x)}.$$

On the other hand, in the eighth entry the function $u(x)$ must satisfy the inequality $|u(x)| < 1$. This new list is important because it will allow us to solve differential equations like $y'(x) = -ky(x)$, $y'(x) = -ky^2(x)$, and the like.

Thanks to the Evaluation Theorem, calculating the definite integral $\int_a^b f(x)dx$ reduces to the task of calculating $\int f(x)dx$. Let us illustrate this fact: Suppose it is necessary to calculate $\int_{-1}^2 \frac{dx}{x^2+16}$. We notice that

$$I = \int \frac{1}{x^2+16}dx = \int \frac{1}{16((\frac{x}{4})^2+1)}dx.$$

Defining $u(x) = \frac{x}{4}$ we get

$$I = \frac{4}{16}\int \frac{1/4}{u^2(x)+1}dx = \frac{1}{4}\int \frac{u'(x)}{u^2(x)+1}dx = \frac{1}{4}\tan^{-1}(u(x)) + C = \frac{1}{4}\tan^{-1}\frac{x}{4} + C.$$

Therefore, $\int_{-1}^2 \frac{dx}{x^2+16} = \frac{1}{4}[\tan^{-1}(\frac{x}{4})]_{-1}^2$. A more systematic study of these techniques will be carried out in section 2.18.

Exercises for section 2.6

Calculate the following five indefinite integrals:

1. $\int x^{3/4}dx$.

2. $\int \cos(2x)dx$.

3. $\int \frac{1}{x^2+9}dx$.

4. $\int x\sin(x^2)dx$.

5. $\int \frac{1}{\cos^2(\frac{x}{3})}dx$.

6. Find a simple formula for the indefinite integral $\int \frac{1}{x^2+a^2}dx$.

7. Evaluate $\int_{-1}^2 (x+1)e^{x^2+2x+3}dx$.

8. Evaluate $\int_0^{\pi/4} \frac{\sin x}{\cos x}dx$.

9. Evaluate $\int_0^{1/3} \frac{1}{\sqrt{1-2x^2}}dx$.

10. Evaluate $\int_0^{\pi/10} \sec^2(4x)dx$.

11. Evaluate $\int \frac{x^3+2x+1}{x^2+1}dx$.

2.7   The Fundamental Theorem of Calculus

The Evaluation Theorem, studied in section 2.5, solves the problem of evaluating $\int_a^b f(x)dx$, f continuous on $[a,b]$, if somehow we know an antiderivative of f. A natural question to ask is whether every continuous function has an antiderivative and, if it exists, what is the nature of the latter. For instance, does $\ln^2(x)$ or e^{-x^2} have an antiderivative? If the answer were to be yes, what are they?

Let $F(x) = \int_3^x \cos u\, du$ for every x. Certainly $F(x)$ is a bona fide function defined over the whole real line. Thanks to the Evaluation Theorem we will have

$$F(x) = [\sin u]_3^x = \sin x - \sin 3.$$

Therefore $F'(x) = \cos x$. Similarly, if $F(x) = \int_0^x \frac{1}{u^2+1} du$ we will have

$$F(x) = [\arctan u]_0^x = \arctan x - \arctan 0.$$

Consequently $F'(x) = \arctan'(x) = \frac{1}{x^2+1}$.

We might then conjecture that if f is a continuous function defined on an interval I, a an arbitrary (fixed) point of I and

$$F(x) = \int_a^x f(u)du \text{ for every } x \text{ in } I,$$

then

$$F'(x) = f(x) \text{ for every } x \text{ in } I$$

where it is to be understood that, at the endpoints of I, we are dealing with lateral derivatives. Using Leibniz's notation, we would write

$$\tfrac{d}{dx} \int_a^x f(u)du = f(x)$$

provided that f is continuous on I. This is the statement of an extremely important theorem, so much so that it is called **The Fundamental Theorem of Calculus** (FTC).

If we have an antiderivative G of f, then ET asserts that $\int_a^x f(u)du = G(x)-G(a)$. Hence

$$\tfrac{d}{dx} \int_a^x f(u)du = \tfrac{d}{dx}(G(x) - G(a)) = \tfrac{d}{dx}G(x) = f(x).$$

Thus, under these circumstances FTC is not quite so interesting. However, the Fundamental Theorem of Calculus becomes particularly important when we have no idea about what could be an antiderivative of f. **FTC implies that every continuous function f has an antiderivative**, namely

$$F(x) = \int_a^x f(u)du.$$

For instance, let $f(x) = e^{-x^2}$. Then $F(x) = \int_0^x e^{-u^2} du$ is an antiderivative of f. The function $F(x)$ is not 'simple,' nonetheless we have that $F'(x) = e^{-x^2}$, so F is increasing everywhere. Moreover, $F''(x) = -2xe^{-x^2}$. Hence F is concave down on $(0, \infty)$ and concave up on $(-\infty, 0)$. Thus, F has a shape that is similar to the well-known function arctan. Using techniques from advanced calculus, one can prove that

$$\lim_{x \to \infty} F(x) = \frac{\sqrt{\pi}}{2} \text{ and } \lim_{x \to -\infty} F(x) = -\frac{\sqrt{\pi}}{2}.$$

One of the most notable features of FTC is its ability to allow the definition of a wide variety of new functions of great importance in mathematics and statistics; the next section will exemplify this use of FTC. However, the Fundamental Theorem of Calculus is not only important on a theoretical level. Rather FTC is a powerful tool, indispensable for the solution of several problems that appear in applied mathematics. Moreover, FTC is the key to the solution of any initial value problem of the type $y'(x) = f(x)$, $y(x_o) = y_o$, where $f(x)$ is a continuous function defined on an interval I (x_o is any fixed point of I, while y_o is an arbitrary real number). Indeed, if there is a solution $y(x)$ then $\int_{x_o}^x y'(u) du = \int_{x_o}^x f(u) du$. But $\int_{x_o}^x y'(u) du = y(x) - y(x_o)$, therefore

$$y(x) = y_o + \int_{x_o}^x f(u) du$$

is the only 'candidate' for a solution of the given IVP. We can check that $y(x_o) = y_o + \int_{x_o}^{x_o} f(u) du = y_o$, while FTC confirms that $y'(x) = f(x)$ for any x in I. Thus, $y(x) = y_o + \int_{x_o}^x f(u) du$ is the only solution of the IVP.

For instance, let us try to solve the IVP $y'(x) = e^{x^2}$, $y(0) = 4$. The solution will be

$$y(x) = 4 + \int_0^x e^{u^2} du.$$

This is a non-elementary function, in a sense to be defined in section 2.30. We notice that $y'(x) = e^{x^2}$ and $y''(x) = 2xe^{x^2}$; thus, $y(x)$ is increasing everywhere, concave down on $(-\infty, 0)$, and concave up on $(0, \infty)$. Using the TMA program we could approximate $y(x)$ at any x, a fact that leads to a graph of $y(x)$ (see exercise 6 at the end of the section). In section 1.16 we were already able to solve differential equations of the type $y'(x) = f(x)$ when it was easy to find an antiderivative of f. Thanks to FTC, we have now found a systematic procedure that works always for any continuous function f, even when there is no simple antiderivative of f.

It is to be noted that, using FTC, the Evaluation Theorem can be proved in a few steps (see exercise 7 at the end of the section). In other words, it is a corollary of FTC. We hope that by now the reader will understand why we did not reserve the word 'fundamental' when referring to the Evaluation Theorem.

A restricted proof

We will provide a geometrical proof of FTC, with the added hypotheses that the function is increasing and positive , besides being continuous. The reader can consult the enrichment note, at the end of this section, for a proof without the added hypotheses.

Let us assume that $f(x)$ is **continuous, positive, and increasing** on an interval I. Choose a point a in I and define

$$F(x) = \int_a^x f(t)dt$$

for every x in I. We will show that $F'(x_o) = f(x_o)$ for any fixed number x_o in the interior of I, in other words

$$\lim_{h \to 0} \frac{F(x_o+h)-F(x_o)}{h} = f(x_o).$$

If $h > 0$ we have

$$F(x_o + h) - F(x_o) = \int_a^{x_o+h} f(t)dt + \int_{x_o}^a f(t)dt = \int_{x_o}^{x_o+h} f(t)dt.$$

Since f is increasing and positive, the area of the rectangle of width h and height $f(x_o)$ is less than the area under f between x_o and $x_o + h$ (Figure 2.2), that is,

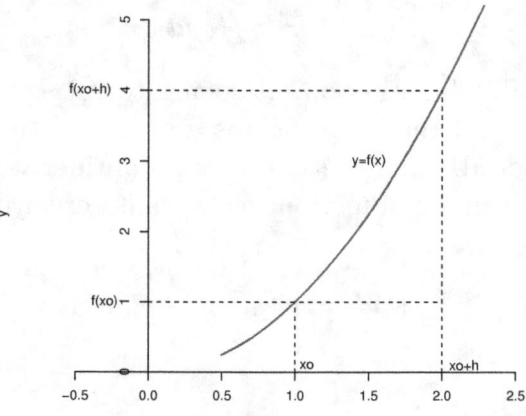

Figure 2.2: An increasing, continuous, positive function

$$f(x_o)h < \int_{x_o}^{x_o+h} f(t)dt.$$

A similar argument leads to the inequality

$$\int_{x_o}^{x_o+h} f(t)dt < hf(x_o+h).$$

Putting these inequalities together we get

$$f(x_o) < \tfrac{1}{h} \int_{x_o}^{x_o+h} f(t)dt < f(x_o+h).$$

The continuity of f implies $\lim_{h \to 0+} f(x_o+h) = f(x_o)$. Then the squeeze property of limits (section 1.4) allows us to conclude that

$$\lim_{h \to 0+} \tfrac{F(x_o+h)-F(x_o)}{h} = f(x_o).$$

Following the same line of reasoning we can prove that

$$\lim_{h \to 0+} \tfrac{F(x_o-h)-F(x_o)}{-h} = f(x_o).$$

The last two limits imply what we wanted to show, namely $F'(x_o) = f(x_0)$. If x_o is an endpoint, the same proof works well. Moreover, the proof for the case when f is continuous, positive, and decreasing is entirely similar. QED

A generalization of FTC

Is it possible to generalize FTC? Suppose that

$$G(x) = \int_a^{g(x)} f(t)dt$$

where $g(x)$ is a differentiable function on an open interval J, f is continuous on an open interval I, a is in I, and the range of g is a subset of the domain of f. Then $G'(x) = f(g(x))g'(x)$. Indeed, suppose that F is an antiderivative of f (recall that FTC implies that any continuous function has an antiderivative). The Evaluation Theorem leads to

$$\int_a^{g(x)} f(t)dt = F(g(x)) - F(a)$$

for each x. Hence

$$G'(x) = F'(g(x))g'(x) = f(g(x))g'(x).$$

For instance,

$$\tfrac{d}{dx} \int_1^{x^3} \sin(t^2)dt = \sin((x^3)^2) \cdot 3x^2 = 3x^2 \sin(x^6);$$

$$\tfrac{d}{dx} \int_0^{\cos x} e^{-t^2}dt = e^{-(\cos x)^2}(-\sin x).$$

Solving a differential equation

In section 1.16 we learned how to solve differential equations of the type $y' + ay = b$. Let us now discuss an equation of the type $y' + ay = b(t)$, where $b(t)$ is any continuous function: first multiply both sides by e^{at}; thus, $e^{at}(y'+ay) = e^{at}b(t)$. The Fundamental Theorem of Calculus implies $e^{at}b(t) = (\int_{t_o}^t e^{as}b(s)ds)'$, where t_o is any fixed point of the interval under consideration, while the chain rule leads to $(e^{at}y(t))' = e^{at}(y'+ay)$. Therefore

$$(e^{at}y(t))' = (\int_{t_o}^t e^{as}b(s)ds)'.$$

Consequently, there exists a constant L such that $e^{at}y(t) = \int_{t_o}^t e^{as}b(s)ds + L$ for every t. That is to say,

$$y(t) = e^{-at}\int_{t_o}^t e^{as}b(s)ds + Le^{-at}.$$

The choice of t_o is immaterial. Nonetheless, if the differential equation is accompanied by an initial condition $y(0) = \gamma$, the natural choice will be $t_o = 0$. For instance, the solution to the initial value problem $y' + 5y = cos(t)$, $y(0) = \pi$, is

$$y(t) = e^{-5t}\int_0^t e^{5s}cos(s)ds + \pi e^{-5t}.$$

A graphing calculator leads to $\int_0^t e^{5s}cos(s)ds = (\frac{5cos(t)}{26} + \frac{sint}{26})e^{5t} - \frac{5}{26}$. Therefore

$$y(t) = \frac{5}{26}cos(t) + \frac{sint}{26} + (\pi - \frac{5}{26})e^{-5t}.$$

Exercises for Section 2.7

1. Calculate $\frac{d}{dx}\int_2^{x^2} cos\,t\,dt$.

2. What function $y(x)$ has the property that $y'(x) = e^{-x^2}$ and satisfies the initial condition $y(0) = 0$? Answer the same question when $y(0) = 1$.

3. Suppose $F_1(x) = \int_3^x \sin u^2 du$ and $F_2(x) = \int_5^x \sin u^2 du$. Show that they differ by a constant. What is this constant?

4. Solve the initial value problem $y'(x) = \cos(2x)$, $y(0) = 3$.

5. Solve the initial value problem $y'(x) = xe^x$, $y(0) = 1$ (Hint: $\int xe^x dx = -e^x + xe^x + C$).

6. Sketch a graph $(-2 \le x \le 2)$ of $y(x) = 4 + \int_0^x e^{u^2}du$, which is the solution of the IVP $y'(x) = e^{x^2}$, $y(0) = 4$, using the TMA program to approximate $y(x)$ at several values of x. Keep in mind that $y(x)$ is increasing, concave up on $(0, \infty)$, and concave down on $(-\infty, 0)$.

7. Show that FTC implies ET in a rather straightforward way. In other words, ET happens to be a corollary of FTC (Hint: We wish to show that, given f continuous on an interval I, $\int_a^b f(x)dx = [F(x)]_a^b$, where F is any antiderivative of f. Define the function $G(x) = \int_a^x f(t)dt$, use FTC and recall that if two functions have the same derivative then they differ by a constant).

8. Provide a proof of FTC assuming that $f(x)$ is continuous, positive, and decreasing on an interval I.

9. Given $f(x) = x^2$, $x < 0$, and $f(x) = x$, $x \geq 0$, sketch a graph of $F(x) = \int_0^x f(u)du$ observing that F is 'smooth' (that is, differentiable everywhere) despite the fact that f is not differentiable at $x = 0$.

10. Solve the initial value problem $y'(x) = f(x)$, $y(0) = 0$, where f is a function defined on the whole real line in the following fashion: $f(x) = 0$ for $x < 0$, $f(x) = x$ for x in $[0, 1]$, $f(x) = -x + 2$ for x in $[1, 2]$, and $f(x) = 0$ for $x > 2$. Furthermore, sketch a graph of the solution (Hint: Since f is continuous, the solution will be $y(x) = \int_0^x f(u)du$; thus, our task is to calculate this integral).

11. Let us consider the IVP $y'(x) = f(x)$, $y(0) = 0$, where $f : [0, \infty) \to \Re$ is defined by the rule $f(x) = x$, $0 \leq x < 1$, while $f(x) = 1/2$, $x \geq 1$. The function f has a point of discontinuity at $x = 1$; however, $\lim_{x\to 1-} f(x)$ and $\lim_{x\to 1+} f(x)$ exist as real numbers. This is an example of a **piecewise-continuous function**, a family of functions of considerable importance in chemistry and physics[5]. Is there a function $y(x)$, continuous everywhere, such that $y'(x) = f(x)$ for $x \neq 1$? (Hint: The solution is $y(x) = \int_0^x f(u)du$ due to the fact that FTC can be applied when $x \neq 1$; your task is to calculate the integral considering, separately, the cases $0 \leq x < 1$ and $x \geq 1$.)

12. In relation to the preceding problem, carefully draw the solution $y(x)$ to corroborate that it is continuous everywhere but not differentiable at $x = 1$. Nonetheless, $y'_-(1) = \lim_{x\to 1-} f(x) = 1$ and $y'_+(1) = \lim_{x\to 1+} f(x) = 1/2$.

13. Let us consider a suspension bridge, whose deck has a uniform weight density coefficient α. Assume that the weight of the deck is much larger than the combined weight of the main cables and the suspenders. Question: What is the shape of the main cables? (Hint: Choose the origin of the coordinate system at the lowest point of a main cable. Between 0 and x, the weight of the deck is αx. Thus $T_1 = T_2 \cos\theta$ and $\alpha x = T_2 \sin\theta$, where T_1 is the horizontal tension at the origin and T_2 is the tangential tension at x.)

---

[5]A characteristic of this family is that they are continuous functions, except possibly at a finite number of points wherein the lateral limits exist.

14. An $R - L$ electric circuit under a constant electromotive force E_o is governed by the differential equation

$$RI(t) + LI'(t) = E_o.$$

Solve the differential equation under the assumption that $I(0) = 0$, i.e., no current was flowing through the circuit before we closed it.

15. Solve the preceding problem, with $R = L = 1$, if instead of the constant electromotive force E_o we have to deal with the electromotive force $\cos 2t$.

Enrichment Note: A Proof of FTC

We wish to provide a proof of the Fundamental Theorem of Calculus. For this purpose a result is needed, which is intuitively obvious but nonetheless requires to be proven.

Lemma

Suppose $f : [a, b] \to \Re$ is continuous. Then there exists θ, $a \le \theta \le b$, such that $\int_a^b f = f(\theta)(b - a)$.

Proof. Since f is continuous on $[a, b]$, the Max-Min theorem (section 1.6) implies that there exist r, s in $[a, b]$ such that

$$f(r) \le f(t) \le f(s) \text{ for every } t \text{ in } [a, b].$$

By exercise 10 (section 2.4) we can conclude that

$$\int_a^b f(r)dt \le \int_a^b f(t)dt \le \int_a^b f(s)dt.$$

But $\int_a^b f(r)dt = f(r)(b - a)$ and $\int_a^b f(s)dt = f(s)(b - a)$. Therefore

$$f(r) \le \frac{\int_a^b f(t)dt}{b-a} \le f(s).$$

If $f(r) = \frac{\int_a^b f(t)dt}{b-a}$ or $f(s) = \frac{\int_a^b f(t)dt}{b-a}$, we are done. Thus, suppose

$$f(r) < \frac{\int_a^b f(t)dt}{b-a} < f(s).$$

The Intermediate Value theorem (section 1.6) implies that there exists θ in (r, s) (if $r < s$) or (s, r) (if $s < r$) such that

$$\frac{\int_a^b f(t)dt}{b-a} = f(\theta).$$

Obviously, θ is in $[a, b]$. QED

Theorem

Let f be continuous on an arbitrary interval I. Choose a point a in I and define

$$F(x) = \int_a^x f(t)dt \text{ for every } x \text{ in } I.$$

Then $F'(x_o) = f(x_o)$ for any x_o in I.

Proof[6]. Suppose x_o lies in the interior of I. For $h > 0$, sufficiently small, we have

$$F(x_o + h) - F(x_o) = \int_a^{x_o+h} f - \int_a^{x_o} f = \int_{x_o}^a f + \int_a^{x_o+h} f = \int_{x_o}^{x_o+h} f.$$

The preceding lemma implies that there exists $\theta(h)$, $x_o \leq \theta(h) \leq x_o + h$, such that

$$\int_{x_o}^{x_o+h} f = f(\theta(h))h.$$

Consequently

$$\frac{F(x_o+h)-F(x_o)}{h} = f(\theta(h)).$$

Since $x_o \leq \theta(h) \leq x_o + h$, the squeeze property of limits (section 1.4) implies that $\lim_{h \to 0+} \theta(h) = x_o$. But f is continuous at x_o, therefore

$$\lim_{h \to 0+} f(\theta(h)) = f(\lim_{h \to 0+} \theta(h)) = f(x_o).$$

Hence

$$\lim_{h \to 0+} \frac{F(x_o + h) - F(x_o)}{h} = f(x_o). \tag{2.10}$$

In a similar fashion one can prove that

$$\lim_{h \to 0+} \frac{F(x_o - h) - F(x_o)}{-h} = f(x_o). \tag{2.11}$$

From (2.10) and (2.11) we can conclude that $F'(x_o) = f(x_o)$. Of course, if x_o is a left endpoint we would have $F'_+(x_o) = f(x_o)$, while if x_o is a right endpoint we would have $F'_-(x_o) = f(x_o)$. QED

---

[6]We will present, from a modern perspective, the proof given by Louis Cauchy (1789-1857) in 1823. One of Cauchy's main concerns was to put calculus on a solid foundation.

Remark

In the preceding lemma, the number θ can be chosen between a and b; that is, $a < \theta < b$. Why is this so? Assume that $f(r) = \int_a^b f(t)dt/(b-a)$, where $f(r) \leq f(t)$ for every t in $[a,b]$. Then $f(r)(b-a) = \int_a^b f(t)dt$ and consequently $\int_a^b f(r)dt = \int_a^b f(t)dt$. Therefore $\int_a^b g(t)dt = 0$, where $g(t) = f(t) - f(r)$ for every t in $[a,b]$. Since g is continuous and $g(t) \geq 0$ on $[a,b]$, it follows that $g(t) = 0$ for every t in (a,b) (see exercise 11 of section 2.4). Then choose any θ in (a,b), say $\theta = (a+b)/2$. We will have that $g(\theta) = 0$, which in turn implies that $f(\theta) = f(r)$. Thus, $\int_a^b f = f(\theta)(b-a)$. The case $f(s) = \int_a^b f(t)dt/(b-a)$ can be treated in an entirely analogous way.

This stronger version of the lemma is often called the Intermediate Value Theorem of Integral Calculus (ITIC). It should be noted that it is possible to prove FTC without using the lemma (Rosenlicht, 1968). Under these circumstances, ITIC happens to stem from FTC. Indeed, let F be an antiderivative of f. Then $\int_a^b f = F(b) - F(a)$. The Mean Value Theorem implies that, for some θ in (a,b), $F(b) - F(a) = F'(\theta)(b-a) = f(\theta)(b-a)$. Consequently $\int_a^b f = f(\theta)(b-a)$.

2.8 A New Approach to the Natural Logarithm

The time has come to look at the natural logarithm from the perspective of integral calculus. Such an approach will allow us to widen our mathematical horizon and provide a rigorous definition of logarithms and exponential functions. Since $\int \frac{1}{x}dx = \ln x + C$, $x > 0$, we can assert that

$$\int_1^x \frac{1}{t}dt = [\ln t]_1^x, \ x > 0.$$

Consequently, for any $x > 0$

$$\ln x = \int_1^x \frac{1}{t}dt. \tag{2.12}$$

We have reached (2.12) through a long process that started when we accepted, in section 1.3, that a^x is a well-defined function for each $a > 0$. This approach has the drawback that it relies on the acceptance of some important limits. From a pedagogical standpoint it made a lot of sense to trust our intuition about the exponential function and use technology to calculate the derivative of e^x in section 1.9. However, it is time to put everything on a solid foundation. We will start from (2.12) as the *definition* of the function ln and we will show thereafter all the well-known properties of the natural logarithm.

Since $\ln' x = 1/x > 0$ on $(0, \infty)$ we can conclude that ln is increasing on its domain of definition, in particular it is one-to-one. The fact that ln is increasing implies that $\ln x < \ln 1 = 0$ for $0 < x < 1$ and $\ln x > \ln 1 = 0$ for $x > 1$. Moreover, ln is concave down because $\ln'' x = -1/x^2 < 0$ and, obviously, $\ln$ is continuous because differentiability implies continuity. We have now enough information to sketch a graph of $\ln$ (Figure 2.3).

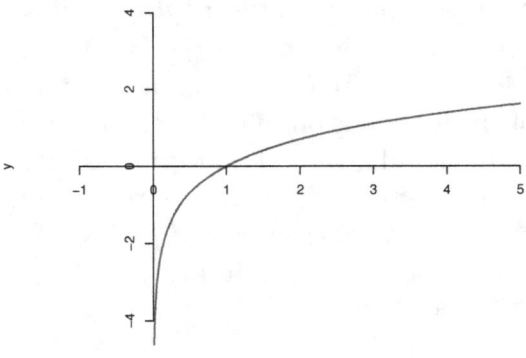

Figure 2.3: The natural logarithm function

The graph of $\ln$ should not be confused with the graph of the function $x \to 1/x$, $x > 0$. When $x > 1$, the area under this function between 1 and x happens to be the number $\ln x$. However, when $0 < x < 1$, the area under $x \to 1/x$ between x and 1 is the number $-\ln x$ because

$$\textstyle\int_x^1 \frac{1}{t}dt = -\int_1^x \frac{1}{t}dt = -\ln x.$$

The word 'logarithm' has been used with regard to the function $\ln$, so we might expect, on the basis of our previous experience with the common logarithm $\log_{10}$, that

$$\ln(xy) = \ln x + \ln y$$

for any $x, y > 0$. A proof goes as follows: Fix an arbitrary $y > 0$ and define the function $h(x) = \ln(xy)$, $x > 0$. The chain rule implies

$$h'(x) = \textstyle\frac{1}{xy}y = \frac{1}{x}.$$

Thus, there exists a constant C such that $h(x) = \ln x + C$ for every $x > 0$. In particular $h(1) = \ln 1 + C$, hence $\ln y = C$. We have then succeeded in proving that $\ln(xy) = \ln x + \ln y$ for any $x > 0$. Although y is fixed, it was arbitrarily chosen;

therefore $\ln(xy) = \ln x + \ln y$ for any $x, y > 0$. Using induction we can then prove that $\ln(x^n) = n \ln x$ for any natural number n.

Two other characteristic properties of a logarithm can be justified right away, namely $\ln x^{-1} = -\ln x$ and $\ln \frac{x}{y} = \ln x - \ln y$. Indeed $xx^{-1} = 1$, thus $\ln x + \ln x^{-1} = \ln 1 = 0$. On the other hand, since $\frac{x}{y} = x\frac{1}{y}$ we will have $\ln(\frac{x}{y}) = \ln x + \ln \frac{1}{y} = \ln x - \ln y$.

Is the function $\ln : (0, \infty) \to \Re$ an onto function? In other words, given any real number y, does there exist $x > 0$ such that $\ln x = y$? If $y = 0$ there is nothing to worry about since $\ln 1 = 0$. Assuming $y > 0$ choose a natural number n such that $n > \frac{y}{\ln 2}$ and define $x_1 = 2^n$, then $\ln x_1 = n \ln 2 > y$, that is, $\ln 1 = 0 < y < \ln x_1$. The Intermediate Value Theorem (IVT), for continuous functions, implies that there exists x, $1 < x < x_1$, such that $\ln x = y$. The case $y < 0$ can be dealt in an entirely similar fashion: choose a natural number m, $m > \frac{-y}{\ln 2}$, and define $x_2 = 2^{-m}$, then $\ln x_2 = -m \ln 2 < y < 0 = \ln 1$. Thanks to IVT we can conclude that there exists x, $x_2 < x < 1$, such that $\ln x = y$.

Having proven that $\ln$ is onto we know, in particular, that there exists a number x such that $\ln x = 1$. This number, which is unique because $\ln$ is one-to-one, is denoted by the symbol e^*. That is to say, e^* is the only real number such that $\ln e^* = 1$. Can we approximate the value of e^*? A first, rather 'crude' approximation, asserts that $2 < e^* < 3$. Just note that the definition of $\ln x$ as the area under $1/x$ between 1 and x leads to $\ln 2 < 1$ (a straightforward verification) and $1 < \ln 3$ (subdivide the interval $[1, 3]$ in eight intervals of common length $1/4$). Then IVT implies that there exists x, $2 < x < 3$, such that $\ln x = 1$. But this x is precisely the number e^* because $\ln$ is one-to-one.

In the next section we will prove that $e^* = e$, where e is the by now well-known limit of the sequence $(1 + \frac{1}{n})^n$. The number e, which happens to be one of the five most important numbers in mathematics (the others are $0, 1, \pi$, and the complex number i), can be retrieved directly from a calculator: if we push the button for e^1 we get on the screen the number 2.71828, the same approximation[7] we obtained in section 2.1 when working with $n = 10^6$.

Let us emphasize that $\int \frac{1}{t} dt = \ln t + C$, $t > 0$, because $\ln' t = \frac{1}{t}$. Moreover

$$\int \frac{u'(t)}{u(t)} dt = \ln u(t) + C \tag{2.13}$$

provided that $u(t) > 0$ on the interval where it is defined and $u'(t)$ is continuous. Why is this so? Using the chain rule we have $\frac{d}{dt} \ln u(t) = \frac{u'(t)}{u(t)}$. Let us remember that

---

[7]The number of decimals that one gets depends on the particular settings of the calculator.

in section 2.6 we got acquainted with the equality (2.13) for the first time. As we will see soon, very many applications of calculus to the natural sciences make use of (2.13).

Remark

If we use the $\int$ command from a graphing calculator (TI-89 or similar ones) to calculate $\int \frac{1}{x} dx$, we get the answer $\ln |x|$, in agreement with what we found in section 2.6. We have to keep in mind that the process of integration takes place either on $(0, \infty)$ or $(-\infty, 0)$. For instance

$$\int_{-2}^{-1} \frac{1}{x} dx = [\ln(-x)]_{-2}^{-1} = \ln 1 - \ln 2.$$

The fact that the answer is a negative quantity should come as no surprise because the graph of $x \to \frac{1}{x}$ lies below the x-axis when $x < 0$; the area between the x-axis and this function happens to be $-(\ln 1 - \ln 2)$.

Similarly, a graphing calculator provides the answer $\int \frac{u'(x)}{u(x)} dx = \ln |u(x)|$ whenever we choose to work with a specific function $u(x)$, corroborating what we found in section 2.6. For instance,

$$\int_1^2 \frac{-2x-3}{-x^2-3x-4} dx = [\ln |-x^2-3x-4|]_1^2 = \ln |-4-6-4| - \ln |-1-3-4| = \ln 14 - \ln 8 = \ln \frac{7}{4}$$

Exercises for section 2.8

1. Calculate $\int_0^3 \frac{x\,dx}{x^2+2}$.

2. Calculate $\int_0^{\pi/4} \frac{\sin x}{\cos x}$.

3. Calculate $\frac{d}{dx} \ln(x^2 + 1)$. Find the interval where the function $x \to \ln(x^2 + 1)$ is increasing or decreasing. Analyze its concavity and sketch a graph of $y = \ln(x^2 + 1)$.

4. Find $\frac{d}{dx} \cos(\ln x)$, $x > 0$.

5. Find $\frac{d}{dx} \ln(\cos x)$, $-\frac{\pi}{2} < x < \frac{\pi}{2}$.

6. Use FTC to find $\frac{d}{dx} \int_0^x \ln(t^2 + 3) dt$.

7. By graphing $y = \ln x$ and $y = x - 1$ we can conjecture that $\ln x \leq x - 1$ for every $x > 0$. Using property (iv) from section 2.4, it is possible to provide a proof. Indeed, assume $x > 1$. Then $\frac{1}{t} \leq 1$ for every t in $[1, x]$. Hence $\int_1^x \frac{1}{t} dt \leq \int_1^x dt$, which in turn leads to $\ln x \leq x - 1$. Show that the same inequality holds if $0 < x \leq 1$ (Hint: Note that, if x is a fixed positive number less than 1, we will have $\frac{1}{t} \geq 1$ for every t in $[x, 1]$).

8. Use mathematical induction to prove that $\ln x^n = n \ln x$, n an arbitrary natural number.

9. Calculate $\int \frac{\arctan x}{x^2 + 1} dx$.

10. Calculate $\int_1^2 \frac{dx}{(x^2 + 1)\arctan x}$.

11. If $u(x) < 0$, and $u'(x)$ is continuous, we will have

$$\int \frac{u'(x)}{u(x)} dt = \ln(-u(x)) + C.$$

Use this formula to calculate $\int_{-3}^{-1} \frac{1}{(x^2+1)\arctan x} dx$.

2.9   The Exponential Function Revisited

We have just learned that $\ln : (0, \infty) \to \Re$ is one-to-one and onto. Its inverse $\ln^{-1}$ is then also one-to-one and onto, and is customarily denoted exp. Thus, $\exp : \Re \to (0, \infty)$ has the property that $\exp(x) = y$ if and only if $x = \ln y$ for any x and any $y > 0$. For instance, $\exp(0) = 1$ because $\ln 1 = 0$ and $\exp(1) = e^*$ because $\ln(e^*) = 1$. Furthermore, the fact that $\exp = \ln^{-1}$ implies

$$\ln(\exp(x)) = x \text{ for any } x, \text{ and}$$

$$\exp(\ln(x)) = x \text{ for any } x > 0.$$

What happens to be the derivative of exp? Thanks to the theorem about the derivative of the inverse (section 1.11), the function exp is differentiable on the whole real line because $\ln$ is differentiable on $(0, \infty)$ and $\ln'(x) \neq 0$ for every $x > 0$ (actually, $\ln'(x) > 0$ on $(0, \infty)$). Since $\ln(\exp(x)) = x$ for any x, the chain rule implies $\frac{1}{\exp(x)} \frac{d}{dx} \exp(x) = 1$; consequently

$$\frac{d}{dx} \exp(x) = \exp(x) \text{ for any } x.$$

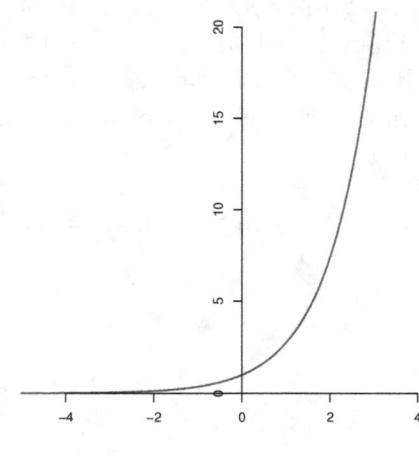

Figure 2.4: The exponential function

That is to say, the derivative of the exponential function is itself! We note that $\frac{d}{dx}\exp(x) > 0$ and $\frac{d^2}{dx^2}\exp(x) > 0$; thus, as expected, exp is increasing and concave up (Figure 2.4).

Right away we can prove that

$$\exp(x_1 + x_2) = \exp(x_1)\exp(x_2).$$

Indeed, let $y_1 = \exp(x_1)$ and $y_2 = \exp(x_2)$. Then $\ln y_1 = x_1$ and $\ln y_2 = x_2$. But $\ln(y_1 y_2) = \ln y_1 + \ln y_2$, consequently $y_1 y_2 = \exp(\ln y_1 + \ln y_2) = \exp(x_1 x_2)$, which in turn leads to $\exp(x_1)\exp(x_2) = \exp(x_1 + x_2)$.

Next we can prove easily two other properties of exp, namely

$$\exp(-x) = \tfrac{1}{\exp(x)} \quad \text{and}$$

$$\exp(x_1 - x_2) = \tfrac{\exp(x_1)}{\exp(x_2)}.$$

It should be noted that the identity $\exp(x_1 + x_2) = \exp(x_1)\exp(x_2)$ can be generalized, in the sense that $\exp(x_1 + ... + x_n) = \exp(x_1) \times ... \times \exp(x_n)$; in particular $\exp(n) = (\exp(1))^n$, where n is any natural number. Moreover, one can check that the linearization of exp at $x = 0$ is $L(x) = x + 1$. Hence

$$\exp(x) \approx x + 1$$

whenever x is very close to zero.

Now it is time to prove that

$$\exp(x) = \lim_{n \to \infty} (1 + \frac{x}{n})^n \tag{2.14}$$

This expression is trivially true for $x = 0$, so let us suppose that $x \neq 0$. Then

$$1 = \ln'(1) = \lim_{h \to 0} \frac{\ln(1+h) - \ln(1)}{h} = \lim_{n \to \infty} \frac{\ln(1 + \frac{x}{n})}{\frac{x}{n}} = \frac{1}{x} \lim_{n \to \infty} \ln(1 + \frac{x}{n})^n.$$

Thus

$$x = \lim_{n \to \infty} \ln(1 + \frac{x}{n})^n.$$

Consequently

$$\exp(x) = \exp(\lim_{n \to \infty} \ln(1 + \frac{x}{n})^n).$$

But the function exp is continuous everywhere. Therefore

$$\exp(x) = \lim_{n \to \infty} \exp(\ln(1 + \frac{x}{n})^n).$$

Since exp and ln are inverse functions, we can conclude that (2.14) is true.

In particular, from (2.14) we get

$$\exp(1) = \lim_{n \to \infty} (1 + \frac{1}{n})^n.$$

That is to say, the sequence $(1 + \frac{1}{n})^n$ converges and does so to the number $\exp(1)$. In section 2.1 the limit of this sequence was denoted by the letter e, hence $\exp(1) = e$ (keep in mind that the limit of any convergent sequence is unique!). However, $\exp(1) = e^*$. Consequently $e = e^*$.

Having $\exp(x)$ expressed as the limit of a sequence (formula (2.14)) is quite important from a theoretical point of view: in advanced calculus it is frequently used to define $\exp(x)$, once the existence of the limit has been proven. On the other hand, the approximation

$$\exp(x) \approx (1 + \frac{x}{n})^n$$

when n is large, has many applications.

It is convenient to write

$$e^x = \exp(x).$$

This definition makes sense because $e^1 = e = \exp(1)$, $e^n = e \times ... \times e = \exp(1) \times ... \times \exp(1) = \exp(1 + ... + 1) = \exp(n)$, and $e^{-n} = \frac{1}{e^n} = \frac{1}{(\exp(1))^n} = \frac{1}{\exp(n)} = \exp(-n)$, in concordance with the common usage of the symbol for the power of a positive or a negative natural number. Having defined $e^x = \exp(x)$ we can assert that

$$e^{x_1+x_2} = e^{x_1}e^{x_2},$$

$$e^{-x} = \tfrac{1}{e^x},$$

$$\tfrac{d}{dx}e^x = e^x,$$

$$\int e^x dx = e^x + C.$$

$$\int e^{u(x)}u'(x)dx = e^{u(x)} + C$$

The graph of the function exp, that is to say, the graph of the function $x \to e^x$, suggests that $\lim_{x\to\infty} e^x = \infty$. We would have to show that given any $M > 0$ there exists p such that if $x > p$ then $e^x > M$. Indeed, given $M > 0$ choose $p = \ln M$; if $x > \ln M$, since the exponential function is increasing we can conclude that $e^x > e^{\ln M} = M$. Having finished the proof, we observe that

$$\lim_{x\to\infty} e^{-x} = \lim_{x\to\infty} \tfrac{1}{e^x} = \tfrac{1}{\lim_{x\to\infty} e^x} = 0.$$

In a similar way, we can prove that for any $b > 0$

$$\lim_{x\to\infty} e^{bx} = \infty \text{ and } \lim_{x\to\infty} e^{-bx} = 0.$$

We should point out that given the differential equation $x'(t) = -kx(t)$, where it is assumed that $t \geq 0$ and $x(t) > 0$, we get $x'(t)/x(t) = -k$ (k is a positive parameter). Consequently,

$$\int \tfrac{x'(t)}{x(t)}dt = \int -kdt.$$

Thus, thanks to (2.13), we get

$$\ln x(t) = -kt + C$$

for some constant C. Hence $x(t) = He^{-kt}$, where $H = e^C$. In particular, $x(0) = H$. Therefore $x(t) = x(0)e^{-kt}$, exactly the same result that was obtained in section 1.15 when $\alpha = -k$.

We have come full circle. In chapter 1 we started with the function e^x, accepted that $(e^x)' = e^x$ and proceeded to define ln as the inverse of e^x. Then we proved that $\ln' x = 1/x$ and consequently $\int(1/x)dx = \ln x + C$. In this chapter we *defined* $\ln x$,

$$\ln x = \int_1^x \tfrac{1}{t}dt$$

and proved all the usual properties of a logarithm. Thereafter exp was defined as $\ln^{-1}$. We could have defined exp, in chapter 1, as the solution of the differential equation $y' = y$ that satisfies $y(0) = 1$, and from there deduce all the usual properties of exp. But this approach is rather sophisticated (Helfgott, 2005).

After all these mathematical developments, it is time to discuss an application to biology.

Growth of fish

Fish tend to grow while they are alive, but the rate of growth is not constant. If L denotes the final length of a mature individual of a certain species of fish, we might put forward the hypothesis that the rate of growth is proportional to the difference between L and $x(t)$ (the length at any time t). That is to say,

$$x'(t) = k(L - x(t)) \tag{2.15}$$

where k is a positive parameter. Dividing by $L - x(t)$ and integrating we get

$$\int \tfrac{-x'(t)}{L-x(t)}dt = \int -kdt.$$

Applying (2.13) we get

$$\ln(L - x(t)) = -kt + C \tag{2.16}$$

for a certain constant C. Thus, if we have t on the horizontal axis and $\ln(L - x(t))$ on the vertical axis, the points should cluster around a straight line; otherwise the model (2.15) is unacceptable. A linear regression analysis will provide the least squares regression line, whose slope is then used as an estimate of k. Thereafter we can find how length varies with time. Indeed, $\ln(L - x(0)) = C$, hence

$$\ln \frac{L - x(0)}{L - x(t)} = kt. \tag{2.17}$$

Then

$$\tfrac{L-x(t)}{L-x(0)} = e^{-kt}.$$

Consequently

$$x(t) = L - (L - x(0))e^{-kt}.$$

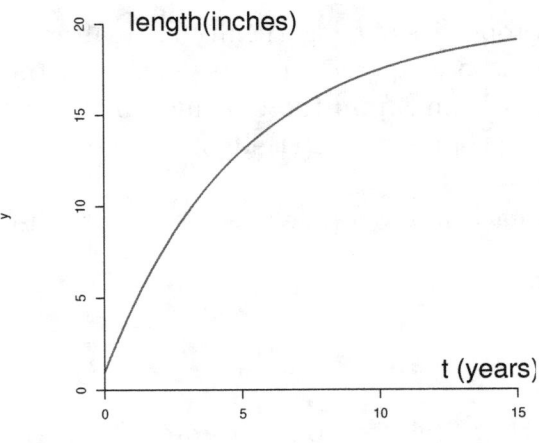

Figure 2.5: Growth of a species of fish

Table 2.1: Growth of *Coregomus clupeaformis*

t (years)	x(t) (cm)	$\ln(71 - x(t))$
1	15	4.0254
2	23	3.8712
4	42	3.3673
6	53	2.8904
8	58	2.5649
10	64	1.9459

As expected, $\lim_{x \to \infty} x(t) = L$ (Figure 2.5).

Let us see whether data related to *Coregomus clupeaformis* (North American lake whitefish) is in agreement with mathematical consequences of the model $x'(t) = k(L - x(t))$ (first two columns of Table 2.1, extracted from Altman and Dittmer, 1964), where $L = 71$ cm has been obtained from observations made by biologists.

We plan to compare data with (2.16) rather than (2.17) because $x(0)$ is not known. With this purpose in mind, we calculate $y = \ln(71 - x(t))$ (third column of Table 2.1). A linear regression analysis provides the line of best fit $\hat{y} = -0.228138t + 4.289559$ (Figure 2.6). Since the correlation coefficient is $r = -0.997212$, a number very close to -1, we accept the model and proceed to estimate k as the number 0.228138. Using (2.16), with the estimation 4.289559 for C, we get $x(t) = 71 - e^{4.289559}e^{-0.228138t} = 71 - 72.934297e^{-0.228138t}$.

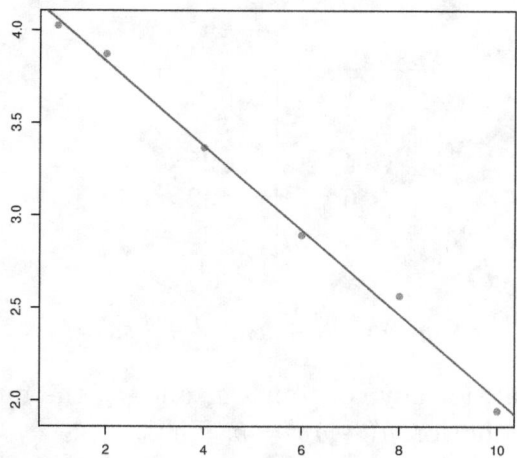

Figure 2.6: Line of best fit related to Table 2.1

A second order differential equation

While analyzing a problem on enzyme kinetics (section 4.6) we will need to solve a differential equation of the type $y''(t) + ay'(t) = b$, where a and b are positive constants. Defining $v = y'$ we get $v' + av = b$. This is a first-order differential equation that we encountered in section 1.16 for the first time. Multiplying both sides of it by e^{at} we get

$$e^{at}(v' + av) = be^{at}.$$

Hence $(e^{at}v(t))' = (\frac{b}{a}e^{at})'$. Then there exists a constant c_1 such that

$$e^{at}v(t) = \frac{b}{a}e^{at} + c_1 \quad \text{for every } t.$$

Therefore

$$y'(t) = v(t) = \frac{b}{a} + c_1 e^{-at},$$

which leads to

$$y'(t) = (\frac{b}{a}t - \frac{c_1}{a}e^{-at})'.$$

Consequently, there exists a constant c_2 such that, for every t,

$$y(t) = \frac{b}{a}t - \frac{c_1}{a}e^{-at} + c_2.$$

Defining $d_1 = c_2$ and $d_2 = -\frac{c_1}{a}$ we get, for every t,

Table 2.2: Growth of *Pomoxis annularis*

t (years)	$x(t)$(cm)
1	7
2	3.15
4	25
6	32
8	38

$$y(t) = d_1 + d_2 e^{-at} + \frac{b}{a}t.$$

The reader can verify that, for any constants d_1 and d_2, the function $d_1 + d_2 e^{-at} + \frac{b}{a}t$ satisfies the differential equation $y'' + ay' = b$. Thus,

$$d_1 + d_2 e^{-at} + \frac{b}{a}t$$

provides the *general solution* of the given differential equation. Notice that in the general solution there are two arbitrary constants, namely d_1 and d_2. If the differential equation $y'' + ay' = b$ comes with the initial conditions $y(0) = \alpha$, $y'(0) = \beta$, the constants d_1 and d_2 will adopt a numerical value in terms of α, β, a, and b. For instance, the solution of $y'' + y' = 1$, $y(0) = 0$, $y'(0) = 2$, is found by noticing that $y(t) = d_1 + d_2 e^{-t} + t$. Then $y'(t) = -d_2 e^{-t} + 1$. But $0 = y(0) = d_1 + d_2$, $2 = y'(0) = -d_2 + 1$. Consequently $d_2 = -1$ and $d_1 = 1$. In summary, the solution of the given IVP happens to be $y(t) = 1 - e^{-t} + t$.

Exercises for section 2.9

1. Consider the function e^{x^2}. Where is it decreasing or increasing? Analyze its concavity. Do the same with e^{-x^2}. Sketch both functions.

2. Calculate $\int xe^{x^2+5}dx$, $\int x^2 e^{-x^3+7}dx$.

3. The growth of *Pomoxis annularis*, a type of fish, is displayed in Table 2.2 (Altman and Dittmer, 1964). Is the model $x'(t) = k(40 - x(t))$ in agreement with data? If the answer is positive, estimate k (40 cm is the maximum length).

4. A patient takes a pill every 6 hours. The drug is degraded and excreted by the kidneys following the differential equation $c'(t) = -kc(t)$ over time. The symbol $c(t)$ denotes the concentration of the drug in the blood stream. Assuming that the initial concentration is c_o, find the concentration right after the second and third pills. What is the concentration right after the nth pill?

5. Next suppose that the pill is ingested every T hours. In the long run, what is the concentration of the drug in the blood? Draw a diagram describing the variation of the concentration across time, noting that there is a sudden jump after the ingestion of each pill (Hint: This problem involves a geometric series, a topic that was discussed in section 2.1).

6. Let us define the function $h(x) = xe^x$. Sketch a graph of $h(x)$ and show that it is increasing on $[-1, \infty)$. Thus, the function is one-to-one on $[-1, \infty)$. From the graph of $h(x)$ we might conclude that its range is $[-1/e, \infty)$.

7. Since $h : [-1, \infty) \to [-1/e, \infty)$ is one-to-one and onto, its inverse h^{-1} exists,

$$h^{-1} : [-1/e, \infty) \to [-1, \infty).$$

The function h^{-1} is commonly denoted W and is called **Lambert W function** in honor of Johann Lambert (1728-1777), who did pioneering work with regard to h and its inverse. Thus, $W : [-1/e, \infty) \to [-1, \infty)$ and we have two basic identities:

$$x = W(x)e^{W(x)} \qquad x \geq -1/e;$$

$$x = W(xe^x) \qquad x \geq -1.$$

Why is this so?

8. Lambert's W function is particularly useful to solve algebraic equations of the type

$$(a + bx)e^{cx} = d$$

where a, b, c, d are given real numbers. Since the unknown x is inside an exponential and, at the same time, outside of it, renders it impossible to solve for x explicitly in the usual way. But the solution can be found in terms of the W function, which is stored in many present-day mathematical software (Mathematica, Maple, etc.). Solve the following equations: $xe^x = 5$, $(3 + x)e^x = 5$, $(3 + 2x)e^x = 5$, $(3 + 2x)e^{7x} = 5$.

9. Solve the equation $(a + bx)e^{cx} = d$.

10. Find the general solution of $y'' + y' = e^{3t}$.

11. Solve the IVP $y'' + 2y' = 7$, $y(0) = 1$, $y'(0) = 0$.

12. Systems of differential equations of the type $x'(t) = \alpha x(t) - \beta y(t)$, $y'(t) = -\gamma x(t) + \delta y(t)$ ($\alpha, \beta, \gamma, \delta$ positive parameters) are used to study the competition between two species for a common resource. Assume that $x(0) = 40$, $y(0) = 50$, and the model under study is

$$x' = x - y,$$

$$y' = -x + y.$$

Which species becomes extinct? What happens to the other species? (Hint: Take the derivative to the first differential equation and use both equations to arrive at $x'' = 2x'$. Then solve this second order linear differential equation, keeping in mind that $x(0) = 40$ and $x'(0) = 40 - 50 = -10$.)

13. Solving the differential equation $y'' + ay' + by = 0$ requires some care. Letting r_1, r_2 be the two roots of the polynomial $p(x) = x^2 + ax + b$ we can assert that $r_1 + r_2 = -a$ and $r_1 r_2 = b$. Thus, the equation becomes $y'' - (r_1 + r_2)y' + r_1 r_2 y = 0$. That is, $(y' - r_1 y)' - r_2(y' - r_1 y) = 0$. The original problem has been reduced to solving two first order linear differential equations, namely $v' - r_2 v = 0$ and $y' - r_1 y = v$. Apply this technique in order to solve $y'' + 2y' + y = 0$ with initial conditions $y(0) = 0$, $y'(0) = 1$.

2.10 A More General Exponential

Next let us define the exponential function with an arbitrary fixed base $a > 0$: for any real number x,

$$a^x = e^{x \ln a}. \tag{2.18}$$

This definition is consistent with our previous usage of the exponential function. For instance, when $a = e$ we have $e^{x \ln e} = e^x$. Right away we note that a^x is always positive, $\ln a^x = x \ln a$, and

$$\tfrac{d}{dx} a^x = e^{x \ln a} \ln a = a^x \ln a.$$

Consequently

$$\textstyle\int a^x dx = \tfrac{a^x}{\ln a} + C$$

and

$$\textstyle\int a^{u(x)} u'(x) dx = \tfrac{a^{u(x)}}{\ln a} + C.$$

Since $\ln a > 0$ whenever $a > 1$, we can conclude that $\frac{d}{dx}a^x > 0$ and consequently a^x is increasing provided that $a > 1$. By the same token, if $0 < a < 1$ it follows that a^x is decreasing because $\ln a < 0$. Moreover, whether $a > 1$ or $0 < a < 1$ we have

$$\frac{d^2}{dx^2}a^x = a^x(\ln a)^2 > 0.$$

Hence, under both circumstances, the function a^x is concave up (Figure 2.7).

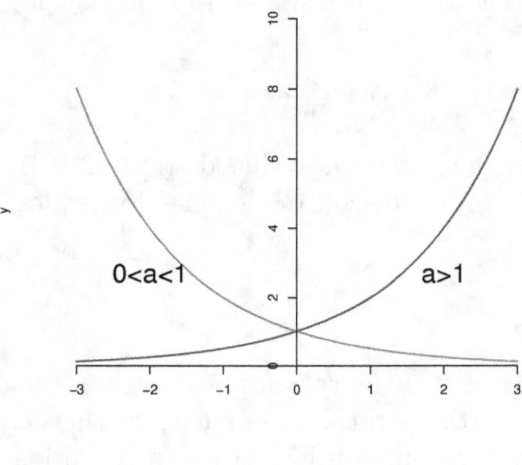

Figure 2.7: Exponential functions

On the other hand,

$$a^{x+y} = e^{(x+y)\ln a} = e^{x\ln a + y\ln a} = e^{x\ln a}e^{y\ln a} = a^x a^y$$

and

$$(a^x)^y = a^{xy}.$$

To justify the last equality we write $z = (a^x)^y$; then $\ln z = y\ln a^x = yx\ln a = \ln a^{xy}$. Since $\ln$ is one-to-one we can conclude that $z = a^{xy}$, consequently $(a^x)^y = a^{xy}$.

Let us suppose that $a > 1$. The function $\exp_a$, where $\exp_a = a^x$, has as domain the whole real line and adopts only positive values, and it is one-to-one because it is increasing. It is also onto because given any $y > 0$ there exists u such that $e^u = y$, thus

$$e^{\frac{u}{\ln a}\ln a} = y,$$

that is, $a^{\frac{u}{\ln a}} = y$. Having shown that $\exp_a$ is one-to-one and onto, we can assert that the inverse of $\exp_a$ exists, and that it has domain $(0, \infty)$ and range $\Re$ (the whole real line). We will denote it $\log_a$. Since $\log_a = \exp_a^{-1}$, the following properties are true:

$$a^{\log_a x} = x, \text{ for every } x > 0, \text{ and}$$

$$\log_a(a^x) = x, \text{ for every } x.$$

From the equality $a^{\log_a x} = x$ we get $\ln a^{\log_a x} = \ln x$, hence $(\log_a x)\ln a = \ln x$. Consequently

$$\log_a x = \frac{\ln x}{\ln a}, \quad x > 0. \tag{2.19}$$

All the usual properties of $\log_a$ can be deduced from (2.19). That is to say, $\log_a$ 'inherits' all the properties of $\ln$: $\log_a(xy) = \log_a x + \log_a y$, $\log_a(\frac{x}{y}) = \log_a x - \log_a y$, and $\log_a x^n = n \log_a x$. Moreover

$$\frac{d}{dx}\log_a x = \frac{d}{dx}\frac{\ln x}{\ln a} = \frac{1}{x\ln a}.$$

We see that the derivative of $\log_a$ is more complicated than the derivative of $\ln$. That is why we prefer to work with $\ln$ rather than $\log_a$ in the context of calculus. Of course, for $a = 10$ we have the common logarithm function $\log_{10}$, which was studied in chapter 1.

Exercises for Section 2.10

1. Calculate $\frac{d}{dx} 3^{x^2+2x}$.

2. Calculate $\frac{d}{dx} \log_2(x + 1)$.

3. Calculate $\int_{-1}^{2} x 2^{x^2} dx$.

4. Verify that $\log_a$ inherits all the properties of $\ln$.

5. Find the set of all antiderivatives of the function $f(x) = \frac{\log_3(x)}{x}$.

6. Show that $\lim_{x\to\infty} a^x = \infty$ $(a > 1)$ while $\lim_{x\to\infty} a^x = 0$ $(0 < a < 1)$.

2.11   Power Functions

What happens if the variable is not in the exponent but in the base? We are accustomed to work with polynomial functions x^n, defined all over the real line, or x^{-n} (defined everywhere, except at the origin). In chapter 1 we dealt also with the function $x^{1/2}$, defined for $x \geq 0$, and the functions $x^{1/3}$, $x^{2/3}$, which are defined on the whole real line. Now we are interested in giving a meaning to x^r, where r is an arbitrary fixed real number, say $r = \sqrt{2}$. To achieve this goal, we need to restrict the domain to the interval $(0, \infty)$. Since $e^{n \ln x} = e^{\ln x^n} = x^n$ for any natural number n, it makes sense to define

$$x^r = e^{r \ln x} \tag{2.20}$$

for every $x > 0$. Right away we observe that

$$\ln x^r = \ln e^{r \ln x} = r \ln x$$

while

$$\frac{d}{dx} x^r = \frac{r}{x} e^{r \ln x} = \frac{r}{x} x^r = r x^{r-1}. \tag{2.21}$$

Thus, $\frac{d}{dx} x^r > 0$ whenever $r > 0$ and $\frac{d}{dx} x^r < 0$ whenever $r < 0$ (for $r = 0$ we have the constant function 1). We can then conclude that the function $x \to x^r$ is increasing when $r > 0$ and decreasing when $r < 0$. Moreover,

$$\frac{d^2}{dx^2} x^r = r(r-1) x^{r-2}.$$

Consequently, the function $x \to x^r$ is concave up when $r > 1$ or $r < 0$, and concave down when $0 < r < 1$ (Figure 2.8). We are dealing with power functions.

From (2.21) we get

$$\int x^r dx = \frac{x^{r+1}}{r+1} + C, \qquad r \neq -1.$$

On the other hand, it is not hard to prove that $x^r x^s = x^{r+s}$, $x^{rs} = (x^r)^s$, and $x^r / x^s = x^{r-s}$.

Having defined x^r for any r, and $x > 0$, we should feel confident about dealing with allometric models, that is to say expressions of the form $y = A x^r$ relating two variables. The next example will illustrate, once more, this type of model and will give us another opportunity to utilize linear regression.

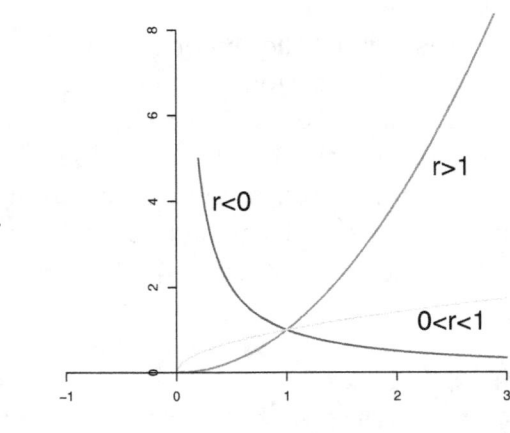

Figure 2.8: Power functions

Ducks

Weight and length of wings of ten randomly selected species of ducks of the Anatidae family have been collected. The first two columns of Table 2.3 (Lislevand et al., 2007) provides the necessary data. We postulate an allometric relationship of the type $y = Ax^\alpha$ between weight (x) and length (y). The relationship $y = Ax^\alpha$ is equivalent to $\ln y = \ln A + \alpha \ln x$, thus we expect that the points $(\ln x, \ln y)$ should cluster around a straight line. Columns three and four of Table 2.3 will allow us to test this prediction. The least squares regression line happens to be $\hat{\ln} y = 0.370816 \ln x + 3.095427$ (Figure 2.9). Since the correlation coefficient is 0.986068, we accept the allometric model and proceed to estimate A and α. Indeed, $\alpha = 0.370816$ and $\ln A = 3.095427$, which in turn leads to $A = e^{3.095427} = 22.096672$. Therefore,

$$y = 22.096672 x^{0.370816}$$

For instance, on the basis of the allometric model that we just built, it is to be expected that the length of the wings of a 1700 gram duck from the Anatidae family will be $y = 22.096672(1700)^{0.370816} = 348.52105$ cm. Of course, this is an approximation to reality.

Table 2.3: Anatidae family of ducks

x (grams)	y (cm)	ln x	ln y
322	187	5.7746	5.2311
1025	275	6.9324	5.6168
6200	511.7	8.7323	6.2377
282	186.5	5.6419	5.2284
2075	392	7.6377	5.9713
404	197	6.0014	5.2832
1261	334	7.1397	5.8111
2770	443	7.9266	6.0936
3814	506	8.2464	6.2265
1210	279	7.0984	5.6312

Exercises for Section 2.11

1. Assume that $y = Ax^\alpha$ describes an allometric relationship and that (x_1, y_1), (x_2, y_2) are two points on $y = Ax^\alpha$. Find A and α.

2. Find the family of all antiderivatives of $x^{2/3}$.

3. In section 1.7 we considered the function $x^{1/3}$, defined over the whole real line. But, as a power function, it is defined only for $x > 0$. The power function $x^{1/3}$ has a continuous extension $f(x)$, where

$$f(x) = e^{\frac{1}{3}\ln x}, \ x > 0, \ f(0) = 0, \ \text{and} \ f(x) = -e^{\frac{1}{3}\ln(-x)}, \ x < 0.$$

We have in mind $f(x)$ when considering the function $x^{1/3}$, defined over the whole real line. Question: what is the 'natural' continuous extension of the power function $x^{2/3}$?

4. Does $\frac{d}{dx}x^{4/7}$ exist for all x? What is its value? Answer the same question for $x^{3/11}$.

5. The egg mass and the corresponding heart rate of nine species of birds (at 80% of incubation period) is displayed in Table 2.4 (Tazawa et al., 2001). Test the validity of the model $y = Ax^\alpha$ and estimate the values of A and α.

6. Assume that the relative growth rates of the leaf area $L(t)$ and the stem diameter $S(t)$ of a plant are proportional at any time while they are growing, that is,

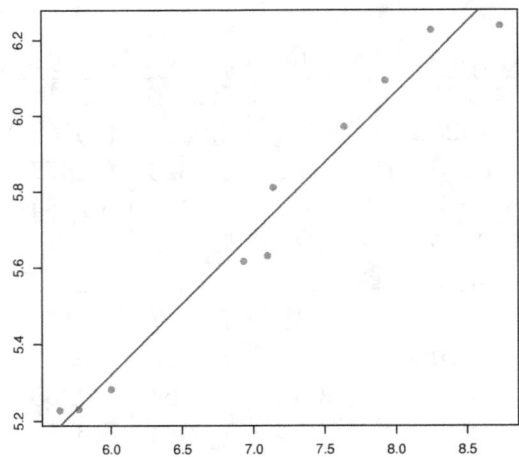

Figure 2.9: The line of best fit for data in table 2.3

$\frac{L'(t)}{L(t)} = k\frac{S'(t)}{S(t)}$. Show that, at any time t, $L(t)$ is proportional to $S(t)^k$. That is to say, the leaf area and the stem diameter are linked through an allometric relationship.

7. Stevens' differential equation (Burghes and Borrie, 1981) asserts that

$$\frac{dR}{dI} = k\frac{R}{I},$$

where R stands for response magnitude, I the intensity of a stimulus, and k a parameter. Show that R and I are related by a power rule. How would you use linear regression to check compatibility between data and Stevens' law, and estimate k?

8. Calculate $\lim_{x\to 0}(1 + x)^{1/x}$ and $\lim_{x\to\infty}(1 + \frac{1}{x})^x$.

2.12   Basic Chemical Kinetics

Tertiary butyl bromure is put in contact with water and begins to decompose. Table 2.5 (Barrow, 1979) provides information about the decomposition process at 50°. According to the law of mass action, a possible model for the chemical reaction $(CH_3)_3CBr + H_2O \rightarrow (CH_3)_3COH + HBr$ is given by $c'(t) = -kc(t)$, where $c(t)$ denotes the concentration of $(CH_3)_3CBr$ as a function of time (k is the parameter of

Table 2.4: Heart rate of nine species of birds

Species	egg mass (grams)	heart rate (beats/minute)
King quail	6	341
Japanese quail	10.7	319
Chicken	64.9	287
Duck	79	247
Turkey	82.9	246
Peafowl	111.3	262
Goose	158.3	224
Emu	634	192
Ostrich	1331	173

Table 2.5: Decomposition of tertiary butyl bromure at $50^\circ C$

t (hours)	c(t) (moles/lt)	$\ln c(t)$
0	0.1056	-2.2481
9	0.0961	-2.3424
18	0.0856	-2.4581
27	0.0767	-2.5679
40	0.0645	-2.7411
54	0.0536	-2.9262

the reaction). This model will be valid insofar as the mathematical consequences of it are in agreement with data.

Right away we get $\int \frac{c'(t)}{c(t)} dt = \int -k dt$, thus there exists a constant H such that

$$\ln c(t) = -kt + H$$

The last expression implies that if time is on the horizontal axis and $y = \ln c(t)$ is on the vertical axis, the experimental values should gather around a straight line. A linear regression analysis, using the first and third column of Table 2.5, provides the 'line of best fit,' namely $\hat{y} = -0.012661x - 2.234956$, with correlation coefficient -0.999423 (we used a TI-89 graphing calculator, but any similar calculator will do the job). Thus, we accept the model and adopt the value 0.012661 as an estimation of k. The chemical reaction is said to obey first order kinetics.

Having completed the validation process, we may well ask: How does the concentration vary with time? From the differential equation $c'(t) = -kc(t)$ we get

$$\int_0^t \frac{c'(s)}{c(s)} ds = \int_0^t -k ds.$$

Hence $[\ln c(s)]_0^t = -kt$, which in turn implies $\ln \frac{c(0)}{c(t)} = -kt$. That is (Figure 2.10),

$$c(t) = c(0)e^{-kt}$$

As expected, $\lim_{t \to \infty} c(t) = c(0) \lim_{t \to \infty} e^{-kt} = 0$.

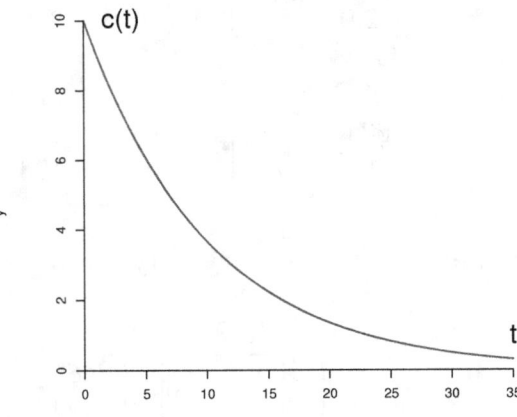

Figure 2.10: Variation of concentration across time

There is another path that can be followed, of an 'algorithmic nature.' In Leibniz's notation the equation becomes $\frac{dc}{dt} = -kc$. Therefore $\int \frac{dc}{c} = \int -k\,dt$, which in turn implies $\ln c(t) = -kt + L$ for a certain constant L. In particular $\ln c(0) = L$, consequently $\ln c(t) = -kt + \ln c(0)$; thus, $\ln \frac{c(0)}{c(t)} = kt$. In other words, we 'separated' the independent variable t and the dependent variable c and thereafter we integrated both sides of the equality. More will be said about this approach in section 2.18.

It may well happen that the model $\frac{dc}{dt} = -kc$ is not valid, in the sense that mathematical consequences of it are not in agreement with data. Then we should try the next simplest model, namely $\frac{dc}{dt} = -kc^2$. Separating variables and integrating we obtain $\int \frac{dc}{c^2} = \int -k\,dt$, thus $-c^{-1}(t) = -kt + L$ for a certain constant L. But $-\frac{1}{c(0)} = L$, consequently

$$\frac{1}{c(t)} = kt + \frac{1}{c(0)}$$

(the same result was reached in section 1.16, using only concepts from differential calculus). We would have to perform a linear regression analysis, with $\frac{1}{c(t)}$ on the

Table 2.6: Decomposition of Ammonium Cyanide

Initial concentration (gmol/lt)	half-life (hours)
0.05	37
0.1	19.2
0.2	9.5

Table 2.7: Decomposition of Hydrogen Peroxide

time (hours)	concentration (moles/lt)
5	37.1
10	29.8
20	19.6
30	12.3
50	5

vertical axis and t on the horizontal axis, in order to test the validity of this new model.

Exercises for Section 2.12

1. Assume that the chemical reaction $A \rightarrow$ products, with kinetic parameter k, obeys first order kinetics; that is to say, $c'(t) = -kc(t)$, where $c(t)$ denotes the concentration at any instant t. Find the time it takes to reach the concentration $c(0)/2$.

2. Find the time it takes to reach the concentration $c(0)/2$ if we assume a second order model, that is to say, $c'(t) = -k(c(t))^2$.

3. Show that if the chemical reaction $A \rightarrow$ products is of second order, the equality $\frac{\tau_1}{\tau_2} = \frac{c_2}{c_1}$ holds (τ_i is the half-life that corresponds to the initial concentration c_i). Data about the decomposition of ammonium cyanide (Jenson and Jeffreys, 1971) can be found in Table 2.6. Is there a reasonable agreement between this data and the above-mentioned equality?

4. Data about the decomposition of hydrogen peroxide can be found in table 2.7 (Barrow, 1979). Is a first order model $c'(t) = -kc(t)$ in agreement with data? If your answer is yes, estimate the parameter k.

Table 2.8: Decomposition of tertiary butyl bromure at 25^oC

time (hours)	concentration (moles/lt)
0	0.1039
3.15	0.0896
4.10	0.0859
6.2	0.0776
8.2	0.0701
10	0.0639
13.5	0.0529
18.3	0.0353
26	0.0270
30.8	0.0207
37.3	0.0142

5. Table 2.8 provides data about the decomposition of tertiary butyl bromure (TBB) at 25^oC (Barrow, 1979). We claim that TBB at this temperature also obeys first order kinetics. Determine whether this claim is in agreement with data. If your answer is positive, estimate k (the kinetic parameter of the reaction).

2.13   Radioactive Decay

At the beginning of the 20th century, physicists put forward the equation $y'(t) = -ky(t)$ to explain radioactive phenomena ($y(t)$ denotes the amount of radioactive substance that has not disintegrated, often measured in milligrams). Of course, the model makes no sense at the level of individual atoms; however, at the macroscopic level it looks plausible because the rate at which a radioactive substance disintegrates is proportional to the amount that has not disintegrated yet.

As usual, the model will be valid provided that the mathematical consequences that stem from it are in agreement with data. The differential equation under consideration is identical to the equation of first order kinetics studied in the previous section. Thus,

$$\ln \frac{y(0)}{y(t)} = kt \qquad (2.22)$$

or, its equivalent expression

$$y(t) = y(0)e^{-kt}. \qquad (2.23)$$

The concept of half-life plays a crucial role in the study of radioactivity. It is the time, τ, it takes the substance to lose half of its weight. From (2.22) we get

$$\ln \frac{y(0)}{\frac{y(0)}{2}} = k\tau.$$

Therefore

$$\tau = \frac{\ln 2}{k}. \tag{2.24}$$

Remarkably, τ does not depend on the initial amount of radioactive material. For instance, it takes the same time if we start with 7 mg and lose 3.5 mg as if we were to start with 10 mg and lose 5 mg through disintegration. Let us discuss two typical problems involving radioactivity.

(i) Radioactive Iodine-131 has a half-life of 8 days. Suppose that we start with 12 mg of I-131. How long will it take to lose 60% of its mass? We need to find t such that $y(t) = 0.4y(0)$. From (2.24) we get $k = \ln 2/8$; thus, (2.22) implies

$$t = \frac{1}{k} \ln \frac{y(0)}{0.4y(0)}.$$

Hence $t = \frac{8}{\ln 2} \ln \frac{1}{0.4} \approx 10.58$ days.

(ii) Carbon-14, a radioactive isotope of carbon, has a half-life of approximately 5,700 years. Suppose a piece of charcoal found in a cave has only 20% of the amount of Carbon-14 of the tree from which charcoal was made. How long ago was the charcoal made? We have $k = \ln 2/5700$. From (2.22) we get

$$t = \frac{5700}{\ln 2} \ln \frac{x(0)}{0.2x(0)} = \frac{5700}{\ln 2} \ln \frac{1}{0.2} = \frac{5700}{\ln 2} \ln 5 \approx 13,235 \text{ years.}$$

Exercises for section 2.13

1. Suppose that a radioactive compound loses 70% of its mass in 15 days. What is its half-life?

2. An artificial radioactive compound, with a half-life of 10 days, is used for medical purposes. It can be effective only if it has not lost more than 60% of its original mass. For how long can it be profitably used?

3. We weigh a radioactive substance twice. The first measurement indicates the presence of 10 mg. The second measurement, made 5 hours later, gives the value of 8 mg. How much will be left 13 hours after the second measurement?

4. Two days after an artificial radioactive substance was manufactured, there were 20 mg of it. One day later there were 18 mg. How much was there at the beginning?

186 CHAPTER 2. INTEGRAL CALCULUS

5. A bone from a bison was found to contain 35% of the amount of C^{14} of a living bison[8]. How long ago did the bison die?

6. The diagram $A \to B \to C$ describes a radioactive process, wherein a radioactive substance A decays into a radioactive substance B that in turn decays into a stable substance C. The process is governed by the following system of differential equations:

$$A'(t) = -k_1 A(t), \quad B'(t) = k_1 A(t) - k_2 B(t), \quad C'(t) = k_2 B(t).$$

Assuming that the initial conditions are $A(0) = a$, $B(0) = 0$, and $C(0) = 0$, find $A(t)$, $B(t)$, and $C(t)$ at any time t.

2.14 Newton's Law of Cooling

If a flask, whose contents are at $T(0)$ degrees, is put in contact with water that circulates at the constant temperature of T_M degrees ($T(0) > T_M$), the temperature inside the flask descends abruptly at the beginning and gradually thereafter. Based on repeated experiments we might put forward the hypothesis that the rate at which the temperature descends is proportional to the difference between $T(t)$ and T_M, that is to say,

$$T'(t) = -k(T(t) - T_M) \tag{2.25}$$

for a certain parameter $k > 0$. Based on our previous experience with first order kinetics and radioactive decay, we divide both sides of (2.25) by $T(t) - T_M$ and integrate. Indeed,

$$\frac{T'(t)}{T(t) - T_M} = -k$$

leads to

$$\int_0^t \frac{T'(s)}{T(s) - T_M} ds = \int_0^t -k ds.$$

But an antiderivative of $\frac{T'(s)}{T(s) - T_M}$ is $\ln(T(s) - T_M)$ (recall (2.13)). Consequently $[\ln(T(s) - T_M)]_0^t = -kt$, which in turn leads to

$$\ln \frac{T(t) - T_M}{T(0) - T_M} = -kt \tag{2.26}$$

or, what is the same,

$$\ln \frac{T(0) - T_M}{T(t) - T_M} = kt. \tag{2.27}$$

---

[8]Tacitly we are assuming that the physiology of bisons has not changed in the last 10,000 years or so.

It is to be noted that (2.26) is equivalent to $\frac{T(t)-T_M}{T(0)-T_M} = e^{-kt}$, hence

$$T(t) = T_M + (T(0) - T_M)e^{-kt}.\qquad(2.28)$$

As expected, $\lim_{t\to\infty} T(t) = T_M$.

Let us solve some practical problems. For instance, suppose that a solution at 70° C is kept in a flask, which is put in contact with circulating water at 45^oC (see Figure 2.11). After two minutes, the temperature drops to 62^oC. How long will it take, since the beginning of the cooling process, to bring down the temperature to 50^oC ? Let $\bar{t}$ be the unknown. From (2.27) we have

$$\ln \frac{70-45}{62-45} = 2k.$$

Hence $k = \frac{1}{2}\ln\frac{25}{17}$. Once more, we apply (2.27) and get

$$\ln \frac{70-45}{50-45} = k\bar{t}.$$

Therefore

$$\bar{t} = \frac{1}{k}\ln\frac{25}{5} = \frac{2}{\ln\frac{25}{17}}\ln 5 \approx 8.35 \text{ minutes.}$$

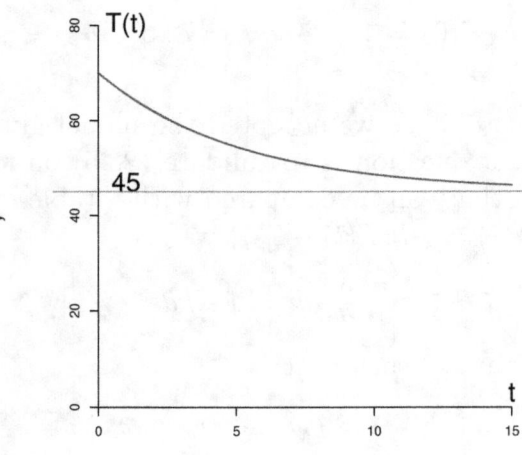

Figure 2.11: Variation of temperature across time

Another example will shed more light on the methodology to be followed in order to solve problems about cooling processes when the temperature of the surrounding medium (often water) is kept constant. Suppose that the flask of the previous example is put in contact with a cooling device, through which water circulates at the constant temperature of 45° C. A first measurement indicates that $T(0) = 82^\circ C$, while a second measurement points out that $T(3) = 76^\circ C$. What is the predicted value for $T(6)$? From (2.27) we get

$$3k = \ln \frac{82-45}{76-45}.$$

Hence $k = \frac{1}{3} \ln \frac{37}{31}$. Using (2.28) we reach the conclusion that

$$T(6) = 45 + (82 - 45)e^{-6k} \approx 70.97^\circ C.$$

It could happen that a chemical reaction is taking place inside the flask and that, at $t = 0$, it is suspended by some established procedure. Right away it is put in contact with circulating water kept at the constant temperature of 45°C. Two measurements indicate that $T(4) = 60^\circ$C and $T(9) = 52^\circ$C. What is the temperature at which the reaction was taking place? In other words, we need to calculate $T(0)$. Applying (2.28) twice, we get

$$15 = 60 - 45 = (T(0) - 45)e^{-4k}, \tag{2.29}$$

$$7 = 52 - 45 = (T(0) - 45)e^{-9k}. \tag{2.30}$$

Dividing (2.29) by (2.30), we arrive at $15/7 = e^{5k}$; consequently $k = \frac{1}{5} \ln \frac{5}{7}$. Finally, from (2.29), we reach the answer:

$$T(0) = 45 + 15e^{4k} \approx 72.6^\circ C.$$

In the three previous problems we accepted the model given by the differential equation (2.25). A different question is to validate (2.25) on the basis of data. We have to draw a prediction that can be compared with a table of experimental values $(t, T(t))$. Starting with $T'(t) = -k(T(t) - T_M)$ we get

$$\int \frac{T'(t)}{T(t)-T_M} dt = \int -k dt.$$

Hence, there exists a constant C such that

$$\ln(T(t) - T_M) = -kt + C.$$

This last expression predicts that the experimental values will gather around a straight line with negative slope if t is on the horizontal axis and $\ln(T(t)-T_M)$ is on the vertical axis. Moreover, if the correlation coefficient is close to -1, we accept the model (2.25) and proceed to estimate k as the slope of the least squares linear regression line. Also, we estimate C from the independent term of the regression line and are thus able to write

Table 2.9: Temperature inside a cup of water

t (minutes)	T(t) (Fahrenheit degrees)	$\ln(T(t) - 66)$
5	155.6	4.4954
10	144.7	4.3656
15	134.8	4.2312
20	126.3	4.0993
25	119.9	3.9871
30	115	3.8918
35	110.6	3.7977
40	105.9	3.6864
45	101.8	3.5779
50	98.6	3.4843
55	95.9	3.3979
60	93.5	3.3142
65	91.5	3.2387
70	89.7	3.1655
75	88.2	3.1001
80	86.9	3.0397
85	84.9	2.9392

$$T(t) = T_M + e^C e^{-kt}$$

Let us analyze an experiment that closely resembles the conditions under which Newton's law of cooling is expected to hold. Inside a room, kept at the constant temperature of $66°F$, a cup of water at $166.9°F$ was left to cool. Sixteen temperature measurements, at intervals of 5 minutes, were made using a probe (first two columns of Table 2.9). Letting $y = \ln(T(t) - 66)$, we build the third column of Table 2.9. A linear regression analysis provides the least squares regression line, namely $\hat{y} = -0.019076t + 4.494406$. Since the correlation coefficient happens to be -0.994696, very close to -1, we accept the validity of the model and use the estimates $k = 0.019076$, $C = 4.494406$, which in turn leads to $T(t) = 66 + e^{4.494406}e^{-0.019076}$; that is,

$$T(t) = 66 + 89.514981e^{-0.019076t}.$$

This expression for $T(t)$ provides a way to predict the temperature at any time between $t = 0$ and $t = 85$ minutes.

Table 2.10: Cooling of a cup of coffee

time	0	5	10	15	20	25	30	35	40
T(t)	131.2	126.9	121.5	116.5	111.3	106.4	102.5	99	96
time	45	50	55	60	65	70	75	80	85
T(t)	93.5	91.3	89.5	87.9	86.3	84.3	82.5	81.1	79.7

Exercises for section 2.14

1. A flask, whose interior temperature is 80°F, is put in contact with a cooling device through which water is circulating at the constant temperature of 40°F. After 5 minutes, the temperature inside the flask drops to 71°F. How long will it take to have a flask temperature of 50°F?

2. The temperature inside a flask is 75°F. We put it in contact with running water kept at the constant temperature of 45°F. After 2 minutes the temperature inside the flask drops to 70°F. Calculate the temperature 5 minutes after the cooling process started.

3. The temperature inside a house is 67°F. A thermometer, which was inside the house, is brought outside. After 3 minutes the thermometer indicates a temperature of 62°F and 3 minutes afterwards the temperature drops to 60°F. Accepting Newton's law of cooling, what is the outside temperature?

4. We wish to determine the temperature at which a reaction is taking place inside a flask, but cannot make a direct measurement while the reaction is on. We stop the reaction and, right afterwards, we put the flask in contact with water circulating at 40°F. After 5 minutes the temperature drops to 68°F and, 2 minutes after the first measurement, the temperature further drops to 56°F. Calculate $T(0)$, where $T(t)$ is the temperature inside the flask at any time t.

5. A cup of coffee is taken out from a microwave and left to cool inside a room kept at the constant temperature of 66°F. Using a scientific thermometer, we make measurements at 5-minute intervals (Table 2.10). Through a linear regression analysis, test Newton's law of cooling and thereafter estimate the parameter k.

6. A corpse of a person is found at 11 a.m. inside a room kept at a constant temperature of 68°F. The temperature of the body at the time it was found was 80°F. At 12 noon, the temperature of the corpse has dropped to 70°F. Taking into consideration that the normal temperature of a living person is 98.6°F, when did the person die?

Table 2.11: Growth of Sunflowers I

t (days)	x(t) (cm)	ln $x(t)$
7	17.93	2.8865
14	36.36	3.5935
21	67.76	4.216
28	98.1	4.586
35	131	4.8752

2.15 Exponential Growth

Suppose that the height of a sunflower plant is given by Table 2.11 (first two columns). Let $x(t)$ denote the height at any time. On the basis of our experience with radioactive decay we might put forward, as a possible model, the differential equation

$$\frac{dx}{dt} = ax(t).$$

Let us try to obtain mathematical consequences of it that can be compared with data. We observe that

$$\int \frac{dx}{x} = \int a\, dt.$$

Thus $\ln x(t) = at + C$ for a certain constant C. Next we try to find out whether this consequence of our model is in agreement with data. With this purpose in mind we build the third column of Table 2.11 and use linear regression to obtain the line of best fit $\hat{y} = 0.070999t + 2.54047$. The correlation coefficient happens to be $r = 0.982991$. The value of r, close to 1, tells us that there is reasonable agreement between a mathematical consequence of the model and data. Our next step is to estimate a and C, namely $a \approx 0.070999$ and $C \approx 2.54047$. Therefore $x(t) = e^C e^{at} = 12.685632 e^{0.070999t}$. This curve (Figure 2.12) 'represents' data adequately between days 7 and 35 but, as we will discuss in the next section, it ceases to do so if we enlarge our observations up to day 84.

In general, the initial value problem $x'(t) = ax(t)$, $x(0) = x_o$, will lead to the unique solution $x(t) = x_o e^{at}$. Indeed,

$$\int_0^t \frac{x'(u)}{x(u)} du = a \int_0^t du.$$

Therefore $\ln \frac{x(t)}{x(0)} = at$, which in turn implies $x(t) = x(0)e^{at}$. The presence of the exponential function is what leads to the expression 'exponential growth.' It should be noted that, in close analogy to the equation for radioactive decay, we can find the time (τ) it takes to double $x_o = x(0)$: from the definition of doubling time we get $2x_o = x_o e^{a\tau}$, hence $\tau = \ln 2/a$.

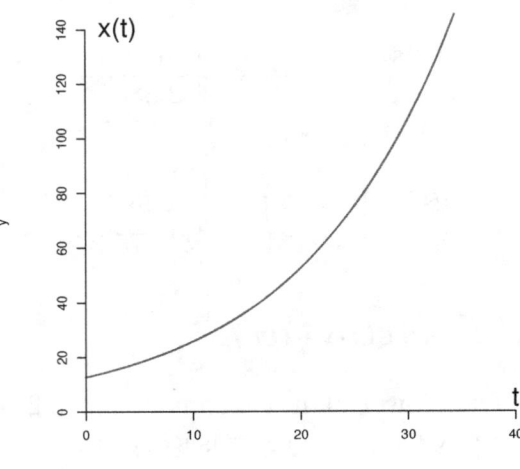

Figure 2.12: Exponential growth

Under ideal conditions, *Escherichia coli* bacteria grow exponentially. If tempera-ture is at an optimum and there is an unlimited supply of nutrients, and cell density is low, this type of bacterium will undergo cell division every 20 minutes approximately. Suppose that we have 100 bacteria in a special soup, full of nutrients. How many will be there after 5 hours? Since $\tau = 20$ we will have $x(300) = 100e^{(\ln 2/20)300} = 3{,}276{,}800$. It is an explosive growth indeed! From the context of the problem we are assuming that $x(t)$ denotes the number of bacteria at any time t, a differentiable function defined on the real line although strictly speaking the number of bacteria is a 'discrete' function of time. However, the continuous representation of $x(t)$ is usually a good approximation to the real phenomenon at hand. To be more pre-cise, we should emphasize that $x_o e^{at}$ is a continuous curve that passes through the points $(0, x_o)$, $(\tau, 2x_o)$, $(2\tau, 4x_o)$, $(3\tau, 8x_o)$, ... If we use the formula $x_o e^{at}$ to calculate the number of bacteria at any time t, we will only find an estimate. That is to say, whenever t/τ is a natural number, $x_o e^{at}$ (where $a = \ln 2/\tau$) provides an exact answer. However, when t/τ is not a natural number, $x_o e^{at}$ provides an approximation.

In section 2.1 we mentioned that the growth of bacteria is governed by the formula $P_o 2^{t/\tau}$, where τ is the time it takes a bacterium to subdivide and P_o is the initial number of bacteria. Is this formula different from $P_o e^{at}$? Not really, because

$$P_o 2^{t/\tau} = P_o (e^{\ln 2})^{t/\tau} = P_o (e^{\ln 2/\tau})^t = P_o e^{at}$$

where $a = \ln 2/\tau$.

Table 2.12: Population of Houston (Texas)

year	Population
1900	44,633
1910	78,800
1920	138,276
1930	292,352
1940	384,514
1950	596,163
1960	938,219
1970	1,232,802
1980	1,595,138

To bring this section to an end, let us discuss one additional problem. Working with a certain type of bacterium, under ideal conditions, we find that 100 bacteria become 600 bacteria after 2 hours. We wish to estimate its doubling time. Assuming exponential growth we will have that $600 = 120e^{120k}$, therefore $\ln 5 = 120k$. Consequently $\tau = \frac{\ln 2}{\ln 5/120}$, i.e. $\tau \approx 51.68$ minutes.

Exercises for section 2.15

1. Two hundred bacteria are put in a soup full of nutrients. Three hours later, there are 9000 bacteria. Assuming exponential growth, estimate the number of bacteria after 10 hours have elapsed since the beginning of the experiment.

2. We found the function $x(t) = 12.685632e^{0.070999t}$, a nice fit to the data on growth of sunflowers (Table 2.11). Estimate the time it would take a sunflower to reach a height of 80 cm.

3. The population of Houston (Texas) between 1900 and 1990 has increased dramatically. According to the U.S. Census Bureau, it has gone from 44,633 in 1900 to 1,595,138 in 1980 (Table 2.12). Use linear regression to find an exponential function that is a good fit to the given data.

4. Estimate the population of Houston in 1955, using for this purpose the exponential function from the previous problem.

5. What is the doubling time of a population that increases exponentially, say $x(t) = x(0)e^{at}$ for a certain parameter a, and it is known that $x(t_1) = N_1$ and $x(t_2) = N_2$?

Table 2.13: Growth of sunflowers II

t (days)	x(t) (cm)	$\ln x(t)$	$\ln \frac{x(t)}{260-x(t)}$
7	17.93	2.8865	-2.6027517
14	36.36	3.5935	-1.8165683
21	67.76	4.216	-1.0427725
28	98.10	4.586	-0.50099149
35	131	4.8752	0.01538492
42	169.5	5.1329	0.62750308
49	205.5	5.3254	1.3272453
56	228.3	5.4307	1.9743439
63	247.1	5.5098	2.9525658
70	250.5	5.5235	3.2721671
77	253.8	5.5365	3.7119973
84	254.5	5.5393	3.8345527

6. Suppose that we have two strains of bacteria (x and y) that undergo exponential growth, say $x'(t) = \alpha x(t)$ and $y'(t) = \beta y(t)$. Let $z(t) = \frac{x(t)}{x(t)+y(t)}$, the fraction of the first strain. Find $z(t)$ in terms of $z(0)$, α, and β.

7. A population $P(t)$ could have an explosive growth, in the sense that $P'(t) = (\alpha t + \beta)P(t)$ for some positive parameters α and β. Under these circumstances, find an expression that could be used to test the model with data on population across time, using linear regression for this purpose. How would you estimate the parameters? (Hint: Divide by $P(t)$ and integrate the given differential equation between 0 and t).

2.16   Logistic Growth

Let us expand Table 2.11 to include seven more observations with regard to the growth of sunflowers (Reed and Holland, 1919). If we were to accept exponential growth, we would have to compare t with $\ln x(t)$ (see Table 2.13) through a linear regression process. We get $r = 0.899564$, with the least squares regression line $\hat{y} = 0.031118t + 3.430399$. We note that the model is not very good anymore (r has dropped from 0.982991, as seen in the previous section, to 0.899564.) A plot of the data on sunflowers will show that the rate of growth slows and seems to level off. To take into account these circumstances, Reed and Holland considered the differential equation

$$\frac{dx}{dt} = kx(t)(L - x(t))$$

that governs autocatalytic reactions in chemistry (L is the initial amount of substance to be transformed, $x(t)$ is the amount of product formed at time t, which in turn catalyzes the reaction, and $x(0)$ has to be positive for the reaction to get started). Let us solve the preceding differential equation keeping in mind that during the whole process both $x(t)$ and $L - x(t)$ are positive. Right away we have:

$$\int \frac{dx}{x(L-x)} = \int k\, dt.$$

But

$$\frac{1}{x(L-x)} = \frac{1}{L}\frac{1}{x} + \frac{1}{L}\frac{1}{L-x}.$$

Therefore

$$\frac{1}{L}\int \frac{dx}{x} - \frac{1}{L}\int \frac{-dx}{L-x} = \int k\, dt.$$

Consequently, there exists a constant C such that

$$\frac{1}{L}\ln x - \frac{1}{L}\ln(L-x) = kt + C.$$

Then

$$\ln \frac{x(t)}{L - x(t)} = kLt + H \tag{2.31}$$

where $H = CL$. Thus $\frac{x(t)}{L-x(t)} = e^H e^{kLt}$, that is,

$$x(t) = \frac{L}{1 + e^{-H}e^{-kLt}} = \frac{Le^H}{e^H + e^{-kLt}} = \frac{Lx_o}{x_o + (L - x_o)e^{-kLt}}, \tag{2.32}$$

where the last equality is obtained from the fact that $e^H = \frac{x_o}{L-x_o}$. Note that $\lim_{t\to\infty} x(t) = L$ because, in the long run, all the substance will have been transformed into product.

Reed and Holland adapted the differential equation of autocatalysis to the problem of growth of sunflowers. In this context $x(t)$ will denote height and L the limiting height, beyond which the plant does not grow anymore. Following Lomen and Lovelock (1999) we choose $L = 260$ as an approximate limiting height for the problem under discussion. A careful plot of the data makes this choice a reasonable one.

We should keep in mind that (2.31) predicts that a line will appear if we put $\ln \frac{x(t)}{260-x(t)}$ on the vertical axis and time on the horizontal axis. The reader might surmise that we plan to apply linear regression with t as the first variable and $y = \ln \frac{x(t)}{260-x(t)}$ as the second variable (Table 2.13).

The least squares regression line happens to be $\hat{y} = 0.087221t - 2.989167$ and the correlation coefficient $r = 0.994812$. The value of r is pretty close to 1, so we accept the model with $kL = 0.087221$ and $H = -2.989167$. Therefore

$$x(t) = \frac{260e^{-2.989167}}{e^{-2.989167} + e^{-0.087221t}} = \frac{13.085629}{0.05032934 + e^{-0.087221t}}.$$

Applying this expression for $x(t)$ (Figure 2.13) we can compare the estimated values of height with observed heights (Table 2.14)

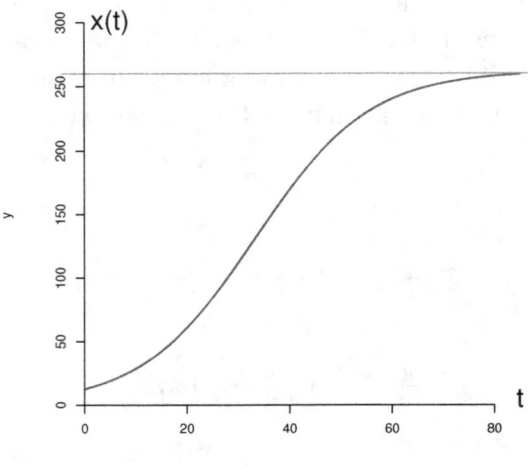

Figure 2.13: Logistic growth of sunflowers

Bacterial growth under constraints

After having discussed the growth of sunflowers, it is time to analyze bacterial growth when resources such as food are limited and overcrowding is detrimental to the multiplication of bacteria. Often it is observed that the population at first increases, reaches a point of high growth, and thereafter tends to a limiting value. To explain this type of behavior we have to modify the model $N'(t) = \alpha N(t) - \beta N(t)$, $\alpha > \beta$, where α is the coefficient of birth and β is the coefficient of death. As we know by now, this model leads to exponential growth, namely $N(t) = N_o e^{at}$ where $a = \alpha - \beta$. Let us rather assume that α is not a constant but $\alpha = \gamma - \delta N(t)$ for some positive constants γ and δ ($\gamma > \beta$). In other words, α decreases linearly. Then $N'(t) = (\gamma - \delta N(t))N(t) - \beta N(t)$, that is to say, $N'(t) = aN(t) - bN^2(t)$ where $a = \gamma - \beta$ and $b = \delta$. But this equation, known as the 'logistic differential equation,' or simply the **logistic equation**, can be rewritten as

$$N'(t) = bN(t)(\frac{a}{b} - N(t)). \tag{2.33}$$

This is precisely the differential equation for autocatalytic reactions and the growth of sunflowers, with k corresponding to b and L corresponding to a/b. Therefore (2.25) implies that

$$N(t) = \frac{\frac{a}{b}N_o}{N_o + (\frac{a}{b} - N_o)e^{-b(a/b)t}} = \frac{aN_o}{bN_o + (a - bN_o)e^{-at}}. \tag{2.34}$$

We observe that

Table 2.14: Comparison between estimated and observed values

t (days)	$x(t)$ (observed)	$x(t)$ (estimated)
7	17.93	22.052582
14	36.36	37.903355
21	67.76	62.17045
28	98.10	95.307053
35	131	134.13053
42	169	172.23023
49	205.5	203.643
56	228.3	226.03056
63	247.1	240.38153
70	250.5	248.96566
77	253.8	253.88926
84	254.5	256.64551

Table 2.15: Growth of a bacterial colony

days	area (cm^2)	$\ln(\frac{N(t)}{50-N(t)})$
0	0.24	-5.3343278
1	2.78	-2.8323666
2	13.53	-0.99158056
3	36.30	0.97442191
4	47.5	2.944439
5	49.4	4.410776

$$\lim_{t \to \infty} N(t) = \frac{a}{b}.$$

Let us analyze the growth of a bacterial colony (Thornton, 1922) subject to the constraints of food and space (first two columns of Table 2.15).

Is the model of logistic growth, namely $N'(t) = aN(t) - bN^2(t)$, in agreement with data? From (2.31) we know that this model implies that

$$\ln(\frac{N(t)}{\frac{a}{b} - N(t)}) = at + H \tag{2.35}$$

for a certain constant H. Looking at data it seems reasonable to choose $a/b = 50$ and apply linear regression with time on the horizontal axis and $y = \ln \frac{N(t)}{50-N(t)}$ on the vertical axis (Table 2.15).

Using technology we get the correlation coefficient $r = 0.997587$ and the least squares regression line $\hat{y} = 1.94368t - 4.996624$. Due to the fact that r is close to 1 we accept the validity of the model of logistic growth and choose to estimate a by the value 1.94368. Then $b = 1.94368/50 = 0.0388736$ is the accepted estimation of the parameter b. From (2.34) we get

$$N(t) = \frac{1.94368 \star 0.24}{0.0388736 \star 0.24 + (1.94368 - 0.0388736 \star 0.24)e^{-1.94368t}} = \frac{0.4664932}{0.00932966 + (1.94368 - 0.4664932)e^{-1.94368t}},$$

whose graph can be seen in Figure 2.14.

Growth of fish in a lake

Suppose that growth of fish in an artificial lake follows a logistic model, say

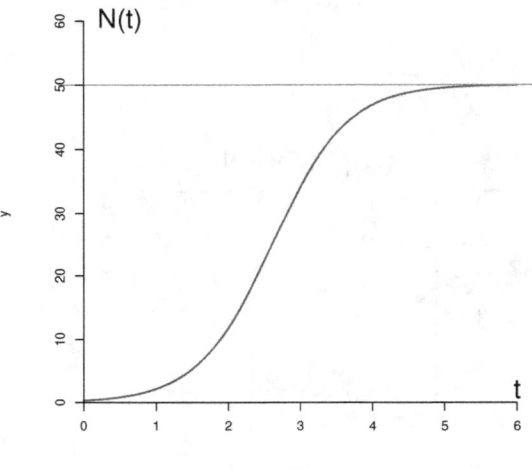

Figure 2.14: Logistic growth of bacteria

$$N'(t) = bN(t)(K - N(t))$$

where $K = \frac{a}{b}$ (a and b are the parameters of the model). From what we have learned so far about logistic models, we know that $\lim_{t \to \infty} N(t) = \frac{a}{b} = K$. Thus, we can interpret K as the 'carrying capacity.' In other words, K happens to be the limit number of fish that the lake can sustain. Assume that $K = 3000$, $N(0) = 150$ and $N(4) = 600$ (t is measured in months). What is the value of $N(t)$ for any t? From (2.34) we get

$$N(t) = \frac{3000(150)}{150+(3000-150)e^{-at}} = \frac{3000}{1+(\frac{3000}{150}-1)e^{-at}}.$$

Therefore $N(t) = \frac{3000}{1+19e^{-at}}$. But

$$600 = N(4) = \frac{3000}{1+19e^{-4a}}.$$

Consequently $e^{-4a} = \frac{4}{19}$, which in turn leads to $a = 0.3895361$. We can conclude that

$$N(t) = \frac{3000}{1+19e^{-0.38953615t}}.$$

Of course, it is possible to solve the same problem without recourse to (2.34). Indeed, from $N'(t) = bN(t)(3000 - N(t))$ we get

$$\frac{N'(t)}{N(t)(3000-N(t))} = b.$$

But

$$\frac{1}{N(3000-N)} = \frac{p}{N} + \frac{q}{3000-N}$$

for some numbers p, q. Since $1 = p(3000 - N) + qN$, it follows that $3000p = 1$ and $q - p = 0$. Consequently[9] $p = q = 1/3000$. Multiplying by N', and integrating, we get

$$\frac{1}{3000} \int \frac{N'(t)}{N(t)} dt - \frac{1}{3000} \int \frac{-N'(t)}{3000-N(t)} dt = \int b \, dt.$$

Thus,

$$\frac{1}{3000} \ln N(t) - \frac{1}{3000} \ln(3000 - N) = bt + C$$

for some constant C. That is to say,

$$\frac{1}{3000} \ln \frac{N(t)}{3000-N(t)} = bt + C.$$

In particular, making $t = 0$, we arrive at the equality

$$\frac{1}{3000} \ln \frac{150}{3000-150} = C.$$

Therefore

$$\frac{1}{3000} \ln \frac{2850N(t)}{150(3000-N(t))} = bt.$$

After simplifying and applying the exponential function, we reach the expression

$$N(t) = \frac{3000}{1+19e^{-3000bt}}.$$

But $N(4) = 600$, so

$$600 = \frac{3000}{1+19e^{-12,000b}}.$$

A little bit of algebraic manipulation, or the use of the *solve* command from our graphing calculator, leads to $b = 0.00012985$. Consequently

$$N(t) = \frac{3000}{1+19e^{-0.38955t}}.$$

---

[9]We are using the algebraic method of partial fractions.

A differential equation

It is time to consider an example of a differential equation of the type $y'(t) = ay(t) - by^2(t)$, which does not necessarily stem from a particular application. Let us analyze the equation $y' = 2y - y^2$. Right away we notice that the constant functions $y(t) = 0$ and $y(t) = 2$, for all t, are solutions. A theorem from advanced mathematics asserts that no two solutions of $y' = ay - by^2$ can intersect each other[10]. Thus, all solutions of $y' = 2y - y^2$ are either between the constant functions 0 and 2 or above 2. It all depends on whether $0 < y(0) < 2$ or $y(0) > 2$.

Let us first consider the case $0 < y(0) < 2$, say $y(0) = 1$. The solution, $y(t)$, will satisfy the inequalities $0 < y(t) < 2$ for all t. Then

$$\frac{y'(t)}{y(t)(2-y(t))} = 1.$$

Using the method of partial fractions (or the *expand* command from a graphing calculator) we obtain

$$\frac{1}{y(2-y)} = \frac{1}{2}\frac{1}{y} + \frac{1}{2}\frac{1}{2-y}.$$

Therefore

$$\frac{y'(t)}{y(t)(2-y(t))} = \frac{1}{2}\frac{y'(t)}{y(t)} + \frac{1}{2}\frac{y'(t)}{2-y(t)}.$$

Consequently,

$$\frac{1}{2}\int_0^t \frac{y'(u)}{y(u)}du + \frac{1}{2}\int_0^t \frac{y'(u)}{2-y(u)}du = \int_0^t 1 du.$$

Hence $\frac{1}{2}[\ln y(u)]_0^t + \frac{1}{2}[\ln(2-y(u))]_0^t = t$, that is, $\frac{1}{2}\ln y(t) + \frac{1}{2}\ln(2-y(t)) = t$ or, what is the same,

$$\ln \frac{y(t)}{2-y(t)} = 2t.$$

Applying the exponential function to both sides, we get

$$\frac{y(t)}{2-y(t)} = e^{2t}.$$

Multiplying and reordering leads to

$$y(t) = \frac{2}{1+e^{-2t}}.$$

Next, let us suppose that $y(0) = 3$. We already know that, since $y(0) > 2$, the inequality $y(t) > 2$ holds for all t. Then

$$\frac{y'(t)}{y(t)(2-y(t))} = \frac{1}{2}\frac{y'(t)}{y(t)} - \frac{1}{2}\frac{y'(t)}{y(t)-2}$$

---

[10]Actually, no two solutions of a differential equation of the type $y' = f(y)$, where the derivative of f is continuous, can intersect each other; see enrichment note at the end of the section.

leads to

$$\tfrac{1}{2}\int_0^t \tfrac{y'(u)}{y(u)}\,du - \tfrac{1}{2}\int_0^t \tfrac{y'(u)}{y(u)-2}\,du = \int_0^t 1\,du$$

Thus, $\tfrac{1}{2}[\ln y(u)]_0^t - \tfrac{1}{2}[\ln(y(u)-2]_0^t = t$, that is,

$$\tfrac{1}{2}\ln \tfrac{y(t)}{3} - \tfrac{1}{2}\ln(y(t)-2) = t.$$

Consequently $\ln \tfrac{y(t)}{3y(t)-6} = 2t$. Applying exp to both sides, and reordering terms, we finally arrive at

$$y(t) = \tfrac{6}{3-e^{-2t}}.$$

A graph of both solutions, when $y(0) = 1$ and when $y(0) = 3$, can be seen in Figure 2.15.

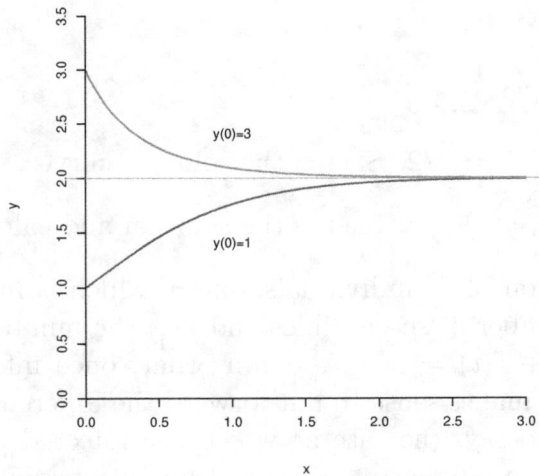

Figure 2.15: Two solutions of the equation $y' = 2y - y^2$

Of course, since we know that any non-constant solution of $y' = 2y - y^2$ does not cross either the constant function 0 or the constant function 2, it is possible to solve the differential equation without worrying about the initial condition until the very end. Indeed,

$$\tfrac{1}{2}\int \tfrac{y'(t)}{y(t)}\,dt - \tfrac{1}{2}\int \tfrac{y'(t)}{y(t)-2}\,dt = \int dt.$$

Hence, there exists a constant C such that

$$\tfrac{1}{2}\ln|y(t)| - \tfrac{1}{2}\ln|y(t)-2| = t + C \text{ for all } t.$$

Therefore

$$\tfrac{1}{2}\ln\left|\tfrac{y(t)}{y(t)-2}\right| = t + C \text{ for all } t.$$

Making $t = 0$ we get $\tfrac{1}{2}\ln\left|\tfrac{y_o}{y_o-2}\right| = C$. Therefore

$$\tfrac{1}{2}\ln\left|\tfrac{(y_o-2)y(t)}{y_o(y(t)-2)}\right| = t,$$

which in turn leads to

$$\left|\tfrac{(y_o-2)y(t)}{y_o(y(t)-2)}\right| = e^{2t}.$$

If $y_o = 1$ we get $\tfrac{y(t)}{2-y(t)} = e^{2t}$, while if $y_o = 3$ we get $\tfrac{y(t)}{3(y(t)-2)} = e^{2t}$. From these expressions we get the corresponding solutions.

Our discussion, with regard to $y' = 2y - y^2$, is valid for any differential equation of the type $y' = ay - by^2$. Besides the constant solutions 0 and a/b, every other solution is obtained through a process of integration that involves the algebraic method of partial fractions.

Exercises for section 2.16

1. Solve $y' = y - y^2$, $y(0) = 1/2$. Sketch the solution curve.

2. Solve $y' = 3y - 2y^2$, $y(0) = 4$. Sketch the solution and calculate $\lim_{t\to\infty} y(t)$.

3. There is a population of N individuals, one of which is initially infected. Let $S(t)$ denote the number of susceptibles and $I(t)$ the number of infected at any time t; thus, $S(t) + I(t) = N$. Assuming that, once infected, an individual remains infected, it makes sense to put forward the differential equation $I'(t) = kS(t)I(t)$; that is to say, the rate at which the infected population increases is proportional to the number of susceptibles and infected (keep in mind that all individuals are either susceptibles or infected, there is no other category[11]). Therefore $I'(t) = k(N - I(t))I(t)$, $I(0) = 1$ or, what is the same, $I'(t) = kNI(t) - kI^2(t)$. Find $I(t)$ and calculate $\lim_{t\to\infty} I(t)$.

4. Suppose that a new measurement were to be added to the data on growth of a bacterial colony (Table 2.15), say 50.8 cm^2 at $t = 6$ days (with carrying capacity of 51 cm^2) . What correlation coefficient do you get? Use linear regression to find the corresponding least squares regression line. Estimate $N(3.5)$, the size of the bacterial colony after three and a half days.

---

[11]We are dealing with a simplified model for the spread of an epidemic. Why? Some people might be immune from the very beginning, others might die or recover from the disease and become immune. Neither is the latency period being considered.

5. The logistic differential equation, namely $N'(t) = aN(t) - bN^2(t)$, can be written as $N'(t) = aN(t)(1 - \frac{N(t)}{K})$ where $K = \frac{a}{b}$ is the carrying capacity. The model has to be modified if 'harvesting' takes place at a constant rate h:

$$N'(t) = aN(t)(1 - \tfrac{N(t)}{K}) - h.$$

For instance, we can think of fish being harvested from a farm, under the absence of predators. Show that $N'(t) < 0$ if $h > \frac{aK}{4}$. That is to say, if the constant rate of harvesting goes beyond $\frac{aK}{4}$, the population will become extinct eventually (Hint: Note that $N'(t) = -\frac{a}{K}(N^2 - kN + \frac{Kh}{a})$; then the problem is reduced to an analysis of the vertical parabola $N^2 - KN + \frac{Kh}{a}$.)

6. Solve the **logistic differential equation with harvesting**, when $a = 0.1$, $K = 100$, $h = 0.099$. Sketch the solution curves for $N_o = 10$, $N_o = 120$, $N_o = 0.5$, and calculate $\lim_{t \to \infty} N(t)$ in each case.

7. Several models of plant growth have been developed (Thornley and Johnson, 1990). The simplest one assumes that the rate of growth of the dry mass of a plant, $W'(t)$, is only proportional to the amount of dry mass, that is, $W'(t) = \alpha W(t)$; α is a positive parameter. Evidently, we will have that $W(t) = W_o e^{\alpha t}$. Next consider the possibility that the rate of growth of the dry mass is proportional to the amount of substrate present ($S(t)$, the amount of nutrients), but is independent of the amount of dry mass, that is, $W'(t) = kS(t)$, where k is a positive parameter. Taking into account that, at any time t, $W(t) + S(t) = W_o + S_o = a$, the differential equation becomes $W'(t) = k(a - W(t))$. Find its solution.

8. A third model of plant growth considers the possibility that the rate of growth of W is proportional both to the amount of substrate and the dry mass. That is to say, $W'(t) = \alpha W(t)S(t)$. Since $S(t) + W(t) = a$, where $a = S_o + W_o$, it follows that $W'(t) = \alpha W(t)(a - W(t))$. This is a model of logistic growth. Find the solution of the differential equation.

9. The last model that we will consider assumes that $W'(t) = \alpha W(t)$, where α diminishes exponentially with time, say $\alpha(t) = \alpha_o e^{-\beta t}$, where $0 < \beta < \alpha_o$. Then $W'(t) = \alpha_o e^{-\beta t} W(t)$. This equation is known as 'Gompertz growth equation.' Before solving it, find the point of inflection of $W(t)$. Thereafter, use a process of integration to solve Gompertz's equation. Furthermore, sketch a typical solution for each of the preceding four models.

10. Verify that, at the inflection point of Gompertz's equation, the dry mass adopts the value $\frac{1}{e} \times$ limiting mass; quite different from the logistic model, wherein at the inflection point the dry mass is one half the limiting value.

11. Verify that, at the beginning, the solution of Gompertz's equation does behave as an equation of exponential growth (Hint: Recall the linearization of e^x at $x = 0$).

Enrichment Note: More About the Logistic Equation

The growth of sunflowers and bacteria, which we discussed in section 2.16, is typical of what is known as logistic growth. Let us analyze in greater detail the underlying differential equation . With this purpose in mind, we will consider the equation

$$N'(t) = aN(t) - bN^2(t)$$

where $a, b > 0$. The constant function $N(t) = a/b$ is a solution since $N'(t) = 0$ and $a(a/b) - b(a^2/b^2) = 0$. Thus, the constant function $t \to a/b$ satisfies the differential equation $N' = aN - bN^2$ and the initial condition $N(0) = a/b$. Obviously, the null function is a solution of the differential equation too. A deep theorem, which we will not prove, asserts that the solutions of any differential equation $y' = f(y)$, where f' is continuous, do not cross each other. It is a consequence of the fact that the initial value problem $y' = f(y)$, $y(t_o) = y_o$, admits only one solution provided that f' exists and is continuous in a neighborhood of y_o. Thus, if $N_o > a/b$ or $0 < N_o < a/b$, any solution of $N' = aN - bN^2$, $N(0) = N_o$ will not cross the horizontal line $y = a/b$ or the horizontal line $y = 0$. To be more precise, either $a - bN(t) > 0$ and $N(t) > 0$ for all t (when $0 < N_o < a/b$) or $a - bN(t) < 0$ for all t (when $N_o > a/b$).

Assuming that $N_o \neq \frac{a}{b}$, $N_o > 0$, let us try to find the solution of

$$\tfrac{dN}{dt} = aN - bN^2, \ N(0) = N_o.$$

We can proceed readily, taking into account our previous study of the differential equation that governs the process of autocatalysis. Indeed

$$\int \tfrac{dN}{aN - bN^2} = \int dt.$$

But $\frac{1}{N(a-bN)} = \frac{p}{N} + \frac{q}{a-bN}$ for certain numbers p, q. Hence $1 = p(a - bN) + qN$, which in turn implies $ap = 1$, $-bp + q = 0$. Therefore $p = 1/a$, $q = b/a$. Consequently

$$\tfrac{1}{N(a-bN)} = \tfrac{1}{a}\tfrac{1}{N} + \tfrac{b}{a}\tfrac{1}{a-bN}.$$

Thus

$$\tfrac{1}{a} \int \tfrac{dN}{N} - \tfrac{1}{a} \int \tfrac{-bdN}{a-bN} = \int dt.$$

Therefore, there exists a constant C such that $\frac{1}{a}\ln N - \frac{1}{a}\ln|a - bN| = t + C$. Hence $\ln\left|\frac{N}{a-bN}\right| = at + aC$, that is, $\left|\frac{N}{a-bN}\right| = He^{at}$, where $H = e^{aC}$. Finally, we can conclude that $\frac{N}{a-bN} = Fe^{at}$ where $F = \pm H$. Making $t = 0$ we get $\frac{N_o}{a-bN_o} = F$, consequently $\frac{N}{a-bN} = \frac{N_o}{a-bN_o}e^{at}$. A little bit of algebra leads to

$$N(t) = \frac{aN_o}{bN_o + (a - bN_o)e^{-at}}. \tag{2.36}$$

The reader may observe that from (2.36), which is identical to (2.34), neither 0 nor a/b, two solutions of the differential equation, can be obtained.

Before going any further let us note that the maximum rate of growth of N will take place when $N = \frac{a}{2b}$. Why is this so? Let $g(N) = aN - bN^2$, which happens to be a parabola that opens downwards (N on the horizontal axis and $g(N)$ on the vertical axis) with vertex at $N = \frac{a}{2b}$. Thus, the maximum rate of growth takes place when $N = \frac{a}{2b}$. The next question, to be expected, is the following: For what value of t do we have $N(t) = \frac{a}{2b}$? Evidently $N_o < \frac{a}{b}$ (recall that the solutions of the differential equation under consideration cannot cross each other), thus $a - bN_o > 0$. We have to solve the algebraic equation

$$\frac{a}{2b} = \frac{aN_o}{bN_o + (a - bN_o)e^{-at}}. \tag{2.37}$$

Suppose t is a solution of it. Then $abN_o + a(a - bN_o)e^{-at} = 2abN_o$, which in turn leads to $e^{-at} = \frac{bN_o}{a-bN_o}$. Applying logarithms to both sides we get

$$t = \frac{1}{a}\ln\frac{a - bN_o}{bN_o}. \tag{2.38}$$

Soon we will see that $N(t)$ has an inflection point at t given by (2.38).

From (2.36) we observe that $\lim_{t\to\infty} N(t) = a/b$. Moreover, as expected, $N(t)$ given by (2.36) does not adopt the value a/b at any time because it would imply that

$$\frac{a}{b} = \frac{aN_o}{bN_o + (a-bN_o)e^{-at}}$$

for a certain t, hence $(a - bN_o)e^{-at} = 0$. Since $a - bN_o \neq 0$ we would have $e^{-at} = 0$, a clear impossibility.

Suppose $N_o < \frac{a}{b}$. We must have $N(t) < \frac{a}{b}$ for every t because otherwise the Intermediate Value Theorem for continuous functions (IVT) would imply that $N(t)$ adopts the value $\frac{a}{b}$ at some t. Similarly, $N_o > \frac{a}{b}$ implies that $N(t) > \frac{a}{b}$ for every t. Using the chain rule we get

$$N'(t) = \frac{-aN_o[-a(a-bN_o)e^{-at}]}{[bN_o + (a-bN_o)e^{-at}]^2}.$$

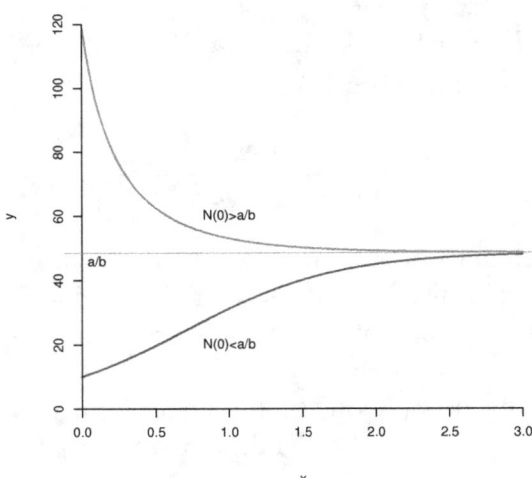

Figure 2.16: Two solutions of the logistic equation

Therefore, if $N_o < \frac{a}{b}$ then $N'(t) > 0$, while $N_o > \frac{a}{b}$ implies $N'(t) < 0$. Consequently, $N(t)$ is increasing if $N_o < \frac{a}{b}$ while $N(t)$ is decreasing if $N_o > \frac{a}{b}$. With a little bit of patience we can calculate $N''(t)$ and prove that $N''(t) > 0$ when $N_o > \frac{a}{b}$; thus, $N(t)$ is concave up when $N_o > \frac{a}{b}$. On the other hand, if $N_o < \frac{a}{b}$ we can prove that $N(t)$ has a point of inflection at $t = \frac{1}{a} \ln \frac{a - bN_o}{bN_o}$, as predicted before. All these characteristics of $N(t)$ are molded in Figure 2.16. From what we have discussed so far, it should be clear why a/b is usually called the 'carrying capacity.'

Interestingly enough, when t is small, logistic growth resembles exponential growth. Why is this so? From (2.36) we get

$$N(t) = \frac{N_o}{\frac{N_o}{a/b} + (1 - \frac{N_o}{a/b})e^{-at}}.$$

If N_o is much smaller than $\frac{a}{b}$ we will have that $\frac{N_o}{a/b}$ is much smaller than 1. Hence $1 - \frac{N_o}{a/b} \approx 1$. Let us pay attention to $\frac{N_o}{a/b} + e^{-at}$. At the beginning e^{-at} is only slightly less than 1, while $\frac{N_o}{a/b}$ is much smaller than 1. Therefore $\frac{N_o}{a/b} + e^{-at} \approx e^{-at}$. Consequently, at the beginning $N(t) \approx \frac{N_o}{e^{-at}} = N_o e^{at}$.

The equation

$$N'(t) = aN(t) - bN^2(t), \tag{2.39}$$

a model of logistic growth, where $N(t)$ denotes a population subject to competition for resources among its members (the term $-bN^2(t)$ takes into consideration this factor), has been studied extensively. We solved the differential equation of logistic growth as an SVE by writing $\frac{N'(t)}{N(t)(a-bN(t))} = 1$, and then integrating. But it is simpler to proceed as follows: Define $v(t) = N^{-1}(t)$; thus, $v'(t) = -N^{-2}(t)N'(t)$. Multiplying the original differential equation by $-N^{-2}(t)$ we get

$$v'(t) + av(t) = b.$$

Then $e^{at}(v'(t) + av(t)) = be^{at}$, which in turn leads to $(e^{at}v(t))' = (\frac{b}{a}e^{at})'$. Therefore $e^{at}v(t) = \frac{b}{a}e^{at} + L$ for some L, that is, $v(t) = \frac{a}{b} + Le^{-at}$. In particular $\frac{1}{N(0)} = v(0) = \frac{b}{a} + L$, thus $L = \frac{1}{N_o} - \frac{b}{a}$ where $N_o = N(0)$. Replacing values we get

$$\frac{1}{N(t)} = \frac{b}{a} + (\frac{1}{N_o} - \frac{b}{a})e^{-at}.$$

After a little bit of algebraic work we reach the expression

$$N(t) = \frac{aN_o}{bN_o + (a - bN_o)e^{-at}}.$$

This is precisely the solution we found in section 2.17 (formula 2.34).

We observe that $\lim_{t \to \infty} N(t) = \frac{a}{b}$. In other words, the model predicts that the population cannot go beyond a/b. In contrast, the simpler model $N'(t) = aN(t)$ implies unlimited growth; actually, this will be explosive growth because the solution of $N'(t) = aN(t)$, $N(0) = N_0$, is $N(t) = N_o e^{at}$. In a real-life setting, say protozoa in a Petri dish full of a nutrient soup, we should expect logistic growth rather than exponential growth because the amount of nutrient is limited; relatively soon, competition for it will appear.

If $N_o < a/b$, it can be proven that $N'(t) > 0$ for every t; thus, $N(t)$ is increasing and tends to the limiting value a/b. Analyzing the second derivative, under the same assumption $N_o < a/b$, we would be able to conclude that $N(t)$ has the shape of an elongated S, a typical behavior of logistic growth. On the other hand, if $N_o > a/b$, it can be proven that $N'(t) < 0$ and $N''(t) > 0$. Consequently, $N(t)$ is decreasing, concave up, and tends to the limiting value a/b. All these characteristics of logistic growth were extensively discussed in section 2.17.

The equation $N'(t) = aN(t) - bN^2(t)$ is a special case of the family of differential equations $y'(t) + ay(t) = by^k(t)$, k an arbitrary fixed positive real number, known as the family of Bernoulli equations[12]. The 'transformation' $v(t) = y^{1-k}(t)$ converts the given equation into a linear first order differential equation that can be solved through the usual method, after we multiply both sides of Bernoulli's equation by $y^{-k}(t)$ and keep in mind the chain rule.

A numerical example may shed further light on the technique involved in the solution of Bernoulli's equation. Let us consider the initial value problem

---

[12]Named in honor of Jacob Bernoulli (1654-1705). He and his younger brother John were disciples of Leibniz.

$$y' + y = y^2, \quad y(0) = 3.$$

Multiplying both sides of the differential equation by y^{-2} we get $y'y^{-2} + y^{-1} = 1$, which turns into $-v' + v = 1$ once we define $v = y^{-1}$. Therefore $e^{-t}(v' - v) = -e^{-t}$, that is,

$$(e^{-t}v)' = (e^{-t})'.$$

Thus, there exists a constant C such that $e^{-t}v(t) = e^{-t} + C$. That is to say,

$$v(t) = 1 + Ce^t.$$

In particular, $1/3 = v(0) = 1 + C$. Hence, $\frac{1}{y(t)} = v(t) = 1 - \frac{2}{3}e^t$. Finally, we reach the conclusion that

$$y(t) = \frac{1}{1 - \frac{2}{3}e^t}, \quad t < \ln \frac{3}{2}.$$

2.17   Problems about Mixing

Suppose that a tank of volume V (in liters) receives b liter/minute of pure water and loses b liter/minute of solution through an opening at the bottom. At time $t = 0$ there are x_o mg of solute, homogeneously distributed across the tank and, at the constant rate of r mg/minute, solute is added to the tank and distributed uniformly using an agitator. Let $x(t)$ denote the amount of solute at any time t, and let Δt denote a very small interval of time. In Δt minutes, $r\Delta t$ milligrams of solute enter the tank while $b\Delta t$ liters of solution leave the tank. But $x(t)$ does not change significantly in the interval of time Δt; thus, in $b\Delta t$ liters of solution there are $b\Delta t \frac{x(t)}{V}$ milligrams of solute. Consequently

$$x(t + \Delta t) - x(t) = r\Delta t - b\Delta t \frac{x(t)}{V}.$$

Therefore

$$\frac{x(t+\Delta t)-x(t)}{\Delta t} = r - b\frac{x(t)}{V}.$$

In the limit, when $\Delta t \to 0$, we get

$$x'(t) = r - \frac{b}{V}x(t). \tag{2.40}$$

Right away, we notice that the constant function $x(t) = \frac{rV}{b}$ is a solution of the differential equation (2.40). Assuming that $x_o = x(0) < \frac{rV}{b}$, since no two solutions of (2.40) can cross each other[13], we must have $x(t) < rV/b$ for all $t \geq 0$, that is, $r - \frac{b}{V}x(t) > 0$ for all t. Then

---

[13]See enrichment note at the end of section 2.16.

$$\int \frac{x'(t)dt}{r - \frac{b}{V}x(t)} = \int dt.$$

Multiplying both sides by $-b/V$ we get

$$\int \frac{\frac{-b}{V}x'(t)dt}{r - \frac{b}{V}x(t)} = \int -\frac{b}{V}dt.$$

Thus, there exists a constant L such that

$$\ln(r - \tfrac{b}{V}x(t)) = -\tfrac{b}{V}t + L \text{ for every } t.$$

Consequently

$$(r - \tfrac{b}{V}x(t)) = He^{-\frac{b}{V}t}, \text{ where } H = e^L.$$

In particular, when $t = 0$ we get $r - \frac{b}{V}x_o = H$. Replacing this value of H in the preceding equation, we reach the equality

$$r - \tfrac{b}{V}x(t) = (r - \tfrac{b}{V}x_o)e^{-\frac{b}{V}t}.$$

Finally,

$$x(t) = \tfrac{rV}{b} + (x_o - \tfrac{rV}{b})e^{-\frac{b}{V}t}.$$

The concentration of solute in the tank is the amount of solute in it divided by V; thus, $c(t) = \frac{x(t)}{V}$. Consequently,

$$c(t) = \frac{r}{b} + (c_o - \frac{r}{b})e^{-\frac{b}{V}t}. \tag{2.41}$$

We note that $\lim_{t\to\infty} c(t) = \frac{r}{b}$. The value $\frac{r}{b}$ is known as the 'equilibrium concentration.' A typical graph of concentration across time can be seen in Figure 2.17.

If $x_o > \frac{r}{b}V$, we must have $x(t) > \frac{r}{b}V$ for all $t \geq 0$, that is, $r - \frac{b}{V}x(t) < 0$ for all $t \geq 0$. Following similar steps to those used in the case $x_o < rV/b$, we can reach the same answer, namely $r - \frac{b}{V}x(t) = (r - \frac{b}{V}x_o)e^{-\frac{b}{V}t}$.

Let us apply what we have learned to a particular problem. Suppose that a drug is introduced into the body of a patient through a vein, at the rate of r mg/minute. We might consider that blood circulates across the body at the unknown rate of b liter/minute. Assuming that the blood is contained in a 'tank' of unknown volume V and that the drug is distributed uniformly across the circulatory system (of course, both are approximations to the real problem), from (2.41) we get

$$\frac{r}{b} - c(t) = \left(\frac{r}{b} - c_o\right)e^{-\frac{b}{V}t}. \tag{2.42}$$

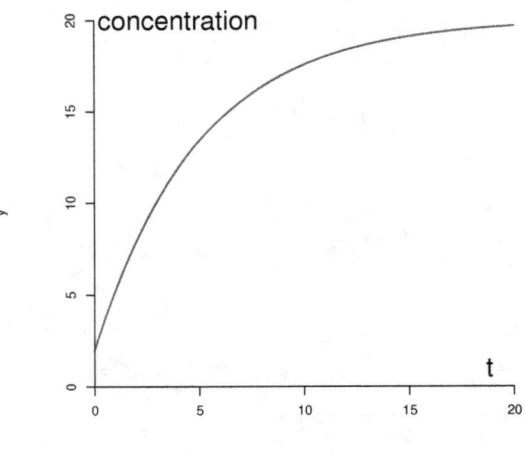

Figure 2.17: Variation of concentration in a tank$(c_o < \frac{r}{b})$

Our goal is to estimate $\frac{b}{V}$, an important task because $-\frac{b}{V}x(t)$ can be interpreted as the rate at which the drug is being metabolized (recall the differential equation (2.40)). With this purpose in mind, we take logarithms to both sides of (2.42) and obtain:

$$\ln(\frac{r}{b} - c(t)) = \ln(\frac{r}{b} - c_o) - \frac{b}{V}t. \tag{2.43}$$

Thereafter we make n measurements $(t_1, c(t_1)), ..., (t_n, c(t_n))$ and measure $\frac{r}{b}$, the equilibrium concentration (it is the concentration, in the long run, of the drug in the blood; see the function defined in (2.41), and apply linear regression with t as the first variable and $\ln(\frac{r}{b} - c(t))$ as the second variable[14]. The slope of the least squares regression line will allow us to estimate b/V.

Many other problems about mixing are governed by differential equations that resemble (2.40). For instance, if we have a tank of volume V that loses solution at the rate of b liter/minute from a valve at the bottom and receives a solution of p mg/liter at the rate of b liter/minute through a valve at the top, and the solution is well-mixed using an agitator, an adequate model is

$$x'(t) = bp - \frac{b}{V}x(t) \tag{2.44}$$

---

[14]Implicitly, we are assuming that the correlation coefficient has been found to be close to -1.

where $x(t)$ denotes the amount of solute at any instant t.

What happens if a tank with volume V loses solution at the rate of b lt/minute and pure water enters at the rate of a lt/minute ($b > a$)? The solution is being diluted and at the same time the volume is diminishing. Since the volume, at any instant t, is $V - (b - a)t$, we can surmise that the system is governed by the differential equation

$$x'(t) = -\frac{b}{V - (b - a)t}x(t). \tag{2.45}$$

Exercises for section 2.17

1. Solve the differential equation (2.44), assuming that $x(0) = x_o$.

2. In the preceding problem, find the equilibrium concentration.

3. Solve the differential equation (2.45), assuming that $x(0) = x_o$

4. A 1-liter tank loses solution at the rate of 0.3 liter/minute and pure water enters at the rate of 0.1 liter/minute. How long will it take to lose half of the original amount of solute?

5. Suppose that a 1-liter tank, with pure water, has two valves, one at the bottom and one at the top. The valves are opened simultaneously to let out solution at the rate of 0.1 liter/minute and to let in, at the same rate, a solution of 1 mg/liter. How long does the system need to be in operation in order to reach a concentration of 0.5 mg/liter?

6. Sketch a graph for the variation of concentration of solute, with regard to the example presented at the beginning of section 2.17, when $c_o > \frac{r}{b}$.

7. If the reader finds that the steps used to solve (2.40) are too long or convoluted, there is an alternative technique already learned in section 1.16: multiply both sides of $x'(t) + \frac{b}{V}x(t) = r$ by $e^{\frac{b}{V}t}$ and reach the equality

$$e^{\frac{b}{V}t}(x'(t) + \frac{b}{V}x(t)) = re^{\frac{b}{V}t}.$$

But the left side is equal to $(e^{\frac{b}{V}t}x(t))'$ and the right side is $(\frac{rV}{b}e^{\frac{b}{V}t})'$. Finish the task at hand by finding an explicit formula for $x(t)$.

8. A tracer is introduced, through the blood stream, into a human organ at the constant concentration of p mgr/lt. The blood stream flows in and out of the organ at the rate of b lt/minute. The organ is capable of excreting the tracer at the constant rate of r mgr/minute. A simple modification of (2.44) leads to

$$x'(t) = bp - \tfrac{b}{V}x(t) - r$$

Solve for $x(t)$ and show that $c_\infty = \frac{pb-r}{b}$, where $c_\infty = \lim_{t\to\infty} c(t)$ is the long-run concentration of tracer in the organ (keep in mind that $c(t) = \frac{x(t)}{V}$ is the concentration of the tracer). It is interesting to note that if c_∞ is measured, and p, b are known, we can calculate r from the formula $r = b(p - c_\infty)$ (Cullen, 1983).

9. A reader might ask whether it is a good idea to use the measurement of r/b and the n measurements $(t_i, c(t_i))$, $1 \le i \le n$, to factor b/V from each of the n expressions

$$\ln(\tfrac{r}{b} - c(t_i)) = \ln(\tfrac{r}{b} - c_o) - \tfrac{b}{V}t_i$$

(see (2.43)) and then estimate b/V as either the mean or the median of the values so obtained. If one or more of the measurements $(t_i, c(t_i))$ were to be an outlier, which strategy would you use? Why?

2.18   Change of Variables

To summarize our knowledge of indefinite integrals, accumulated so far, we have to enlarge a little the list of integrals mentioned in section 2.6. On their respective domains of definition we have:

1. $\int x^r dx = \frac{x^{r+1}}{r+1} + C$, $r \ne -1$

2. $\int \frac{1}{x}dx = \ln|x| + C$, $\int \frac{u'(x)}{u(x)}dx = \ln|u(x)| + C$

3. $\int e^{bx}dx = \frac{1}{b}e^{bx} + C$, $\int a^x dx = \frac{a^x}{\ln a} + C$

4. $\int \sin bx\, dx = -\frac{1}{b}\cos bx + C$

5. $\int \cos bx\, dx = \frac{1}{b}\sin bx + C$

6. $\int \sec^2 bx\, dx = \frac{1}{b}\tan bx + C$

7. $\int \sec bx \tan bx\, dx = \frac{1}{b}\sec bx + C$

8. $\int \frac{1}{x^2+b^2}dx = \frac{1}{b}\arctan \frac{x}{b} + C$

9. $\int \frac{1}{\sqrt{b^2-x^2}}dx = \arcsin \frac{x}{b} + C$

10. $\int \frac{dx}{x\sqrt{x^2-b^2}} = \frac{1}{b}\sec^{-1}(|\frac{x}{b}|) + C$

This is a list of the most important indefinite integrals, and their validity can be tested by taking the derivative of the corresponding function to the right. As expected, some restrictions have to be imposed on b. For instance, in the 9th entry we must have $|x| < |b|$ while in the 10th entry we must have $|x| > |b|$.

Suppose that we have to deal with $\int x \cos x^2 dx$. A trial and error approach allows us to conclude, thanks to the chain rule, that $(\frac{1}{2} \sin x^2)' = x \cos x^2$. Thus $\int x \cos x^2 dx = \frac{1}{2} \sin x^2 + C$, which in turn leads to $\int_a^b x \cos x^2 dx = [\frac{1}{2} \sin x^2]_a^b$. We might observe that given an antiderivative F of a continuous function f and a differentiable function $u(x)$ such that the composition $F(u(x))$ makes sense and $u'(x)$ is continuous,

$$\frac{d}{dx} F(u(x)) = F'(u(x))u'(x) = f(u(x))u'(x).$$

Thus

$$\int f(u(x))u'(x)dx = F(u(x)) + C.$$

Although this formula is important from a theoretical point of view, often we will use a more direct approach keeping in mind that sometimes it is better to remember methods rather than rules. Let us analyze some concrete examples:

(i) $\int \frac{x}{x^2+1}dx = \frac{1}{2} \int \frac{2x}{x^2+1}dx = \frac{1}{2} \ln(x^2 + 1) + C$. Hence

$$\int_a^b \frac{x}{x^2+1}dx = \frac{1}{2}[\ln(x^2 + 1)]_a^b.$$

(ii) $\int 3x \sin x^2 dx = \frac{3}{2} \int (\sin x^2)2xdx = \frac{3}{2}(-\cos x^2) + C$. Therefore

$$\int_a^b 3x \sin x^2 dx = -\frac{3}{2}[\cos x^2]_a^b.$$

(iii) Suppose that we wish to calculate $\int_a^b \sqrt{5 + x^2}xdx$. Defining $u(x) = 5 - x^2$ we get $u'(x) = 2x$, thus

$$\int \sqrt{5 + x^2}xdx = \frac{1}{2} \int \sqrt{u(x)}u'(x)dx = \frac{1}{2}\frac{2}{3}(u(x))^{3/2} + C = \frac{1}{3}(5 + x^2)^{3/2} + C.$$

Consequently

$$\int_a^b \sqrt{5 + x^2}xdx = \frac{1}{3}[(5 + x^2)^{3/2}]_a^b = \frac{1}{3}((5 + b^2)^{3/2} - (5 + a^2)^{3/2}).$$

It should be noted that $\int \sqrt{5 + x^2}xdx$ and many other integrals can be calculated through a time-honored symbolic approach. Indeed, let $u = 5 + x^2$. Then $du = 2xdx$, which in turn leads to $\int \sqrt{5 + x^2}xdx = \frac{1}{2} \int u^{1/2}du = \frac{1}{2}\frac{2}{3}u^{3/2} + C = \frac{1}{3}(5 + x^2)^{3/2} + C$.

Since

$$\int_a^b f(u(x))u'(x)dx = [F(u(x))]_a^b = [F(x)]_{u(a)}^{u(b)} = \int_{u(a)}^{u(b)} f(x)dx$$

we can calculate definite integrals without having to calculate first the corresponding indefinite integral, *but we should not forget that the limits of integration need to be changed.* Going back to the example $\int_a^b \sqrt{5+x^2}xdx$, we define $u = 5 + x^2$. Then $du = 2xdx$, thus

$$\int_a^b \sqrt{5+x^2}xdx = \tfrac{1}{2}\int_{5+a^2}^{5+b^2} u^{1/2}du = \tfrac{1}{2}\tfrac{2}{3}[u^{3/2}]_{5+a^2}^{5+b^2} = \tfrac{1}{3}((5+b^2)^{3/2} - (5+a^2)^{3/2}),$$

exactly the same answer found before! The techniques we have studied so far in this section are generically called 'change of variables techniques' or sometimes 'substitution techniques.'

Let us discuss an example from chemistry. Suppose that a molecule of substance A combines with a molecule of substance B giving birth to a molecule of the product C. Assume that we start with equal concentrations of both reactants, say a moles/lt, and let $x(t)$ denote the concentration of product at any instant t; no product is present at the beginning of the reaction. According to the law of mass action, a probable model is $x'(t) = k(a - x(t))^2$ for a certain parameter k. Dividing by $(a - x(t))^2$ and integrating we obtain

$$\int (a - x(t))^{-2}x'(t)dt = \int kdt.$$

The integral to the right is simply $kt + C_1$, thus we can concentrate our attention on the integral to the left. Letting $u(t) = (a - x(t))$ we will have $u'(t) = -x'(t)$. Then

$$\int (a - x(t))^{-2}x'(t)dt = -\int u(t)^{-2}u'(t)dt = u(t)^{-1} + C_2 = \tfrac{1}{a-x(t)} + C_2.$$

Consequently

$$\tfrac{1}{a-x(t)} = kt + C$$

where $C = C_1 - C_2$ is a certain constant whose value can be determined taking into consideration that $x(0) = 0$. Indeed $1/(a - x(0)) = k*0 + C$, therefore $C = 1/a$. In summary,

$$\tfrac{1}{a-x(t)} = kt + \tfrac{1}{a}.$$

Thus, the model predicts that if t lies on the horizontal axis and $1/(a - x(t))$ on the vertical axis, the experimental values should gather around a straight line. If we were to have data, linear regression would provide an estimate of k once we validate the model. Afterwards we can reach the equality

$$x(t) = a - \tfrac{a}{1+akt},$$

valid at any time during the course of the reaction.

Two additional examples may shed more light on the change of variables technique. Let us try to calculate $\int \frac{dx}{x^2-x+1}$. We have

$$I = \int \frac{dx}{x^2-x+1} = \int \frac{dx}{(x-\frac{1}{2})^2+\frac{3}{4}} = \int \frac{dx}{(x-\frac{1}{2})^2+(\frac{\sqrt{3}}{2})^2}.$$

Making the change of variable $u = x - \frac{1}{2}$ we get

$$I = \int \frac{du}{u^2+(\sqrt{3}/2)^2} = \frac{2}{\sqrt{3}} \int \frac{\sqrt{3}/2}{u^2+(\sqrt{3}/2)^2} = \frac{2}{\sqrt{3}} \arctan \frac{2u}{\sqrt{3}} + C$$

$$= \frac{2}{\sqrt{3}} \arctan \frac{2(x-\frac{1}{2})}{\sqrt{3}} + C.$$

Next let us try to calculate $I = \int \sec^2 x \tan x\, dx$. The change of variable $u = \tan x$ leads to $du = \sec^2 x\, dx$, hence

$$I = \int u\, du = \frac{u^2}{2} + C = \frac{1}{2} \tan^2 x + C.$$

But we could have defined $v = \sec x$, thus $dv = \sec x \tan x\, dx$ and we will have that

$$I = \int v\, dv = \frac{1}{2}v^2 + D = \frac{1}{2} \sec^2 x + D.$$

Is this a different answer from the one found before? Not really because

$$\tfrac{1}{2} \sec^2 x + D = \tfrac{1}{2}(1 + \tan^2 x) + D = \tfrac{1}{2} \tan^2 x + (D + \tfrac{1}{2}) = \tfrac{1}{2} \tan^2 x + C$$

where $C = D + \frac{1}{2}$.

Exercises for section 2.18

In problems 1-12 calculate the given integrals:

1. $\int_{-1}^{3} x \sin(x^2 + 1)dx$.

2. $\int_{-\pi/3}^{\pi/4} \tan x\, dx$

3. $\int_{-\pi/3}^{\pi/4} x \tan(x^2)dx$.

4. $\int_{-2}^{5} \frac{dx}{x^2+x+1}$ (Hint: Complete squares in the denominator).

5. $\int \sec x\, dx$ (Hint: Multiply and divide by $\sec x + \tan x$).

6. $\int_{-1}^{2} \frac{x}{\sqrt{x^2+1}}dx$.

Table 2.16: Experimental Values Linked to Exercise 15

t (min.)	$x(t)$ (moles/lt)
1	0.1042
10	0.1832
30	1.668
120	2.1011
300	2.6458

7. $\int_0^4 \frac{e^x}{e^x+5} dx$.

8. $\int_0^{-1} \sin(\pi t + 7) dt$.

9. $\int_2^6 \frac{1}{3x \ln x} dx$.

10. $\int \frac{dx}{1+e^x}$ (Hint: Observe that $\frac{1}{1+e^x} = \frac{e^x}{e^x + e^{2x}}$).

11. $\int \cos^2 x dx$ and $\int \sin^2 x dx$ (Hint: Recall that $\cos^2 x = \frac{1+\cos 2x}{2}$ and $\sin^2 x = \frac{1-\cos 2x}{2}$).

12. $\int \frac{1}{x^3} \sqrt{\frac{1}{x^2} + 5} dx$.

13. Solve the first-order initial value problem $\frac{dy}{dx} = \cos^2(\pi - x)$, $y(0) = 1$.

14. Solve the second-order initial value theorem $\frac{d^2 y}{dx^2} = \sin(2x - \pi)$, $y(0) = 1$, $y'(0) = -1$.

15. Assume that the initial value problem $x'(t) = k(3 - x(t))^2$, $x(0) = 0$, governs the kinetics of the chemical reaction $A + B \rightarrow C$, where $x(t)$ is the concentration of C at any instant t, k is the parameter of the reaction, and 3 moles/lt is the initial concentration of each reactant. Solve the initial value problem and thereafter use Table 2.16 to validate the model. Furthermore, estimate k.

16. Calculate $\int \frac{\sqrt{x}}{x+2} dx$ (Hint: Define $x = u^2$.)

17. Calculate $\int \frac{dx}{\sqrt{x}+3}$.

18. Calculate $\int \sqrt{3x + 1} dx$ (Hint: Define $\sqrt{3x + 1} = u$, i.e., $x = \frac{1}{3}u^2 - \frac{1}{3}$.)

2.19   Integration by Parts

How could we calculate the integral $\int_0^3 xe^x dx$? No change of variables seems to help, so we might just start guessing and checking. After some false starts, we might conclude that $(xe^x - e^x)' = xe^x$, thus $\int_0^3 xe^x dx = [xe^x - e^x]_0^3$.

Calculating $\int_1^\pi x \sin x dx$ is a greater challenge since it is not easy to find an antiderivative of $x \rightarrow x \sin x$, so some systematic procedure is needed. Indeed, given differentiable functions f and g we note that $(f(x)g(x))' = f'(x)g(x) + f(x)g'(x)$. Assuming that the derivatives of f and g are also continuous, the Evaluation Theorem allows us to conclude that

$$\int_a^b (f'(x)g(x) + f(x)g'(x))dx = [f(x)g(x)]_a^b.$$

Hence

$$\int_a^b f(x)g'(x)dx = [f(x)g(x)]_a^b - \int_a^b f'(x)g(x)dx.$$

This is the famous formula for 'integration by parts.'

Let us see how it works when we try to calculate $\int_1^\pi x \sin x dx$. Define $f(x) = x$ and $g'(x) = \sin x$. Then $f'(x) = 1$ and $g(x) = -\cos x$. Therefore $\int_1^\pi x \sin x dx = [-x \cos x]_1^\pi - \int_1^\pi - \cos x dx = [-x \cos x + \sin x]_1^\pi$. One can check that indeed $(-x \cos x + \sin x)' = x \sin x$, in other words $\int x \sin x dx = -x \cos x + \sin x + C$. If we had chosen $f(x) = \sin x$ and $g'(x) = x$ then integration by parts would lead to the integral $\int_1^\pi x^2 \cos x dx$, undoubtedly more difficult to calculate than the original one. Whenever we wish to apply integration by parts, we have to decide who is $f(x)$ and who is $g'(x)$; luckily there are only two possibilities. Note also that it is enough to choose just one antiderivative of $g'(x)$, it is not necessary to write $g(x) + C$ (see exercise 20. at the end of the section).

We know that $\int \frac{1}{x} dx = \ln x + C$, but so far we have not dealt with the integral $\int \ln x dx$. Surprisingly, integration by parts provides the answer. Defining $f(x) = \ln x$ and $g'(x) = 1$ we get $f'(x) = 1/x$ and $g(x) = x$. Therefore

$$\int_a^b \ln x dx = [x \ln x]_a^b - \int_a^b dx = [x \ln x - x]_a^b$$

which in turn allows us to conclude that $\int \ln x dx = x \ln x - x + C$.

Although the integration by parts formula was stated for definite integrals, it is also valid for indefinite integrals. Indeed

$$\int f(x)g'(x)dx = f(x)g(x) - \int f'(x)g(x)dx.$$

For instance, let us try to express $\int \arcsin x dx$ in terms of elementary functions. Defining $f(x) = \arcsin x$ and $g'(x) = 1$ we get $f'(x) = 1/\sqrt{1-x^2}$ and $g(x) = x$. Therefore

$$\int \arcsin x dx = x \arcsin x + \tfrac{1}{2} \int \frac{-2x}{\sqrt{1-x^2}} dx = x \arcsin x + \tfrac{1}{2} \int u^{-1/2} du$$

where we made the change of variables $u = 1 - x^2$. Therefore

$$\int \arcsin x dx = x \arcsin x + u^{1/2} + C = x \arcsin x + \sqrt{1-x^2} + C.$$

Sometimes it is necessary to apply a change of variable before using integration by parts. For instance, let us calculate $\int_0^\pi \cos \sqrt{x} dx$. We start by defining $u = \sqrt{x}$; thus $u^2 = x$, which in turn leads to $2u du = dx$. Consequently

$$\int \cos \sqrt{x} dx = 2 \int u \cos u du.$$

Defining $f(u) = u$, $g'(u) = \cos u$ we get $f'(u) = 1$ and $g(u) = \sin u$. Hence

$$\int u \cos u du = u \sin u - \int \sin u du = u \sin u + \cos u + C = \sqrt{x} \sin \sqrt{x} + \cos \sqrt{x} + C.$$

Finally,

$$\int_0^\pi \cos \sqrt{x} dx = 2[\sqrt{x} \sin \sqrt{x} + \cos \sqrt{x}]_0^\pi = 2(\sqrt{\pi} \sin \sqrt{\pi} + \cos \sqrt{\pi} - 1).$$

If the reader finds that the preceding argument is not rigorous enough, here is an alternative: recall the basic formula for change of variables analyzed in the previous section, namely

$$\int_a^b f(u(x))u'(x) dx = \int_{u(a)}^{u(b)} f(x) dx.$$

Define $f(x) = \cos \sqrt{x}$ and $u(x) = x^2$. Since we wish to calculate $\int_0^\pi \cos \sqrt{x} dx$, it is necessary to have $u(a) = 0$ and $u(b) = \pi$. Therefore $a = 0$ and $b = \sqrt{\pi}$, which in turn leads to

$$I = \int_0^\pi \cos \sqrt{x} dx = \int_0^{\sqrt{\pi}} \cos \sqrt{x^2} 2x dx = 2 \int_0^{\sqrt{\pi}} x \cos x dx.$$

Using integration by parts, we get $\int x \cos x dx = x \sin x + \cos x$. Therefore $I = 2[x \sin x + \cos x]_0^{\sqrt{\pi}}$.

Exercises for section 2.19

Calculate the following fourteen integrals:

1. $\int x \cos(3x) dx$.

2. $\int (\ln x)^2 dx$.

3. $\int \arctan(3x) dx$.

4. $\int_a^b e^{2x} \sin x\, dx$ (Hint: Apply integration by parts two times.)

5. $\int_0^1 x \arctan x\, dx$.

6. $\int_1^\pi x^2 \ln x\, dx$.

7. $\int e^{3x} \cos 2x\, dx$.

8. $\int e^{-x} \sin x\, dx$.

9. $\int x \cos \frac{x}{2} dx$.

10. $\int_0^3 e^{\sqrt{x}} dx$.

11. $\int \cos(\ln x) dx$.

12. $\int_1^4 \tan^{-1}(\sqrt{y}) dy$.

13. $\int_2^r \frac{\ln x}{\sqrt{x}} dx$, $r > 2$.

14. $\int \frac{\sin^{-1}(\sqrt{x})}{\sqrt{x}} dx$

15. In section 2.5 we found that if $D(r)$ is the population density as a function of the distance to the center of a city, the total population who lives within R miles of the center is $\int_0^R 2\pi r D(r) dr$. Assuming that $D(r) = 100,000 e^{-0.15r}$, calculate the above-mentioned total population when $R = 10$.

16. Solve $\int \frac{dx}{(x^2+1)^2}$ (Hint: Integrate by parts $\int \frac{dx}{x^2+1}$ letting $f(x) = \frac{1}{x^2+1}$ and $g'(x) = 1$.)

17. Solve $\int \sec^3 \theta\, d\theta$ (Hint: Define $f(\theta) = \sec\theta$, $g'(\theta) = \sec^2\theta$.)

18. Solve $y'' + y' = t$, $y(0) = 2$, $y'(0) = 1$.

19. Solve $y'' + 2y' = \sin t$, $y(0) = 1$, $y'(0) = 0$.

20. To solve the integral $\int_a^b xe^x dx$, we define $f(x) = x$ and $g'(x) = e^x$. Then $f'(x) = 1$ and $g(x) = e^x$. But $g(x)$ could be chosen as $e^x + C$, where C is any constant. Then

$$\int_a^b xe^x dx = [x(e^x + C)]_a^b - \int_a^b (e^x + C)dx =$$

$$= [xe^x]_a^b + [Cx]_a^b - \int_a^b e^x dx - \int_a^b C dx =$$

$$= [xe^x]_a^b + [Cx]_a^b - [e^x]_a^b - [Cx]_a^b = [xe^x - e^x]_a^b.$$

We note that the constant C cancels in the process! Thus, the choice $C = 0$, at the very beginning, makes a lot of sense. It is now your turn to show that, in general, if in the formula for integration by parts one chooses $g(x) + C$ instead of $g(x)$ nothing paradoxical will occur (Hint: Keep in mind that ET implies that $\int_a^b f'(x)dx = [f(x)]_a^b$ provided that f' is continuous).

2.20   Trigonometric Integrals

Our main concern in this section will be to study integrals of the type $\int \cos^m x \sin^n x \, dx$, where m and n are positive integers. Let us first deal with the case when m or n are odd. Suppose we wish to solve the indefinite integral $\int \cos^3 x \sin^2 x dx$, a typical example of the above-mentioned category of trigonometric integrals. Using the basic identity $\sin^2 x + \cos^2 x = 1$ we get

$$\int \cos^3 x \sin^2 x dx = \int \cos^2 x \sin^2 x \cos x dx = \int (1 - \sin^2 x) \sin^2 x \cos x dx.$$

Letting $u = \sin x$ it follows that $du = \cos x dx$. Hence

$$\int \cos^3 x \sin^2 x dx = \int (1 - u^2)u^2 du = \int u^2 du - \int u^4 du = \tfrac{1}{3}(\sin x)^3 - \tfrac{1}{5}(\sin x)^5 + C.$$

In a similar fashion we can solve the integral $\int \cos^4 x \sin^5 x dx$. Indeed,

$$\int \cos^4 x \sin^5 x dx = \int \cos^4 x \sin^4 x \sin x dx = -\int \cos^4 x (1 - \cos^2 x)^2 (-\sin x)dx.$$

Letting $u = \cos x$ we get $du = -\sin x dx$. Hence

$$\int \cos^4 x \sin^5 x dx = -\int u^4 (1 - u^2)^2 du = -\int u^4 (1 - 2u^2 + u^4)du =$$
$$-\int u^4 du + 2\int u^6 du - \int u^8 du = -\tfrac{1}{5}(\cos x)^5 + \tfrac{2}{7}(\cos x)^7 - \tfrac{1}{9}(\cos x)^9 + C.$$

On the basis of the two previous examples we can observe that there is an established procedure whenever m or n is odd. What happens if m and n are even? The following example will show the path to be followed:

$$\int \cos^2 x \sin^2 x dx = \int (1 - \sin^2 x) \sin^2 x dx = \int \sin^2 x dx - \int \sin^4 x dx.$$

Recall that $\int \sin^2 x dx = \frac{1}{2}x - \frac{1}{2}\sin x \cos x + C$ (exercise 11, section 2.18). The task of calculating $\int \sin^4 x dx$ requires integration by parts. Indeed, defining $f(x) = \sin^3 x$, $g'(x) = \sin x$ one gets $f'(x) = 3\sin^2 x \cos x$ and $g(x) = -\cos x$. Therefore

$$\int \sin^4 x dx = -(\cos x)\sin^3 x + 3\int \sin^2 x \cos^2 x dx =$$
$$-\cos x \sin^3 x + 3\int \sin^2 x (1 - \sin^2 x) dx = -\cos x \sin^3 x + 3\int \sin^2 x dx - 3\int \sin^4 x dx.$$

Hence $4\int \sin^4 x dx = -\cos x \sin^3 x + 3\int \sin^2 x dx$, which in turn leads to

$$\int \sin^4 x dx = -\frac{1}{4}\cos x \sin^3 x + \frac{3}{4}[\frac{1}{2}x - \frac{1}{4}\sin(2x)] + C =$$
$$-\frac{1}{4}\cos x \sin^3 x + \frac{3}{8}x - \frac{3}{8}\sin x \cos x + C.$$

Putting everything together we finally arrive at

$$\int \cos^2 x \sin^2 x dx = \frac{1}{8}x - \frac{1}{8}\sin x \cos x + \frac{1}{4}\cos x \sin^3 x.$$

The last family of trigonometric integrals that we wish to study are of the type $\int \cos(ax)\sin(bx)dx$, $\int \cos ax \cos bx dx$, and $\int \sin ax \sin bx dx$. Luckily, they can be solved using three well-known trigonometric identities:

$$\cos ax \cos bx = \frac{1}{2}(\cos(a+b)x + \cos(a-b)x),$$

$$\sin ax \sin bx = \frac{1}{2}(\cos(a-b)x - \cos(a+b)x),$$

$$\sin ax \cos bx = \frac{1}{2}(\sin(a+b)x + \sin(a-b)x).$$

For instance,

$$\int \sin 2x \cos 3x dx = \frac{1}{2}\int \sin 5x dx + \frac{1}{2}\int \sin(-x)dx = -\frac{1}{10}\cos 5x + \frac{1}{2}\cos x + C,$$

$$\int \sin 7x \sin 5x dx = \frac{1}{2}\int \cos 2x dx - \frac{1}{2}\int \cos 12x dx = \frac{1}{4}\sin 2x dx - \frac{1}{24}\sin 12x + C.$$

Exercises for Section 2.20

Solve the following integrals:

1. $\int \cos^2 x \sin^3 x dx$.

2. $\int \cos^3 x \sin^3 x dx$.

3. $\int \cos^2 x \sin^4 x dx$.

4. $\int \sin 3x \cos x dx$.

5. $\int \sin 4x \sin 5x dx$.

2.21   Partial Fractions

Let us consider the indefinite integral $\int \frac{dx}{x^2+x+1}$. Based on the methodology we learned in section 2.18, let us try to 'complete the squares' in the denominator hoping to find a way to solve it:

$$\int \frac{dx}{x^2+x+1} = \int \frac{dx}{(x+\frac{1}{2})^2-\frac{1}{4}+1} = \int \frac{dx}{(x+\frac{1}{2})^2+\frac{3}{4}} = \int \frac{dx}{(x+\frac{1}{2})^2+\frac{3}{4}} = \int \frac{dx}{(x+\frac{1}{2})^2+(\frac{\sqrt{3}}{2})^2}.$$

Defining $u = x + \frac{1}{2}$ we get $du = dx$. Hence

$$\int \frac{dx}{x^2+x+1} = \int \frac{du}{u^2+(\frac{\sqrt{3}}{2})^2} = \frac{2}{\sqrt{3}} \tan^{-1} \frac{2u}{\sqrt{3}} + C = \frac{2}{\sqrt{3}} \tan^{-1} \frac{2(x+\frac{1}{2})}{\sqrt{3}} + C.$$

Faced with the task of solving the integral $\int \frac{dx}{x^2+3x+2}$ one might think about completing squares:

$$x^2 + 3x + 2 = (x + \tfrac{3}{2})^2 - \tfrac{9}{4} + \tfrac{8}{4} = (x + \tfrac{3}{2})^2 - \tfrac{1}{4}.$$

Then, defining $u = x + \frac{3}{2}$, we get

$$\int \frac{dx}{x^2+3x+2} = \int \frac{dx}{(x+\frac{3}{2})^2-\frac{1}{4}} = \int \frac{du}{u^2-(\frac{1}{2})^2}.$$

But the integral $\int \frac{du}{u^2-a^2}$ is not among the ten integrals listed at the beginning of section 2.18. We might try to use the algebraic method of partial fractions, which was introduced rather informally when the logistic equation was discussed. Indeed, $x^2 + 3x + 2 = (x + 2)(x + 1)$. Thus, there exist constants A and B such that

$$\frac{1}{(x+2)(x+1)} = \frac{A}{x+2} + \frac{B}{x+1}$$

Then $1 = A(x + 1) + B(x + 2)$, which in turn leads to $A + B = 0$ and $A + 2B = 1$. Thus, $A = -1$ and $B = 1$. Therefore,

$$\int \frac{dx}{(x+2)(x+1)} = \int \frac{-dx}{x+2} + \int \frac{dx}{x+1} = -\ln|x+2| + \ln|x+1| + C = \ln\left|\frac{x+1}{x+2}\right| + C.$$

In general, given the integral $\int \frac{dx}{ax^2+bx+c}$ we have to look at the polynomial $p(x) = ax^2 + bx + c$. If $\Delta = b^2 - 4ac < 0$ we need to use the method of completion of squares, which will lead to an expression that involves the function $\tan^{-1} x$. If $\Delta = b^2 - 4ac > 0$, partial fractions is the method of choice and the answer is an expression that involves ln. The case $\Delta = b^2 - 4ac = 0$ is the simplest because it leads to the calculation of an integral of the type $\int \frac{dx}{a(x-r)^2}$. Evidently,

$$\int \frac{dx}{a(x-r)^2} = -\frac{1}{a}(x-r)^{-1} + C = -\frac{1}{a}\frac{1}{x-r} + C.$$

Another type of integral is $\int \frac{rx+s}{ax^2+bx+c}$. Let us analyze $\int \frac{3x+1}{2x^2+x-1} dx$, an integral of the aforementioned case. We observe that if $u(x) = 2x^2 + x - 1$ then $u'(x) = 4x + 1$, therefore

$$\int \frac{3x+1}{2x^2+x-1}dx = \frac{3}{4}\int \frac{4x+\frac{4}{3}}{2x^2+x-1}dx = \frac{3}{4}\int \frac{4x+1+\frac{1}{3}}{2x^2+x-1}dx =$$

$$\frac{3}{4}\int \frac{4x+1}{2x^2+x-1}dx + \frac{1}{4}\int \frac{dx}{2x^2+x-1} = \frac{3}{4}\int \frac{u'(x)}{u(x)}dx + \frac{1}{4}\int \frac{dx}{2x^2+x-1}.$$

Evidently,

$$\int \frac{u'(x)}{u(x)}dx = \ln|u(x)| + C_1 = \ln|2x^2 + x - 1| + C_1.$$

To calculate $\int \frac{dx}{2x^2+x-1}$, partial fractions is the method of choice. Indeed,

$$\int \frac{dx}{2x^2+x-1} = \frac{1}{2}\int \frac{dx}{x^2+\frac{1}{2}x-\frac{1}{2}} = \frac{1}{2}\int \frac{dx}{(x-\frac{1}{2})(x+1)}.$$

According to the method, there exist constants A and B such that

$$\frac{1}{(x-\frac{1}{2})(x+1)} = \frac{A}{x-\frac{1}{2}} + \frac{B}{x+1}.$$

Therefore $A + B = 0$ and $A - \frac{1}{2}B = 1$; thus, $A = \frac{2}{3}$ and $B = -\frac{2}{3}$. Consequently

$$\int \frac{dx}{2x^2+x-1} = \frac{1}{2}\left(\frac{2}{3}\int \frac{dx}{x-\frac{1}{2}} - \frac{2}{3}\int \frac{dx}{x+1}\right)$$

$$= \frac{1}{3}\ln\left|x - \frac{1}{2}\right| - \frac{1}{3}\ln|x + 1| + C = \frac{1}{3}\ln\left|\frac{x-\frac{1}{2}}{x+1}\right|.$$

Putting everything together we can conclude that

$$\int \frac{3x+1}{2x^2+x-1}dx = \frac{3}{4}\ln|2x^2 + x - 1| + \frac{1}{4}\left(\frac{1}{3}\ln\left|\frac{x-\frac{1}{2}}{x+1}\right|\right) + C.$$

Is there a way to solve $\int \frac{dx}{(x^2+1)(x+1)}$? Since $x^2 + 1$ cannot be expressed as the product of two linear factors, the algebraic method of partial fractions asserts that there exist constants A, B, C such that

$$\frac{1}{(x^2+1)(x+1)} = \frac{Ax+B}{x^2+1} + \frac{C}{x+1}.$$

Thus

$$\frac{1}{(x^2+1)(x+1)} = \frac{(Ax+B)(x+1)+C(x^2+1)}{(x^2+1)(x+1)} = \frac{Ax^2+Ax+Bx+B+Cx^2+C}{(x^2+1)(x+1)}.$$

Therefore $A + C = 0$, $A + B = 0$, and $B + C = 1$, which in turn lead to $B = \frac{1}{2}$, $A = -\frac{1}{2}$, and $C = \frac{1}{2}$. Then

$$\int \frac{dx}{(x^2+1)(x+1)} = \frac{1}{2}\int \frac{(-x+1)}{x^2+1}dx + \frac{1}{2}\int \frac{dx}{x+1}$$

$$= -\frac{1}{4}\int \frac{2xdx}{x^2+1} + \frac{1}{2}\frac{dx}{x^2+1} + \frac{1}{2}\int \frac{dx}{x+1}$$

$$= -\frac{1}{4}\ln(x^2 + 1) + \frac{1}{2}\tan^{-1}x + \frac{1}{2}\ln|x + 1| + C.$$

We have not exhausted all the possible cases where the method of partial fractions is the tool of choice. Rather, only a glimpse of the method has been provided. It should be noted that a graphing calculator is of great help when a complicated fraction appears under the integral sign, and it needs to be expressed as the sum of simpler fractions; the *expand* command does the job.

Exercises for Section 2. 21

Solve the following integrals:

1. $\int \frac{dx}{x^2-x-2}$

2. $\int \frac{dx}{x^2+8x+15}$

3. $\int \frac{dx}{x^2-x+1}$

4. $\int \frac{2x+1}{x^2-x-2}dx$

5. $\int \frac{5x+2}{x^2+x+1}dx$

6. $\int \frac{dx}{(x^2+2)(x-1)}$

7. $\int \frac{dx}{(x^2+3)^2}$

8. $\int \frac{dx}{x^3+x}$

9. $\int \frac{dx}{x^2-1}$

10. $\int \frac{3x-1}{x^2+x}dx$

2.22   Trigonometric Substitutions

I. Suppose that we have to solve the integral $\int \frac{dx}{(1+x^2)^{3/2}}$. None of the techniques learned so far seem to work, so we will try something radically new. Let $x = \tan\theta$, $-\frac{\pi}{2} < \theta < \frac{\pi}{2}$. Then $dx = \sec^2 d\theta$, consequently

$$\int \frac{dx}{(1+x^2)^{3/2}} = \int \frac{\sec^2\theta d\theta}{(\sqrt{1+\tan^2\theta})^3} = \int \frac{\sec^2\theta d\theta}{(\sec\theta)^3} = \int \cos\theta d\theta = \sin\theta + C.$$

These calculations make sense because $\sec\theta > 0$ on the interval $(-\pi/2, \pi/2)$. Next comes a crucial step: from the enclosed triangle, built in such a way that $\tan\theta = x$, it follows that $\sin\theta = \frac{x}{\sqrt{1+x^2}}$. Consequently

$$\int \frac{dx}{(1+x^2)^{3/2}} = \frac{x}{\sqrt{1+x^2}} + C.$$

The reader may have noticed that we chose to work with the tangent function because $1 + \tan^2\theta = \sec^2\theta$. An alternative approach, more in line with what we discussed in previous sections but rather lengthy, can be found at the end of this section. Next let us try to solve the indefinite integral $\int \frac{dx}{\sqrt{a^2+x^2}}$, $a > 0$. In analogy to what we did before, let $x = a\tan\theta$, $-\pi/2 < \theta < \pi/2$. Then $dx = a\sec^2\theta d\theta$; thus,

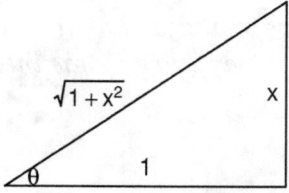

Figure 2.18: Right triangle linked to first example

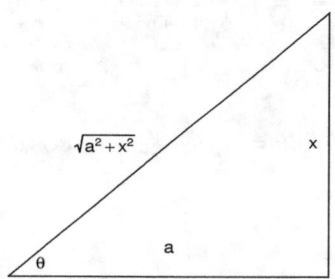

Figure 2.19: Right triangle linked to second example

$$\int \frac{dx}{\sqrt{a^2+x^2}} = \int \frac{a\sec^2\theta d\theta}{\sqrt{a^2+a^2\tan^2\theta}} = \int \sec\theta d\theta = \ln|\sec\theta + \tan\theta| + C.$$

From the triangle in Figure 2.19 we can see that $\sec\theta = \frac{\sqrt{x^2+a^2}}{a}$ and $\tan\theta = \frac{x}{a}$, consequently

$$\int \frac{dx}{\sqrt{a^2+x^2}} = \ln\left|\frac{\sqrt{x^2+a^2}}{a} + \frac{x}{a}\right| + C = \ln\frac{1}{a} + \ln\left|\sqrt{x^2+a^2} + x\right| + C = \ln\left|\sqrt{x^2+a^2} + x\right| + D.$$

A third example will shed more light on the technique being introduced. The indefinite integral $\int \sqrt{x^2+1}\,dx$ seems to be amenable to the idea of defining $x = \tan\theta$, $-\pi/2 < \theta < \pi/2$. Indeed,

$$\int \sqrt{x^2+1}\,dx = \int \sqrt{\tan^2+1}\,\sec^2\theta d\theta = \int \sec^3\theta d\theta.$$

But (see exercise 17, section 2.19)

$$\int \sec^3\theta d\theta = \frac{1}{2}\sec\theta\tan\theta + \frac{1}{2}\ln|\sec\theta + \tan\theta| + C.$$

Using the triangle from Figure 2.18 we finally get the required answer, namely

$$\int \sqrt{x^2+1}\,dx = \tfrac{x}{2}\sqrt{x^2+1} + \ln|\sqrt{x^2+1}+x| + C.$$

By now the reader must have realized that it is a good idea to define $x = a\tan\theta$ whenever the expression $\sqrt{a^2+x^2}$ appears in an integral, either in the numerator or denominator.

II. We know how to solve the indefinite integrals $\int \frac{dx}{\sqrt{1-x^2}}$ and $\int \frac{x\,dx}{\sqrt{1-x^2}}$. But what can be done with $\int \frac{x^2\,dx}{\sqrt{1-x^2}}$? It promises to be more of a challenge. Keeping in mind that $1 - \sin^2\theta = \cos^2\theta$, we might try to define $x = \sin\theta$, $-\frac{\pi}{2} < \theta < \frac{\pi}{2}$, in the hope of finding a path to the solution of the problem. Indeed, $dx = \cos\theta\,d\theta$, so

$$\int \frac{x^2\,dx}{\sqrt{1-x^2}} = \int \frac{\sin^2\theta\cos\theta\,d\theta}{\sqrt{1-\sin^2\theta}} = \int \sin^2\theta\,d\theta.$$

But (see exercise 11, section 2.18)

$$\int \sin^2\theta\,d\theta = \tfrac{1}{2}\theta - \tfrac{1}{2}\sin\theta\cos\theta + C.$$

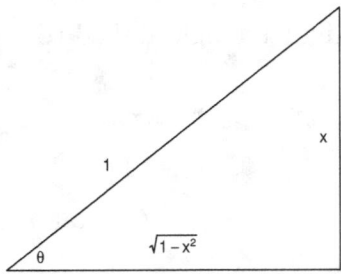

Figure 2.20: Right triangle linked to fourth example

Using the triangle from Figure 2.20 we finally get

$$\int \frac{x^2\,dx}{\sqrt{1-x^2}} = \tfrac{1}{2}\sin^{-1}x - \tfrac{1}{2}x\sqrt{1-x^2} + C.$$

As a new example, and again using $x = \sin\theta$ and the triangle from Figure 2.20, we can see that

$$\int \frac{dx}{(1-x^2)^{3/2}} = \int \frac{\cos\theta\,d\theta}{(\sqrt{1-\sin^2\theta})^3} = \int \frac{\cos\theta\,d\theta}{(\cos\theta)^3} = \int \sec^2\theta\,d\theta = \tan\theta + C = \frac{x}{\sqrt{1-x^2}} + C.$$

Next let us try to analyze the indefinite integral $\int \sqrt{r^2 - x^2}dx$, where $r > 0$. On the basis of our previous experience we might define $x = r\sin\theta$, $-\frac{\pi}{2} < \theta < \frac{\pi}{2}$, in the hope that it serves us well. Indeed,

$$\int \sqrt{r^2 - x^2}dx = \int \sqrt{r^2 - r^2\sin^2\theta}\; r\cos\theta d\theta = \int r^2\cos^2\theta d\theta.$$

However,

$$\int \cos^2\theta d\theta = \tfrac{1}{2}\theta + \tfrac{1}{2}\sin\theta\cos\theta + C.$$

This is a well-known integral (recall exercise 11, section 2.18). Using the triangle from Figure 2.21 we get

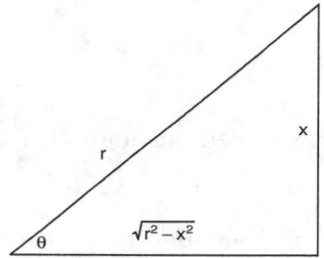

Figure 2.21: Right triangle linked to sixth example

$$\int \sqrt{r^2 - x^2}dx = \tfrac{r^2}{2}\left(\sin^{-1}\tfrac{x}{r} + \tfrac{x}{r}\tfrac{\sqrt{r^2-x^2}}{r}\right) + C = \tfrac{r^2}{2}\sin^{-1}\left(\tfrac{x}{r}\right) + \tfrac{x}{2}\sqrt{r^2 - x^2} + C.$$

For instance, $\int_0^1 \sqrt{1 - x^2}dx = \tfrac{1}{2}\sin^{-1}(1) = \tfrac{1}{2}\tfrac{\pi}{2} = \tfrac{\pi}{4}$. Of course, we expected this result because the integral under scrutiny provides the area of the quarter of a circle of unit radius!

The analysis of the last three integrals shows that a sound strategy is to define $x = a\sin\theta$ *whenever the expression* $\sqrt{a^2 - x^2}$ *appears either in the numerator or the denominator.*

III. In section 1.12 we learned that, whenever $x > 1$, $\frac{d}{dx}\sec^{-1}x = \frac{1}{x\sqrt{x^2-1}}$; thus, $\int \frac{dx}{x\sqrt{x^2-1}} = \sec^{-1}x + C$. If asked to calculate $\int \frac{dx}{\sqrt{x^2-1}}$ we might try to define $x = \sec\theta$, $0 < \theta < \frac{\pi}{2}$. Then $dx = \sec\theta\tan\theta d\theta$, consequently

$$\int \frac{dx}{\sqrt{x^2-1}} = \int \frac{\sec\theta\tan\theta d\theta}{\sqrt{\sec^2\theta-1}} = \int \frac{\sec\theta\tan\theta d\theta}{\tan\theta} = \int \sec\theta d\theta = \ln|\sec\theta + \tan\theta| + C.$$

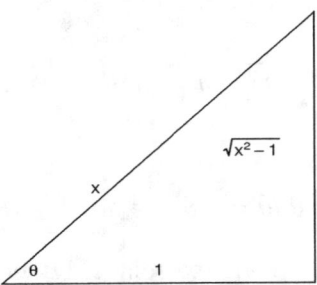

Figure 2.22: Right triangle linked to the seventh example

From the triangle in Figure 2.22 it follows that $\tan\theta = \sqrt{x^2-1}$, hence

$$\int \frac{dx}{\sqrt{x^2-1}} = \ln(x + \sqrt{x^2-1}) + C$$

whenever $x > 1$.

Another integral that can be solved through the same techniques is $\int \frac{x^2 dx}{\sqrt{x^2-1}}$. Indeed[15],

$$\int \frac{x^2 dx}{\sqrt{x^2-1}} = \int \frac{\sec^2\theta \sec\theta\tan\theta d\theta}{\sqrt{\sec^2\theta-1}} = \int \sec^3\theta d\theta = \tfrac{1}{2}\sec\theta\tan\theta + \tfrac{1}{2}\ln|\sec\theta + \tan\theta|.$$

Using the triangle from Figure 2.22 we get the equality $\tan\theta = \sqrt{x^2-1}$. Therefore

$$\int \frac{x^2 dx}{\sqrt{x^2-1}} = \tfrac{1}{2}x\sqrt{x^2-1} + \tfrac{1}{2}\ln(x + \sqrt{x^2-1})$$

whenever $x > 1$.

We should expect that $\int \sqrt{x^2-1}\, dx$ is solvable. Letting $x = \sec\theta$ we get $dx = \sec\theta\tan\theta d\theta$. Hence

$$\int \sqrt{x^2-1}\, dx = \int \sqrt{\sec^2\theta-1}\sec\theta\tan\theta d\theta = \int \tan\theta\sec\theta\tan\theta d\theta = \int \sec\theta\tan^2\theta d\theta =$$
$$\int \sec\theta(\sec^2\theta-1)d\theta = \int \sec^3\theta - \int \sec\theta d\theta.$$

But $\int \sec^3\theta d\theta = \tfrac{1}{2}\sec\theta\tan\theta + \tfrac{1}{2}\ln|\sec\theta + \tan\theta| + C_1$ and $\int \sec\theta = \ln|\sec\theta + \tan\theta| + C_2$. From the corresponding triangle (Figure 2.22) we get the equality $\tan\theta = \sqrt{x^2-1}$. Therefore

$$\int \sqrt{x^2-1}\, dx = \tfrac{1}{2}x\sqrt{x^2-1} - \tfrac{1}{2}\ln|x + \sqrt{x^2-1}| + C$$

The last three problems suggest that when the expression $\sqrt{x^2-a^2}$ appears inside an integral, we should try to make $x = a\sec\theta$ hoping that the new integral is more manageable than the original one.

---

[15]Keep in mind exercise 17, section 2.19.

A new perspective of trigonometric substitution

The first example that we discussed in this section dealt with $\int \frac{dx}{(x^2+1)^{3/2}}$. Let

$$F(x) = \int_0^x f(t)dt \text{ where } f(t) = \frac{1}{(t^2+1)^{3/2}}$$

and define $u(t) = \tan t$, $-\frac{\pi}{2} < t < \frac{\pi}{2}$. Then $u'(t) = \sec^2 t$ and

$$\int_0^x \frac{dt}{(t^2+1)^{3/2}} = \int_{u(a)}^{u(b)} f(t)dt = \int_a^b f(u(t))u'(t)dt.$$

How do we calculate the values of a and b? We must have $u(a) = 0$ and $u(b) = x$, consequently $\tan a = 0$ and $\tan b = x$; thus, $a = 0$ and $b = \arctan x$. Therefore

$$F(x) = \int_0^{\arctan x} \frac{1}{\sec^3(t)} \sec^2(t)dt = \int_0^{\arctan x} \cos t\, dt = [\sin t]_0^{\arctan x}.$$

That is, $F(x) = \sin(\arctan x)$. Finally we realize that $\sin(\arctan x) = \frac{x}{\sqrt{x^2+1}}$ because

$$x = \tan(\arctan x) = \frac{\sin(\arctan x)}{\cos(\arctan x)} = \sin(\arctan x)\sec(\arctan x) =$$
$$\sin(\arctan x)\sqrt{1 + (\tan(\arctan x))^2} = \sin(\arctan x)\sqrt{1 + x^2}.$$

Since $F'(x) = \frac{1}{(x^2+1)^{3/2}}$ we can conclude that

$$\int \frac{dx}{(x^2+1)^{3/2}} = F(x) + C.$$

Thus

$$\int \frac{dx}{(x^2+1)^{3/2}} = \frac{x}{\sqrt{x^2+1}} + C.$$

This approach is more 'rigorous' than the one we used at the beginning of the current section. We could provide similar analyses for all the problems where trigonometric substitutions are involved, but such a level of rigor is not necessary in a first-year calculus course.

Exercises for Section 2.22

Evaluate the following eleven integrals using an appropriate trigonometric substitution:

1. $\int \sqrt{x^2 + 9}\, dx$.

2. $\int \frac{\sqrt{x^2-1}}{x} dx$.

3. $\int \frac{1}{\sqrt{x^2-16}} dx$.

4. $\int \sqrt{9 - x^2}\, dx$.

5. $\int \frac{dx}{(x^2-25)^{3/2}}$.

6. $\int \frac{dx}{x^2\sqrt{16-x^2}}$.

7. $\int \frac{x^2}{\sqrt{4-x^2}}dx$.

8. $\int \frac{\sqrt{x^2-4}}{x}dx$.

9. $\int \frac{x^3}{\sqrt{x^2+4}}dx$.

10. $\int \frac{\sqrt{1-x^2}}{x^2}dx$.

11. $\int \frac{dx}{x\sqrt{1-x^2}}$.

12. The integral $\int \frac{dx}{(x^2+1)^2}$ can be solved through the method of integration by parts (see exercise 16, section 2.19). Use the change of variables $x = \tan\theta$ to find an alternative path to the solution.

2.23   Hyperbolic Functions

Two functions frequently appear in calculus and its applications, namely $\frac{1}{2}(e^x - e^{-x})$ and $\frac{1}{2}(e^x + e^{-x})$. For instance, the latter is the solution to the initial value problem

$$y'' - y = 0, \quad y(0) = 1, y'(0) = 0$$

while the former is the solution to the initial value problem

$$y'' - y = 0, \quad y(0) = 0, y'(0) = 1.$$

Besides, both functions play an important role in the solution of the problem of the hanging chain, a classic example from physics (see section 2.26).

Let us define the functions sinh ('hyperbolic sine') and cosh ('hyperbolic cosine') in the following way:

$$\sinh(x) = \tfrac{1}{2}(e^x - e^{-x})$$

$$\cosh(x) = \tfrac{1}{2}(e^x + e^{-x}).$$

A simple calculation leads to the three basic properties that relate the hyperbolic sine and the hyperbolic cosine[16]:

---

[16]The reader will surely note the striking similarities with the common cosine and sine function.

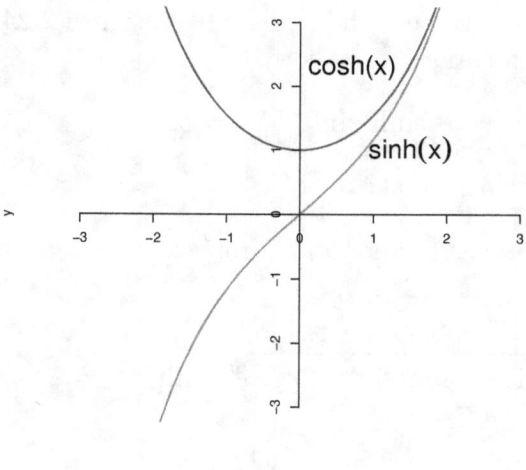

Figure 2.23: The two main hyperbolic functions

$$\cosh^2(x) - \sinh^2(x) = 1$$

$$\tfrac{d}{dx}\sinh(x) = \cosh(x)$$

$$\tfrac{d}{dx}\cosh(x) = \sinh(x).$$

The word 'hyperbolic' comes from the fact that $(\cosh(a), \sinh(a))$, a an arbitrary real number, lies on the hyperbola $x^2 - y^2 = 1$ due to the aforementioned first property.

Building the graphs of sinh and cosh

Since

$$\sinh(x) = \tfrac{1}{2}\tfrac{e^{2x}-1}{e^x},$$

it follows that $\sinh(x) > 0$ if $x > 0$ and $\sinh(x) < 0$ if $x < 0$ (obviously, $\sinh(0) = 0$), while from the very definition of cosh we can assert that $\cosh(0) = 1$ and $\cosh(x) > 1$ whenever $x \neq 0$. On the other hand, since $\sinh'(x) = \cosh x > 0$ for all x, the hyperbolic sine function is increasing over the whole real line. Similarly, since $\cosh'(x) = \sinh(x) > 0$ when $x > 0$ and $\cosh'(x) = \sinh(x) < 0$ when $x < 0$, we can conclude that the hyperbolic cosine function is increasing on $[0, \infty)$ and decreasing on $(-\infty, 0]$.

Moreover, $\sinh''(x) = \cosh'(x) = \sinh(x)$, hence sinh is concave up on $(0, \infty)$ and concave down on $(-\infty, 0)$. By the same token, $\cosh''(x) = \sinh'(x) = \cosh(x)$, thus cosh is concave up everywhere. All these facts are corroborated by the graphs of the hyperbolic sine and hyperbolic cosine (Figure 2.23).

The inverse hyperbolic functions

Since sinh is increasing, its inverse $\sinh^{-1}$ exists (see Figure 2.24). To find its derivative we proceed as usual: for any x

$$\sinh(\sinh^{-1} x) = x.$$

Applying the chain rule we arrive at

$$\cosh(\sinh^{-1} x)\tfrac{d}{dx}\sinh^{-1} x = 1,$$

so

$$\sqrt{1 + (\sinh(\sinh^{-1} x))^2}\tfrac{d}{dx}\sinh^{-1} x = 1.$$

Thus,

$$\tfrac{d}{dx}\sinh^{-1} x = \tfrac{1}{\sqrt{1+x^2}}$$

for every x. Hence

$$\int \tfrac{dx}{\sqrt{1+x^2}} = \sinh^{-1} x + C.$$

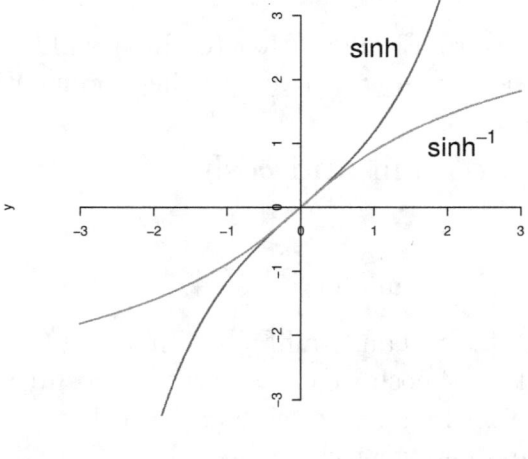

Figure 2.24: The inverse of the hyperbolic sine

In the previous section, using the trigonometric substitution $x = \tan\theta$ we found that

$$\int \tfrac{dx}{\sqrt{1+x^2}} = \ln(\sqrt{x^2 + 1}) + C.$$

Consequently

$$\tfrac{d}{dx}\sinh^{-1}x = \tfrac{d}{dx}\ln(\sqrt{x^2+1}+x).$$

Thus, there exists a constant D such that, for every x,

$$\sinh^{-1}x = \ln(\sqrt{x^2+1}+x)+D.$$

In particular $0 = \sinh(0) = \ln(1)+D$, hence $D = 0$. We have arrived at the surprising result that

$$\sinh^{-1}x = \ln(\sqrt{x^2+1}+x)$$

for every x.

Does $\cosh^{-1}$ exist? We would have to restrict the domain of definition of $\cosh$ since it is not one-to-one. Let us consider $y = \cosh x$, $x \geq 0$. Then $\cosh^{-1}$, with domain $[1,\infty)$ and range $[0,\infty)$, is a well-defined function (see Figure 2.25). Since

$$\cosh(\cosh^{-1}x) = x, \quad x \geq 1,$$

we will have

$$\tfrac{d}{dx}\cosh(\cosh^{-1}x) = 1, \quad x > 1,$$

so

$$\sinh(\cosh^{-1}x)\tfrac{d}{dx}\cosh^{-1}x = 1, \quad x > 1.$$

But

$$\sinh(\cosh^{-1}x) = \sqrt{(\cosh(\cosh^{-1}x))^2-1} = \sqrt{x^2-1},\, x > 1,$$

therefore

$$\tfrac{d}{dx}\cosh^{-1}x = \tfrac{1}{\sqrt{x^2-1}}, \quad x > 1.$$

Then

$$\textstyle\int \frac{dx}{\sqrt{x^2-1}} = \cosh^{-1}x + C$$

whenever the interval of integration lies on $(1,\infty)$.

In section 2.22 we found that

$$\textstyle\int \frac{dx}{\sqrt{x^2-1}} = \ln(x+\sqrt{x^2-1}) + C.$$

Therefore

$$\tfrac{d}{dx}\cosh^{-1}x = \tfrac{d}{dx}\ln(x+\sqrt{x^2-1}).$$

Thus, there exists a constant D such that

$$\cosh^{-1}x = \ln(x+\sqrt{x^2-1}) + D$$

for every $x \geq 1$. In particular $0 = \cosh^{-1}(1) = \ln(1+\sqrt{x^2-1}) + D = 0 + D = D$, consequently $D = 0$. We can then assert that

$$\cosh^{-1}x = \ln(x+\sqrt{x^2-1}), \quad x \geq 1.$$

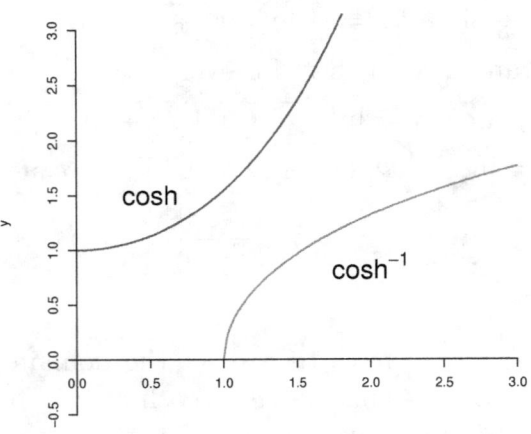

Figure 2.25: The inverse of the hyperbolic cosine

Using a hyperbolic substitution

In section 2.6 it was established that

$$\int \frac{dx}{x\sqrt{x^2-1}} = \sec^{-1}|x| + C$$

whenever the interval of integration lies on $(1, \infty)$ or $(-\infty, -1)$. Interestingly enough, a hyperbolic substitution can be used in order to find another expression for the above-mentioned indefinite integral (Fulling, 2005). Define $x = \cosh u$, $u \geq 0$. Then $dx = \sinh u\, du$, thus

$$\int \frac{dx}{x\sqrt{x^2-1}} = \int \frac{\sinh u\, du}{\cosh u \sinh u} = \int \frac{du}{\cosh u} = \int \frac{2du}{e^u + e^{-u}} = 2\int \frac{e^u\, du}{e^{2u}+1} = 2\int \frac{dv}{v^2+1}$$

where $v = e^u$, which in turn leads to $dv = e^u du$. Thus

$$\int \frac{dx}{x\sqrt{x^2-1}} = 2\tan^{-1} v + C = 2\tan^{-1}(e^u) + C = 2\tan^{-1}\left(e^{\cosh^{-1} x}\right) + C =$$
$$2\tan^{-1}\left(e^{\ln(x+\sqrt{x^2-1})}\right) + C = 2\tan^{-1}\left(x + \sqrt{x^2-1}\right) + C.$$

That is to say,

$$\int \frac{dx}{x\sqrt{x^2-1}} = 2\tan^{-1}\left(x + \sqrt{x^2-1}\right) + C,$$

provided that the interval of integration lies on $(1, \infty)$. Using the chain rule we can verify that, whenever $x < -1$,

$$\frac{d}{dx} 2\tan^{-1}\left(x + \sqrt{x^2-1}\right) = \frac{1}{x\sqrt{x^2-1}}.$$

Therefore, the previous expression for the integral under study is valid if the interval of integration lies on $(1, \infty)$ or $(-\infty, -1)$.

Exercises for Section 2.23

Calculate the derivative of the functions in exercises 1 through 4.

1. $f(x) = x^2 \cosh x$.

2. $f(x) = x \sinh^{-1} x$.

3. $f(x) = \cosh^{-1}(x^2)$.

4. $f(x) = \sinh^{-1}(1/x)$.

5. Evaluate $\int_3^7 \frac{dx}{\sqrt{x^2-1}}$.

6. Evaluate $\int_{-1}^2 \frac{dx}{\sqrt{x^2+1}}$.

7. Define $\tanh x = \frac{\sinh x}{\cosh x} = \frac{e^x - e^{-x}}{e^x + e^{-x}}$. Calculate $\frac{d}{dx} \tanh x$.

8. Show that $\tanh$ is strictly increasing on the whole real line, concave down on $(0, \infty)$ and concave up on $(-\infty, 0)$. Sketch a graph of $\tanh$.

9. Calculate $\frac{d}{dx} \tanh(x^2 + 1)$.

10. Show that $y(x) = A \cosh x + B \sinh x$ (A, B arbitrary constants) satisfies the differential equation $y'' = y$.

11. We could define $\cosh^{-1}(-x)$ on $(-\infty, -1]$. Express $\cosh^{-1}(-x)$ in terms of ln.

2.24 The Use of Technology

In section 2.18 we discussed the rate equation related to the chemical reaction $A + B \rightarrow C$, assuming that at $t = 0$ the concentrations of A and B are the same. What happens if the initial concentrations, say a and b, are different? According to the law of mass action we would have $x'(t) = k(a - x(t))(b - x(t))$ for a certain parameter k, where $x(t)$ denotes the concentration of the product at time t. Thus

$$\int \frac{x'(t)dt}{(a-x(t))(b-x(t))} = \int k \, dt.$$

Using a graphing calculator we get

$$\int (1/((a - x) * (b - x))) = \frac{1}{a-b} \ln \frac{|x-a|}{|x-b|}.$$

But from the nature of the problem we know that $|x - a| = a - x$ and $|x - b| = b - x$. Therefore

$$\frac{1}{a-b}\ln\frac{a-x(t)}{b-x(t)} = kt + C.$$

If we put t on the horizontal axis and $\frac{1}{a-b}\ln\frac{(a-x(t))}{(b-x(t))}$ on the vertical axis, we can predict that the experimental values will gather around a hypothetical line. A linear regression analysis will then provide the line of best fit and an estimate of k. Once the model has been validated, through an analysis of the coefficient of correlation, we could try to find $x(t)$ explicitly – a rather simple algebraic procedure.

Arguably, we could have solved the previous problem without the use of technology. The method of partial fractions provides the necessary path: We know that

$$\frac{1}{(a-x)(b-x)} = \frac{A}{a-x} + \frac{B}{b-x}$$

for certain constants A, B. Therefore

$$\frac{1}{(a-x)(b-x)} = \frac{A(b-x)+B(a-x)}{(a-x)(b-x)}$$

which in turn leads to the system $Ab + aB = 1$, $-A - B = 0$. Then $A = \frac{1}{b-a}$ and $B = \frac{-1}{b-a}$. Consequently

$$\int \frac{x'(t)}{(a-x(t))(b-x(t))}dt = \frac{-1}{b-a}\int \frac{-x'(t)}{a-x(t)}dt + \frac{1}{b-a}\int \frac{-x'(t)}{b-x(t)}dt$$

$$= -\frac{1}{b-a}\ln(a - x(t)) + \frac{1}{b-a}\ln(b - x(t)) = \frac{1}{a-b}\ln\frac{a-x(t)}{b-x(t)}$$

exactly the same answer we found using technology.

Let us revisit the topic of autocatalytic chemical reactions, already analyzed at the beginning of section 2.16. They are of the type $A \to B + C$, wherein one of the products of the reaction, say B, catalyzes the process. Under the absence of catalysis we would have $A'(t) = -kA(t)$ as a possible model, but for an autocatalytic reaction one must investigate the model $A'(t) = -kA(t)B(t)$ where $A(t)$, $B(t)$ denote, respectively, the concentrations of A, B at any instant t. Let $y(t) = a - A(t)$, where $A(0) = a$ (note that $y(t)$ tells us how much of A has been transformed). Obviously $y(0) = 0$ and $B(t) = b + y(t)$, where $B(0) = b > 0$. Thus $(a - y(t))' = -k(a - y(t))(b + y(t))$, which in turn implies $y'(t) = k(a - y(t))(b + y(t))$. We have taken into consideration the fact that, for an autocatalytic reaction to get started, some amount of the product B has to be present.

To analyze a concrete example assume that $a = 3$ and $b = 1$. Then

$$\int \frac{y'(t)}{(3-y(t))(1+y(t))} = \int k\,dt.$$

Using a graphics calculator we get

$$\int (1/((3-y)*(1+y)), y) = \tfrac{1}{4} \ln \left| \tfrac{y+1}{y-3} \right|.$$

Therefore $\tfrac{1}{4} \ln \tfrac{y(t)+1}{3-y(t)} = kt + C$ for some constant C.

After the usual validation process, and the estimation of k and C through linear regression, we can factor out $y(t)$. Indeed

$$y(t) = \tfrac{3H - e^{-4kt}}{H + e^{-4kt}},$$

where $H = e^{4C}$. As expected, $\lim_{t \to \infty} y(t) = 3$.

As in the first example, we could have used the method of partial fractions in order to reach the equality

$$\tfrac{1}{(3-y)(1+y)} = \tfrac{1}{4}\tfrac{1}{3-y} + \tfrac{1}{4}\tfrac{1}{1+y}.$$

Thereafter, the process of integration proceeds smoothly. However, sometimes calculating the partial fraction decomposition of a rational expression can be an onerous task.

The $\int$ option of a graphing calculator is a powerful tool that has somehow diminished the importance of time-honored advanced methods of integration. For instance:

$$\int (x^2/\sqrt{9-x^2}, x) = \tfrac{9}{2} \arcsin \tfrac{x}{3} - \tfrac{1}{2} x \sqrt{9-x^2} + C,$$

$$\int (1/(x^2+3)^2, x, 0, 1) = \tfrac{\pi \sqrt{3}}{108} + \tfrac{1}{24},$$

$$\int (\sin 2x * \cos 7x, x) = -\tfrac{5}{90} \cos(9x) + \tfrac{1}{10} \cos(5x) + C, \text{ and}$$

$$\int (1/((x-3)*(x-5)^2), x) = \tfrac{1}{4} \ln \left| \tfrac{x-3}{x-5} \right| - \tfrac{1}{2}\tfrac{1}{x-5} + C.$$

Table 2.17: Chemical reaction data

t(min.)	$A(t)$ (moles/lt)	$B(t)$ (moles/lt)
0	35.35	18.25
4.75	30.5	13.4
10	27	9.9
20	23.2	6.1
35	20.3	3.2
55	18.6	1.5

Exercises for section 2.24

1. Calculate by hand $\int \cos(ax)dx$, where a is any real number. Why is it that your graphing calculator provides the answer $\int(\cos(ax), x) = \cos(ax)x$, a result that is in disagreement with the answer you get by hand?

2. Calculate the area enclosed by the curves $y = x^2$ and $y = \arctan x$ (Hint: Use technology to find the point of intersection (a, a^2) between both curves, on the first quadrant; thereafter calculate $\int_0^a \arctan x dx - \int_0^a x^2 dx$).

3. Suppose that the rate of an autocatalytic reaction is $y'(t) = k(2-y(t))(1+y(t))$, where $y(t)$ is the amount of reactant that has been transformed (as a function of time). Use technology to find an expression that can be used to test the model.

4. Use technology to calculate $\int_0^6 e^{-0.3t} \sin(1.7t)dt$ (a hand calculation would require applying integration by parts twice, a lengthy endeavor).

5. Table 2.17 provides information with regard to the chemical reaction $A+B \to C$. Use linear regression to estimate the value of k, the parameter of the reaction (Hint: Recall that we showed, with and without technology, that $\frac{1}{a-b} \ln \frac{(a-x(t))}{(b-x(t))} = kt$; but $x(t)$ is the concentration of C, thus $a - x(t) = A(t)$ and $b - x(t) = B(t)$).

6. The indefinite integral $\int \frac{\sin^{-1}(\sqrt{x})}{\sqrt{x}}dx$ can be solved using a graphing calculator, but $\int \frac{\cos^{-1}(\sqrt{x})}{\sqrt{x}}dx$ cannot. Make the change of variables $u = \sqrt{x}$ and then use technology. What answer do you get?

7. Calculate $\int(\sin x)/x, x, 1, 2)$. Find the smallest value of n needed to obtain the same approximation, up to seven decimals, using the TMA program from section 2.3.

8. Find the points of inflection of the function $f(x) = \frac{1}{1+e^{-x}}$. For this purpose, use technology to sketch a graph and make a guess with regard to any possible inflection point. Thereafter proceed to take the first and second derivative to confirm your guess.

9. Using a graphing calculator we get

$$\smallint (1/(x\sqrt{x^2 - b^2}, x) = \tfrac{1}{b}\tan^{-1}(\tfrac{\sqrt{x^2-b^2}}{b}) + C.$$

But at the beginning of section 2.18 we found (10th entry of the list of integrals) that $\int \frac{dx}{x\sqrt{x^2-b^2}} = \frac{1}{b}\sec^{-1}(\frac{x}{b}) + C$. Why do we have this apparent discrepancy? (Hint: Use exercise 14, section 1.15).

10. Use the fMin option of your graphics calculator in order to find the exact value where the function $f(x) = \sqrt{40,000 + x^2} + \frac{500-x}{5}$ adopts its minimum (recall the example about a woman swimming in a lake, which was discussed in section 1.18).

2.25 Separable Variables Differential Equations

Let us start by analyzing the differential equation $y'(t) = ty(t)$. Obviously, the null function 0 is a solution. Any other solution $y(t)$ will never adopt the value zero at any point t because no two solutions of a differential equation of the type $y' = f(t)g(y)$, where f and g' are continuous, can cross each other. It is a consequence of the fact that any initial value problem $y' = f(t)g(y)$, $y(t_o) = y_o$, admits only one solution on an interval that contains t_o[17].

Suppose that we wish to solve the IVP $y'(t) = ty(t)$, $y(0) = 1$. Obviously, the null function is not a solution of it, thus any solution $y(t)$ will never adopt the value zero. Moreover, since $y(0) > 0$, we must have $y(t) > 0$ for every t (otherwise, the Intermediate Value Theorem would imply that $y(t) = 0$ for some t). Therefore

$$\tfrac{y'(t)}{y(t)} = t$$

for every t, which in turn implies

$$\int_0^t \tfrac{y'(u)}{y(u)} du = \int_0^t u\,du.$$

Consequently $[\ln y(u)]_0^t = \frac{t^2}{2}$; thus, $\ln y(t) = \frac{t^2}{2}$. Finally, $y(t) = e^{t^2/2}$. One can check that this function, defined over the whole real line, is a solution of the given IVP. Actually, we have shown that it is the *only* solution!

---

[17]This fact was mentioned before, in sections 2.16 and 2.17, when dealing with $y' = g(y)$.

What happens if our task is to solve the IVP $y'(t) = ty(t)$, $y(0) = -1$? Since $y(0) < 0$, every solution $y(t)$ will satisfy the inequality $y(t) < 0$ for any t. Following the same strategy employed in the previous IVP, we reach the equality $[\ln|y(u)|]_0^t = \frac{t^2}{2}$, that is, $[\ln(-y(u))]_0^t = \frac{t^2}{2}$. Therefore

$$\ln(-y(t)) = \frac{t^2}{2}$$

Hence $y(t) = -e^{t^2/2}$, a function that is defined over the whole real line. Indeed, it can be easily verified that this function is a solution of the IVP under consideration. Its only solution.

Let us solve the differential equation $y'(t) = ty(t)$ from a different perspective. We start with $\int \frac{y'(t)}{y(t)} dt = \int t\, dt$, which in turn leads to $\ln|y(t)| = \frac{t^2}{2} + L$ for a certain constant L. Consequently $y(t) = Ce^{t^2/2}$, where $C = \pm e^L$. It is easy to check that for any constant C the function defined by $y(t) = Ce^{t^2/2}$ is a solution of the given differential equation. Moreover, if we add the initial condition $y(0) = a$, the infinite number of solutions reduces to just one; namely the function $y(t) = ae^{t^2/2}$, defined for any t.

Another way of finding the solution, commonly found in many science books, is as follows. Using Leibniz's notation, the differential equation can be written as $\frac{dy}{dt} = ty$. Next we 'separate the variables' y and t (the latter is the independent variable while the former is the dependent variable) and integrate; in symbols, $\int \frac{dy}{y} = \int t\, dt$. Therefore $\ln|y| = \frac{t^2}{2} + L$ for a certain constant L, the same answer found before. This approach works well whenever we deal with an equation of the type $\frac{dy}{dt} = f(t)g(y)$, generically known as separable variables differential equation (SVE). *We should mention that particular examples of this type of differential equation have been discussed in preceding sections, when either $f(t) = 1$ or $g(y) = 1$.*

Let us consider the differential equation $\frac{dy}{dt} = ty^2$ with initial condition $y(0) = 5$. Then $\int \frac{dy}{y^2} = \int t\, dt$, thus $-y^{-1} = \frac{t^2}{2} + C$ for a certain constant C. Hence

$$y(t) = -\frac{2}{t^2 + 2C}.$$

But $5 = y(0) = -\frac{2}{2C}$, thus $C = -1/5$. Consequently, the solution to the problem is

$$y(t) = -\frac{2}{t^2 - 2/5}.$$

This solution is valid in the open interval $(-\sqrt{2/5}, \sqrt{2/5})$.

By now we must realize what we have to do whenever we are dealing with an arbitrary SVE: just divide by $g(y)$ and then integrate, thus reaching the equality

$$\int \frac{dy}{g(y)} = \int f(t) dt.$$

If we can solve both integrals, either by hand or using appropriate technology, we are on our way to the solution of the differential equation.

A particularly interesting case of an SVE is $\frac{dy}{dt} = g(y)$, which often appears in applications. For instance, the differential equations that govern radioactive decay, exponential growth, logistic growth, mixing, or cooling are of this type. Let us analyze the equation $\frac{dy}{dt} = a - by$, $a, b \neq 0$, with initial condition $y(0) = c$. The constant function $y(t) = a/b$ is the solution of the initial value problem under consideration if $c = a/b$. Assuming that $c < a/b$, we can conclude that $y(t) < a/b$ for every t because $y(t)$ cannot cross the horizontal line a/b (recall that no two solutions, of an SVE, can cross each other). Therefore $a - by(t) > 0$ for every t. Hence

$$\int \frac{-b}{a-by} dy = \int -b \, dt$$

Then there exists a constant D such that $\ln(a - by) = -bt + D$ for every t. In particular, $\ln(a - bc) = D$. Consequently

$$\ln \frac{a-by(t)}{a-bc} = -bt$$

Applying the exponential function to both sides, and after some simple algebraic manipulations, we finally reach the expected answer; namely, $y(t) = \frac{a}{b} - (\frac{a}{b} - c)e^{-bt}$. The case $c > a/b$ can be analyzed in a similar fashion (see exercise 9 at the end of the section); in either case, the same answer is achieved.

There is an alternative path that can be followed when analyzing the differential equation $y' = a - by$, that is, $y' + by = a$. Let us apply the same technique introduced at the end of section 1.16: multiply both sides by e^{bt} and use the chain rule! Then

$$(e^{bt}y(t))' = (\tfrac{a}{b}e^{bt})'$$

Thus, there exists a constant D such that $e^{bt}y(t) = \frac{a}{b}e^{bt} + D$ for every t. Consequently

$$y(t) = \tfrac{a}{b} + De^{-bt}$$

for every t. In particular, $c = y(0) = \frac{a}{b} + D$. Therefore $y(t) = \frac{a}{b} + (c - \frac{a}{b})e^{-bt}$ for every t. The idea of multiplying both sides of the differential equation by an exponential function, as we did above, *does not* work for an arbitrary separable variables equation. For the latter, the technique of choice is 'separate variables and integrate.'

Torricelli's law

Suppose that we have a vessel of volume V, without a top, full of water. A small hole is made on its side, close to the bottom. Torricelli's law (named after Evangelista Torricelli, 1608-1647, a disciple of Galileo), a basic law of hydrodynamics, asserts that, at any instant t,

$$v(t) = \sqrt{2gh(t)}, \tag{2.46}$$

where $v(t)$ is the horizontal velocity of the jet of water coming out of the hole, $h(t)$ is the height of the column of water in the vessel (that is, the depth of the water above the hole), and g is the acceleration due to gravity. Physicists justify the law on the basis of the principle of conservation of energy, namely the potential energy of a tiny mass of water at the surface is converted into kinetic energy of the same volume of water coming out from the hole. Thus $mgh(t) = \frac{1}{2}mv(t)^2$, which in turn leads to Torricelli's law. We have to keep in mind that the law is valid for any non-viscous fluid, provided that the area of the hole is much smaller than the cross-sectional area of the open vessel.

It is interesting to note that the movement of a drop of water of mass m that starts from rest, and is subjected only to the force of gravity, is governed by the differential equation $mv'(t) = mg$. Hence $v(t) = gt$ and consequently $h(t) = \frac{1}{2}gt^2$. Therefore $\sqrt{\frac{2h(t)}{g}} = t$, which in turn leads to

$$v(t) = g\frac{\sqrt{2h(t)}}{\sqrt{g}} = \sqrt{2gh(t)}.$$

Is it possible to find an expression that can be compared with experimental values of time and corresponding height of the column of water? In a very small interval of time Δt, the corresponding difference of volume ΔV of water in the vessel, which is the same as the volume of the jet at the bottom during the interval Δt is given by

$$\Delta V = -av(t)\Delta t$$

where the negative sign reminds us that the volume is diminishing, and a is the area of the hole (since Δt is very small, we can assume that $v(t)$ is practically constant during this interval of time). Using (2.46) we get

$$\frac{\Delta V}{\Delta t} = -a\sqrt{2gh(t)}.$$

Making $\Delta t \to 0$ we reach the differential equation

$$\frac{dV}{dt} = -a\sqrt{2g}h(t)^{1/2}. \tag{2.47}$$

Sometimes this differential equation is referred to as Torricelli's law.

When the vessel is a cylinder of radius r, $V(t) = Ah(t)$ ($A = \pi r^2$ is the area of the cross-section of the cylinder). Therefore

$$\frac{dh}{dt} = -\frac{a}{A}\sqrt{2g}h(t)^{1/2},$$

which is a separable variables differential equation. The usual procedure leads to

Table 2.18: Data collected in the hydrodynamics experiment

t(seconds)	h(t)(cm)	$\sqrt{h(t)}$
0	12	$\sqrt{12}$
42	11	$\sqrt{11}$
93	10	$\sqrt{10}$
146	9	$\sqrt{9}$
206	8	$\sqrt{8}$
269	7	$\sqrt{7}$
342	6	$\sqrt{6}$
419	5	$\sqrt{5}$
516	4	$\sqrt{4}$
639	3	$\sqrt{3}$

$$\int h^{-1/2}dh = \tfrac{-a}{A}\sqrt{2g}\int dt.$$

Hence

$$h(t)^{1/2} = \tfrac{-a}{2A}\sqrt{2g}\,t + C.$$

Therefore, if we make n measurements $(t_i, h(t_i))$, $1 \le i \le n$, the points $(t_i, \sqrt{h(t_i)})$, $1 \le i \le n$, should gather around a line of negative slope. Thus, a prediction of the model takes the form

$$\sqrt{h(t)} = -\alpha t + b.$$

To test this prediction we used a transparent open bottle of cylindrical shape and drew 12 marks on it (1 cm apart)[18]. Thereafter we made a very small hole on the side of the bottle, next to the first mark at the bottom. Then we filled the bottle with water and recorded the time (in seconds) that it took the level of water to reach each of the marks (Table 2.18) up to 3 cm. The time corresponding to the last three marks was discarded because a jet ceased to exist after the level of water went beyond the 3 cm mark.

Next, we built the third column of Table 2.18 and did a linear regression analysis. The least squares regression line was $\hat{y} = -0.00273x + 3.412926$, and the correlation coefficient $r = -0.998196$. Note how close to -1 is the correlation coefficient, thus showing that the prediction is in agreement with the data set.

---

[18]We conducted the experiment employing ideas from Burhes and Borrie (1981, p. 58).

Exercises for Section 2.25

1. Solve the initial value problem $\frac{dy}{dt} = (1 + y^2)t$, $y(0) = 0$.

2. Solve the initial value problem $\frac{dy}{dt} = (y - 2)(y - 3)$, $y(0) = 1$.

3. Solve the differential equation $y'(t) = -2ty$ subject to the initial condition $y(0) = 5$. What is the solution if we change the initial condition to $y(0) = -1$?

4. Solve the initial value problem $y' = \frac{t}{y(t)}$, $y(0) = -1$. Does the IVP $yy' = t$, $y(0) = 0$ have more than one solution?

5. What is the half-life of a chemical reaction governed by the differential equation $c'(t) = -kc^n(t)$? (n is a positive integer bigger than 1, $c(t)$ is the concentration at any instant t, while k is the parameter of the reaction).

6. Suppose that we are dealing with the chemical reaction of the preceding exercise. Let τ_1 and τ_2 be the half-life of the reaction when one starts with the initial concentrations c_1 and c_2, respectively. Calculate n in terms of the four above-mentioned quantities (Hint: Only basic algebra and logarithms are involved in the solution of this problem).

7. At high velocities, the force of resistance of air is not proportional to the velocity but to the square of the velocity. Thus, under these circumstances, the differential equation for the descent of an object under the influence of gravity becomes

$$mv'(t) = mg - k(v(t))^2.$$

   Solve the differential equation, assuming that $v(0) = 0$. Moreover, find $\lim_{t \to \infty} v(t)$.

8. We did the experiment on hydrodynamics a second time (Table 2.19). Perform the necessary linear regression analysis to compare the prediction $\sqrt{h(t)} = -at + b$ with data. You should not be surprised after realizing that the new data is not identical to the one we found on the first try; observational errors introduce variability, the more so because we are not using optical devices or other electronic gadgets, just the bare eye and a simple watch.

9. Solve the IVP $y'(t) = a - by(t)$, $y(0) = c$, assuming that $c > a/b$.

10. Draw a graph of the solution of the preceding differential equation, considering separately the cases $c > \frac{a}{b}$ and $c < \frac{a}{b}$ ($a, b > 0$).

Table 2.19: New data collected in the hydrodynamics experiment

t(seconds)	h(t)(cm)
0	12
46	11
96	10
151	9
213	8
276	7
347	6
425	5
521	4
653	3

2.26   Length of a Curve and the Catenary Problem

Let us consider a flexible chain of uniform density, suspended between its endpoints and hanging under its own weight. What curve does the chain describe? Galileo Galilei (1564-1642) thought that it was a parabola, but there were lingering doubts about it. In 1691, Newton, Leibniz, and John Bernoulli (1667-1748), independently of each other, finally solved the puzzle of the chain ('catenary' in Latin). From a historical point of view, the catenary problem is very important because it was one of the first open problems that was successfully solved using the new calculus techniques. Calculus succeeded where Euclidean or Cartesian geometry, all by themselves, could not.

Before discussing the problem, we need a formula for the length of an arbitrary 'smooth' curve defined by a function $y = f(x)$. To be more precise, let us assume that both f and f' are continuous on an interval $[a, b]$. We subdivide $[a, b]$ in n subintervals of equal length. The length of the segment that joins the points $(x_{i-1}, f(x_{i-1}))$ and $(x_i, f(x_i))$ is

$$\sqrt{(x_i - x_{i-1})^2 + (f(x_i) - f(x_{i-1}))^2}.$$

That is to say,

$$\sqrt{1 + \left(\frac{f(x_i) - f(x_{i-1})}{x_i - x_{i-1}}\right)^2}(x_i - x_{i-1}).$$

By the Mean Value Theorem we can assert that there exists x_i^*, between x_{i-1} and x_i, such that

$$\frac{f(x_i) - f(x_{i-1})}{x_i - x_{i-1}} = f'(x_i^*).$$

Hence, the length of the ith segment is

$$\sqrt{1 + (f'(x_i^*))^2}(x_i - x_{i-1}).$$

An approximation to the length of the curve will be

$$\Sigma_{i=1}^n \sqrt{1 + (f'(x_i^*))^2}(x_i - x_{i-1})$$

which is equal to

$$\Sigma_{i=1}^n \sqrt{1 + (f'(x_i^*))^2}(\tfrac{b-a}{n}).$$

This is a Riemann sum, corresponding to the integral $\int_a^b \sqrt{1 + (f'(x))^2}dx$. Therefore,

$$\lim_{n \to \infty} \Sigma_{i=1}^n \sqrt{1 + (f'(x_i^*))^2}(\tfrac{b-a}{n}) = \int_a^b \sqrt{1 + (f'(x))^2}dx.$$

We can thus adopt the number $\int_a^b \sqrt{1 + (f'(x))^2}dx$ as the length of the curve defined by $y = f(x)$, $a \le x \le b$. Having found the formula, we can go back to the problem of the hanging chain.

Let us choose the origin of the coordinate axes in such a way that it coincides with the lowest point of the chain. The chain, between $(0,0)$ and an arbitrary point (x,y), is in equilibrium (Figure 2.26). The only forces acting on this piece of the chain are T_1, the constant tension at the lowest point of the chain, lying on the horizontal axis and directed toward the left, and the variable tension T_2 at (x,y), which lies on the tangent to the curve at (x,y), plus the weight $\rho L(x)$ (ρ is the uniform density of the chain and $L(x)$ is the length of the chain between $(0,0)$ and (x,y)). Let us recall that

$$L(x) = \int_0^x \sqrt{1 + (f'(u))^2}du,$$

where $y = f(x)$ is the curve adopted by the chain. The force T_2 can be decomposed in its vertical and horizontal components: $T_2 \sin\theta$ and $T_2 \cos\theta$. Since the chain is in equilibrium, we must have $T_1 = T_2 \cos\theta$, $\rho L(x) = T_2 \sin\theta$. Consequently,

$$\tan\theta = \tfrac{\rho}{T_1}L(x).$$

But $\tan\theta = f'(x)$, therefore

$$f'(x) = \tfrac{\rho}{T_1} \int_0^x \sqrt{1 + (f'(u))^2}du.$$

This is the integro-differential equation satisfied by the chain. How can we solve it? The equation of the chain is equivalent to

$$f''(x) = \alpha\sqrt{1 + (f'(x))^2},$$

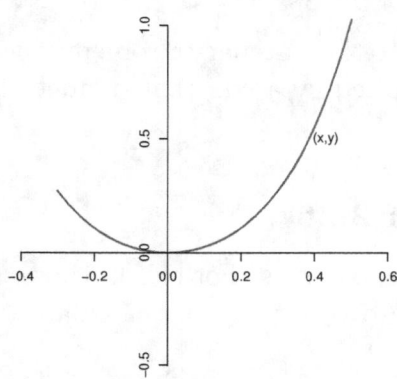

Figure 2.26: The hanging chain

where $\alpha = \rho/T_1$ (observe that $f'(0) = 0$, due to the way that we chose the coordinate axes). Defining $g(x) = f'(x)$ we reach the differential equation

$$g'(x) = \alpha\sqrt{1 + (g(x))^2},$$

which is a separable variables differential equation! As usual, we 'separate' and then integrate:

$$\int \frac{g'(x)\,dx}{\sqrt{1+(g(x))^2}} = \int \alpha\,dx.$$

But

$$\int \frac{dx}{\sqrt{1+x^2}} = \sinh^{-1} x + C.$$

Hence $\sinh^{-1}(g(x)) = \alpha x + C$. In particular, $\sinh^{-1}(g(0)) = \alpha \times 0 + C$. However, $g(0) = f'(0) = 0$; thus, $C = 0$. Consequently

$$\sinh^{-1}(g(x)) = \alpha x,$$

which in turn leads to

$$f'(x) = g(x) = \sinh(\alpha x).$$

Then there exists a constant D such that

$$f(x) = \tfrac{1}{\alpha}\cosh(\alpha x) + D.$$

Making $x = 0$ we get $0 = f(0) = \frac{1}{\alpha} + D$. Finally,

$$f(x) = \frac{1}{\alpha} \cosh(\alpha x) - \frac{1}{\alpha}.$$

We have reached the answer to the Catenary Problem: the chain describes a curve that is a hyperbolic cosine multiplied by a constant, namely $1/\alpha$, and shifted downwards by the amount $1/\alpha$.

Exercises for Section 2.26

Find the length of the following curves. For this purpose, solve by hand the first three exercises. For the last two use your graphing calculator to approximate the value of the corresponding integral.

1. $y = \frac{1}{2}x^2$, $0 \le x \le 1$.

2. $y = \sqrt{1 - x^2}$, $-1 \le x \le 1$.

3. $y = \cosh x$, $0 \le x \le 3$.

4. $y = \sin x$, $0 \le x \le \frac{\pi}{2}$.

5. $y = \frac{1}{x}$, $1 \le x \le 3$.

2.27   Volumes of Solids of Revolution

So far integrals have been mainly used to calculate areas or to interpret diverse problems in terms of areas. Interestingly, integrals are also useful to calculate volumes. It is fitting to say something about this new geometrical application.

From Pre-Calculus mathematics, it is known that the volume of a prism or a cylinder is given by $B \times h$, where B is the area of the base and h is the height. Our goal in this section is to find formulas for several other solids. Suppose we wish to find the volume of a cone of radius r and height h. We can build the cone by revolving, around the x-axis, the segment defined by the function $f(x) = \frac{r}{h}x$, $0 \le x \le h$ (Figure 2.27). Let us subdivide the interval $[0, h]$ in n subintervals of equal length, namely h/n, and let $x_i = i\frac{h}{n}$. We observe that

$$\pi f(x_i)^2 \frac{h}{n}$$

is the volume of the cylinder of radius $f(x_i)$ and height h/n. An approximation to the volume of the cone will be

$$\sum_{i=1}^{n} \pi f(x_i)^2 \frac{h}{n}.$$

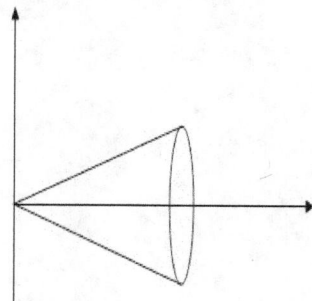

Figure 2.27: Finding the volume of a cone

This is a Riemann sum corresponding to the continuous function $f(x)$, which is a good approximation to the volume of the cone insofar as n is large. Indeed,

$$\lim_{n\to\infty} \Sigma_{i=1}^{n} \pi f(x_i)^2 \tfrac{h}{n} = \pi \int_0^h f(x)^2 dx$$

and we adopt $\pi \int_0^h f(x)^2 dx$ as the volume V of the cone. It follows that

$$V = \pi \int_0^h (\tfrac{r}{h})^2 x dx = \tfrac{\pi r^2}{h^2} \tfrac{h^3}{3} = \tfrac{\pi r^2 h}{3}.$$

The same argument can be applied to find the volume of a solid generated by revolving the curve $y = f(x)$, $a \le x \le b$, around the x-axis. We would reach the formula

$$V = \pi \int_a^b f(x)^2 dx. \tag{2.48}$$

For instance, revolving around the x-axis the curve $y = \sqrt{r^2 - x^2}$, $0 \le x \le r$, we get a hemisphere of radius r. Its volume is

$$V = \pi \int_0^r (r^2 - x^2) dx = \pi r^3 - \tfrac{\pi r^3}{3} = \tfrac{2\pi r^3}{3}.$$

Therefore, the volume of any sphere of radius r happens to be $\tfrac{4\pi r^3}{3}$.

Solids can also be obtained revolving a curve around the y-axis. Suppose we revolve the curve $y = x^2$, $0 \le x \le 1$, around the y-axis, obtaining a bowl-shaped solid (Figure 2.28). We observe that for any y on the range, $0 \le y \le 1$, the corresponding radius of the approximating cylinder is $\sqrt{y}$, therefore

$$V = \pi \int_0^1 (\sqrt{y})^2 dy = \pi \int_0^1 y dy = \tfrac{\pi}{2}.$$

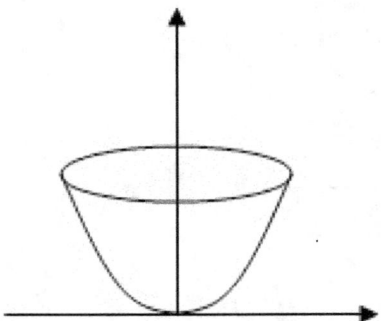

Figure 2.28: A bowl-shaped solid

If the same curve is revolved around the x-axis, we obtain a trumpet-like solid (Figure 2.29) with volume

$$V = \pi \int_0^1 (x^2)^2 dx = \pi \int_0^1 x^4 dx = \tfrac{\pi}{5}.$$

In general, given any increasing continuous function $y = f(x)$, $p \leq x \leq q$, we can revolve it around the y-axis obtaining a solid of revolution whose volume is

$$V = \pi \int_a^b (f^{-1}(y))^2 dy \qquad (2.49)$$

where $a = f(p)$ and $b = f(q)$. For instance, let $y = \arctan x$, $0 \leq x \leq 1$. Since $\arctan(0) = 0$ and $\arctan(1) = \pi/4$, by revolving the curve around the y-axis we can construct a solid with volume

$$V = \pi \int_0^{\pi/4} \tan^2 y \, dy.$$

But $\int \tan^2 y \, dy = \int \sec^2 y \, dy - \int dy = \tan y - y + C$. Therefore

$$V = \pi [\tan y - y]_0^{\pi/4} = \pi (1 - \tfrac{\pi}{4}).$$

A formula, analogous to (2.49), can be obtained when f is decreasing[19].

---

[19]We assume that the function is increasing or decreasing to guarantee that its inverse exists.

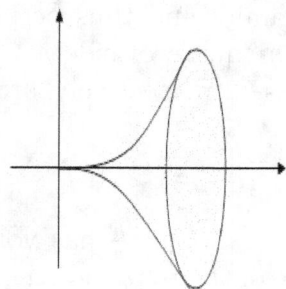

Figure 2.29: A trumpet-like solid

Flow of blood across an artery

In section 2.5 we found that the volume of blood that flows through a cross-section of an artery, in unit time, is given by the formula $\pi k \frac{R^4}{2}$, where R is the radius of the artery. Let us reach the same formula from a different perspective. According to Poiseuille's law, the velocity at which blood circulates is given by $v = k(R^2 - r^2)$, $0 \le r \le R$. We observe that the distance d traveled by every particle of blood, in unit time, is $d = kR^2 - kr^2$, $0 \le r \le R$. Its pictorial representation, having r on the horizontal axis and d on the vertical axis, is a parabola that opens downwards and is shifted up kR^2. The total amount of blood under consideration is the volume of the solid generated by revolving, around the vertical axis, the above-mentioned parabola. From a mathematical point of view, the same volume is generated by revolving, around the y-axis, the quadratic curve $y = kr^2$, $0 \le r \le R$. Since $r = \sqrt{y/k}$, we will have, thanks to (2.49),

$$V = \pi \int_0^{kR^2} (\sqrt{y/k})^2 dy = \tfrac{\pi}{k} \int_0^{kR^2} y\,dy = \tfrac{\pi}{2k}(kR^2)^2 = \tfrac{\pi k}{2} R^4.$$

A bowl-shaped solid is obtained, like the one in Figure 2.28.

A more general formula

Suppose we wish to find the volume of a right pyramid of height h whose base is a square of side b. A pyramid is *not* a solid of revolution, thus we cannot apply the formula (2.48). Let us put the top of the pyramid on the origin of the coordinate system in such a way that the x-axis passes through the center of the base. Then we subdivide the interval $[0, h]$ in n subintervals of equal length h/n and let

$$x_i = i\frac{h}{n}.$$

A plane parallel to the base of the pyramid, and thus perpendicular to the x-axis, cuts the pyramid at x_i giving birth to a square of side s_i. The two right triangles with heights h and x_i, and legs $b/2$ and $s_i/2$ respectively, are similar. Therefore,

$$\frac{x_i}{h} = \frac{\frac{s_i}{2}}{\frac{b}{2}}.$$

Hence, $s_i = \frac{bx_i}{h}$. The prism of side s_i and height $\frac{h}{n}$ has volume $s_i^2 \frac{h}{n}$, thus the sum $\Sigma_i^n s_i^2 \frac{h}{n}$ is an approximation to the volume of the pyramid (the approximation becomes 'good' if n is a large number). That is to say,

$$\Sigma_{i=1}^n \frac{b^2}{h^2} x_i^2 \frac{h}{n}.$$

But this is a Riemann sum corresponding to the integral $\frac{b^2}{h^2} \int_0^h x^2 dx$. Consequently,

$$V = \frac{b^2}{h^2}\frac{h^3}{3} = \frac{1}{3}b^2 h.$$

Next, suppose that we have an arbitrary solid lying between $x = a$ and $x = b$. At any x, $a \le x \le b$, a plane, perpendicular to the x-axis, cuts the solid and produces an area $A(x)$, where $A(x)$ is a continuous function on $[a, b]$. Thereafter, we subdivide $[a, b]$ in n subintervals of length $\frac{b-a}{n}$ and let $x_i = a + i\frac{b-a}{n}$. Then

$$\Sigma_{i=1}^n A(x_i)\frac{b-a}{n}$$

is an approximation to the volume of the solid. It is a Riemann sum corresponding to $\int_a^b A(x)dx$. Therefore

$$\lim_{n\to\infty} \Sigma_{i=1}^n A(x_i)\frac{b-a}{n} = \int_a^b A(x)dx.$$

Hence,

$$V = \int_a^b A(x)dx. \tag{2.50}$$

In the example of the square pyramid we found that $A(x) = \frac{b^2}{h^2}x^2$. In general, the challenge is to find $A(x)$ for each particular solid.

Suppose we are dealing with a solid of revolution between $x = a$ and $x = b$, obtained by revolving the continuous function $y = f(x)$ around the x-axis. Then $A(x) = \pi f(x)^2$, and consequently $V = \pi \int_a^b f(x)^2 dx$. Hence (2.48) is a special case of (2.50).

Exercises for Section 2.27

Calculate the volume of the solid of revolution generated by the following curves and sketch a graph of the corresponding solid.

1. $y = \sqrt{x}$, $0 \leq x \leq 1$. Revolve around the x-axis.

2. $y = \sin x$, $0 \leq x \leq \frac{\pi}{2}$. Revolve around the x-axis.

3. $y = \sin x$, $0 \leq x \leq \frac{\pi}{2}$. Revolve around the y-axis.

4. $y = \cos x$, $0 \leq x \leq 1$. Revolve around the y-axis.

5. $y = e^x$, $0 \leq x \leq 1$. Revolve around the x-axis.

6. $y = \frac{1}{x}$, $\frac{1}{2} \leq x \leq 1$. Revolve around the y-axis.

7. $y = \sqrt{4 - x^2}$, $0 \leq x \leq 2$. Revolve around the x-axis. What answer would you get if we decide to revolve the same curve around the y-axis?

2.28 L'Hôpital's Rules

A question we have not addressed so far is related to 'how fast,' comparatively speaking, two functions approach infinity. For instance, in section 2.29, we will need to calculate $\lim_{x \to \infty} \frac{x}{e^{ax}}$, where $a > 0$. Both the function in the numerator (which happens to be the identity function) and the function in the denominator tend to ∞ when $x \to \infty$. From their graphs we surmise that the identity function increases at a slower pace than e^{ax} as $x \to \infty$; thus, it is to be expected that $\lim_{x \to \infty} \frac{x}{e^{ax}} = 0$. Is there a mathematical technique to deal with this and similar problems about limits of quotients of functions?

Luckily, mathematicians invented a collection of techniques, generically known as 'L'Hôpital's Rules,' to analyze what happens when there is a 'competition' between the growth of two functions. The rules are named in honor of the Marquis de L'Hôpital (1661-1704), who wrote the first ever printed book on the calculus in 1696 (wherein one of the rules can be found). Although a proof of all L'Hôpital's rules is beyond a first course[20], its use is not complicated at all. We will state carefully four of the rules, grouped in two categories, and we will illustrate them through examples.

---

[20]A proof of the fourth rule can be found in Appendix B.6.

Type $\frac{\pm\infty}{\pm\infty}$

Rule 1

Let f and g be differentiable functions defined on an interval (a, ∞). Suppose that $\lim_{x\to\infty} f(x) = \infty$, $\lim_{x\to\infty} g(x) = \infty$. Furthermore, assume that $\lim_{x\to\infty} \frac{f'(x)}{g'(x)}$ exists (either as a real number or ∞). Then

$$\lim_{x\to\infty} \frac{f(x)}{g(x)} = \lim_{x\to\infty} \frac{f'(x)}{g'(x)}$$

For instance, if $b > 0$ we will have

$$\lim_{x\to\infty} \frac{x}{e^{bx}} = \lim_{x\to\infty} \frac{1}{be^{bx}} = 0$$

while

$$\lim_{x\to\infty} \frac{x}{\ln x} = \lim_{x\to\infty} \frac{1}{1/x} = \lim_{x\to\infty} x = \infty.$$

We will say that $f(x)$ **grows faster** than $g(x)$ as $x \to \infty$ if $\lim_{x\to\infty} \frac{f(x)}{g(x)} = \infty$ or, equivalently, if $\lim_{x\to\infty} \frac{g(x)}{f(x)} = 0$. Thus, e^{bx}, $b > 0$, grows faster than the identity function x, which in turn grows faster than $\ln x$

What grows faster as $x \to \infty$, $\ln(\ln x)$ or $\ln x$? We have

$$\lim_{x\to\infty} \frac{\ln(\ln x)}{\ln x} = \lim_{x\to\infty} \frac{\frac{1}{\ln x}\frac{1}{x}}{\frac{1}{x}} = 0.$$

Thus, $\ln x$ grows faster than $\ln(\ln x)$.

The rule can be applied as many times as needed, as the following example shows:

$$\lim_{x\to\infty} \frac{x^2}{e^x} = \lim_{x\to\infty} \frac{2x}{e^x} = \lim_{x\to\infty} \frac{2}{e^x} = 0.$$

In the same fashion, we can show that $\lim_{x\to\infty} \frac{x^n}{e^x} = 0$. Thus, as $x \to \infty$ the exponential function grows faster than any polynomial (Figure 2.30 illustrates this fact when $n = 3$).

We will say that $f(x)$ and $g(x)$ **grow at the same rate** when $x \to \infty$ if there exists a number $L > 0$ such that

$$\lim_{x\to\infty} \frac{f(x)}{g(x)} = L.$$

For instance, $\ln(5x)$ and $\ln(7x)$ grow at the same rate. Indeed,

$$\lim_{x\to\infty} \frac{\ln(5x)}{\ln(7x)} = \lim_{x\to\infty} \frac{\frac{5}{5x}}{\frac{7}{7x}} = 1.$$

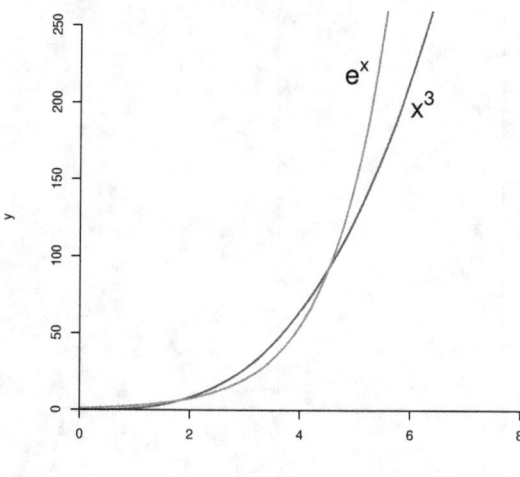

Figure 2.30: The curve $y = e^x$ is eventually above the curve $y = x^3$

Similarly, $\ln(1 + \frac{1}{x})$ and $\frac{1}{x}$ grow at the same rate when $x \to \infty$ (Figure 2.31) because

$$\lim_{x \to \infty} \frac{\ln(1+\frac{1}{x})}{\frac{1}{x}} = \lim_{x \to \infty} \frac{\frac{1}{1+\frac{1}{x}}(-\frac{1}{x^2})}{-\frac{1}{x^2}} = \lim_{x \to \infty} \frac{1}{1+\frac{1}{x}} = 1.$$

In both these examples, the limit of the quotient, when $x \to \infty$, happened to be 1. But it does not have to be so in order to claim that two functions grow at the same rate when $x \to \infty$. For instance,

$$\lim_{x \to \infty} \frac{\ln(x^2+7)}{\ln x} = \lim_{x \to \infty} \frac{\frac{2x}{x^2+7}}{\frac{1}{x}} = \lim_{x \to \infty} \frac{2}{1+\frac{7}{x^2}} = 2.$$

Thus, both functions grow at the same rate when $x \to \infty$ and we should expect that $\ln(x^2 + 7) \approx 2\ln x$ whenever x is big enough. For example, when $x = 500$, $2 \times \ln(500) = 2 \times 6.214608098 = 12.429216$ while $\ln(250,007) = 12.4292442$. By far, rule 1 is the rule of L'Hôpital that appears most often in applications.

Rule 2

Let f and g be defined and continuous on $(c - \delta_o, c + \delta_o)$, and differentiable on this interval with the possible exception at c. Suppose that $\lim_{x \to c} f(x) = \pm\infty$, $\lim_{x \to c} g(x) = \pm\infty$. Furthermore, assume that $\lim_{x \to c} \frac{f'(x)}{g'(x)}$ exists (either as a real number or $\pm\infty$). Then

$$\lim_{x \to c} \frac{f(x)}{g(x)} = \lim_{x \to c} \frac{f'(x)}{g'(x)}.$$

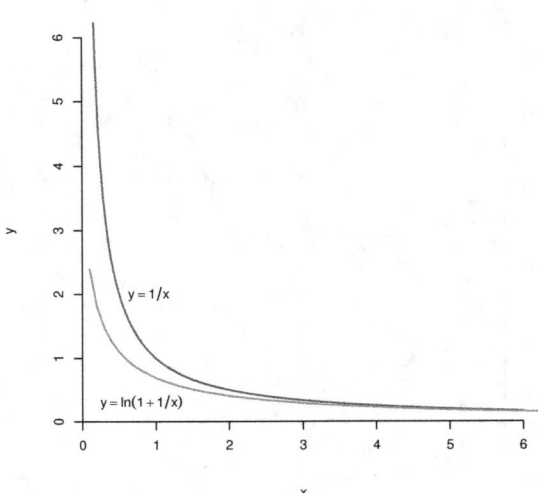

Figure 2.31: $y = \frac{1}{x}$ and $y = \ln(1 + \frac{1}{x})$ grow at the same rate as $x \to \infty$

For instance,

$$\lim_{x \to 0} \frac{1}{x^2 e^{1/x^2}} = \lim_{x \to 0} \frac{1/x^2}{e^{1/x^2}} = \lim_{x \to 0} \frac{-2/x^3}{e^{1/x^2}(-2/x^3)} = \lim_{x \to 0} \frac{1}{e^{1/x^2}} = 0.$$

It should be noted that the second rule is valid if we approach c either from the left or the right, that is, if we have to deal with lateral limits. For instance, what is $\lim_{x \to 0^+} x \ln x$? There is a competition between the identity function, which tends to zero, and the natural logarithm, which tends to $-\infty$. We notice that $x \ln x = \frac{\ln x}{1/x}$, thus we have a quotient of two functions, one of which converges to $-\infty$ as $x \to 0^+$ and the other converges to ∞ as $x \to 0^+$. Under these circumstances, the second rule of L'Hôpital asserts that the original limit is the limit of the quotient of the corresponding derivatives. Therefore

$$\lim_{x \to 0^+} \frac{\ln x}{1/x} = \lim_{x \to 0^+} \frac{1/x}{-1/x^2} = \lim_{x \to 0^+} (-x) = 0.$$

That is to say, the identity function wins and forces the product $x \ln x$ towards zero. It just so happens that often the corresponding limit of the quotient $f'(x)/g'(x)$ is easier to calculate than the limit of the quotient $f(x)/g(x)$; this fact is what makes L'Hôpital's rules so useful!

Type $\frac{0}{0}$

Rule 3

Let f and g be differentiable functions defined on an interval (a, ∞). Suppose that $\lim_{x \to \infty} f(x) = 0$, $\lim_{x \to \infty} g(x) = 0$. Furthermore, assume that $\lim_{x \to \infty} \frac{f'(x)}{g'(x)}$ exists (either as a real number or $\pm\infty$). Then

$$\lim_{x \to \infty} \frac{f(x)}{g(x)} = \lim_{x \to \infty} \frac{f'(x)}{g'(x)}.$$

For example,

$$\lim_{x \to \infty} x \sin \frac{1}{x} = \lim_{x \to \infty} \frac{\sin \frac{1}{x}}{\frac{1}{x}} = \lim_{x \to \infty} \frac{(\cos \frac{1}{x})(-\frac{1}{x^2})}{-\frac{1}{x^2}} = \lim_{x \to \infty} \cos \frac{1}{x} = 1$$

Rule 4

Let f and g be defined and continuous on $(c - \delta_o, c + \delta_o)$, and differentiable on this interval with the possible exception at c. Suppose that $\lim_{x \to c} f(x) = 0$, $\lim_{x \to c} g(x) = 0$, and $\lim_{x \to c} \frac{f'(x)}{g'(x)}$ exists (either as a real number or $\pm\infty$). Then

$$\lim_{x \to c} \frac{f(x)}{g(x)} = \lim_{x \to c} \frac{f'(x)}{g'(x)}.$$

For instance,

$$\lim_{x \to 0} \frac{\sin x}{\cos x - 1} = \lim_{x \to 0} \frac{\cos x}{-\sin x} = -\infty.$$

Sometimes a little bit of algebra is needed before applying L'Hôpital's rules. For instance, suppose that we are asked to calculate

$$\lim_{x \to 0} \left(\frac{1}{\sin x} - \frac{1}{x^2} \right).$$

Since $\lim_{x \to 0} \frac{1}{\sin x} = \infty$ and $\lim_{x \to 0} \frac{1}{x^2} = \infty$, we do not know how to deal with the limit of the difference because it would require calculating the indeterminate expression $\infty - \infty$. However,

$$\frac{1}{\sin x} - \frac{1}{x^2} = \frac{x^2 - \sin x}{x^2 \sin x}.$$

L'Hôpital's fourth rule implies that

$$\lim_{x \to 0} \frac{x^2 - \sin x}{x^2 \sin x} = \lim_{x \to 0} \frac{2x - \cos x}{2x \sin x + x^2 \cos x}.$$

Since $\lim_{x \to 0} (2x - \cos x) = -1$ and $\lim_{x \to 0} (2x \sin x + x^2 \cos x) = 0$, we can conclude that

$$\lim_{x\to 0}\left(\frac{1}{\sin x} - \frac{1}{x^2}\right) = -\infty.$$

On the other hand,

$$\lim_{x\to 0} \frac{\ln(1+x)}{x} = \lim_{x\to 0} \frac{1}{1+x} = 1.$$

This result is interesting because it will allow us to find a new expression for the number e. Indeed, since exp is continuous we have

$$\lim_{x\to 0}(1+x)^{\frac{1}{x}} = \lim_{x\to 0} e^{\frac{1}{x}\ln(1+x)} = e^{\lim_{x\to 0}\frac{1}{x}\ln(1+x)} = e^1 = e.$$

The fact that $\lim_{x\to 0}(1+x)^{\frac{1}{x}} = e$ implies, right away, that

$$\lim_{x\to\infty}\left(1+\frac{1}{x}\right)^x = e.$$

We should keep in mind that the fourth rule is true if we approach c either from the left or the right, that is, if we have to deal with lateral limits. For instance,

$$\lim_{x\to 0^+}\frac{\sin x}{\sqrt{x}} = \lim_{x\to 0^+}\frac{\cos x}{\frac{1}{2\sqrt{x}}} = \lim_{x\to 0^+} 2\sqrt{x}\cos x = 0.$$

Similarly, the first and third rules are also true when $x \to -\infty$ (under these circumstances, we should be dealing with functions defined on an interval of the form $(-\infty, a)$). For example,

$$\lim_{x\to-\infty} xe^x = \lim_{x\to-\infty}\frac{x}{e^{-x}} = -\lim_{x\to-\infty}\frac{1}{e^{-x}} = 0.$$

Although we have not explicitly mentioned that $g(x)$ and $g'(x)$ are different from zero in each of the rules, on their respective domains, we are assuming that this is so because otherwise the quotients $f(x)/g(x)$ and $f'(x)/g'(x)$ would not be defined. By now we hope that the reader must have noticed the incredible versatility of the rules of L'Hôpital.

Note

Assume that f and g are functions defined on an open interval $(c - \delta, c + \delta)$, both differentiable at c, with $g'(c) \neq 0$ and $\lim_{x\to c} f(x) = 0 = \lim_{x\to c} g(x)$. Under these circumstances, one can easily prove that

$$\lim_{x\to c}\frac{f(x)}{g(x)} = \frac{f'(c)}{g'(c)}.$$

Indeed, since $\lim_{x\to c} f(x) = 0$ and f is continuous at c, we can conclude that $f(c) = 0$. Similarly, $g(c) = 0$. Then

$$\lim_{x \to c} \frac{f(x)}{g(x)} = \lim_{x \to c} \frac{\frac{f(x)-f(c)}{x-c}}{\frac{g(x)-g(c)}{x-c}} = \frac{\lim_{x \to c} \frac{f(x)-f(c)}{x-c}}{\lim_{x \to c} \frac{g(x)-g(c)}{x-c}} = \frac{f'(c)}{g'(c)}.$$

For instance,

$$\lim_{x \to 0} \frac{\tan x}{x} = \frac{\sec^2(0)}{1} = 1.$$

It is especially the condition $g'(c) \neq 0$ which sets a striking difference between the fourth rule of L'Hôpital and the proposition discussed in this note. It is a truly remarkable fact that rule 4 implies

$$\lim_{x \to c} \frac{f(x)}{g(x)} = \lim_{x \to c} \frac{f'(x)}{g'(x)}$$

even though it may well happen that $g'(c) = 0$. Thus, we cannot calculate $\lim_{x \to 0} \frac{\sin x}{\cos x - 1}$ using the above-mentioned proposition because the derivative of $\cos x - 1$ at $x = 0$ happens to be zero. Nonetheless, rule 4 immediately leads to the equality

$$\lim_{x \to 0} \frac{\sin x}{\cos x - 1} = \lim_{x \to 0} \frac{\cos x}{-\sin x}$$

and the value of the second limit happens to be $-\infty$ because $-\cos x$ tends to -1 and the denominator tends to 0. On the other hand, since in many examples that we might encounter, in a first course on the calculus, the functions are differentiable everywhere, if a limit can be calculated using the proposition then it can also be calculated using rule 4. For example,

$$\lim_{x \to 0} \frac{e^x - 1}{\tan(5x)} = \frac{e^x}{5\sec^2(5x)}\Big|_{x=0} = \frac{1}{5}$$

and, at the same time,

$$\lim_{x \to 0} \frac{e^x - 1}{\tan(5x)} = \lim_{x \to 0} \frac{e^x}{5\sec^2(5x)} = \frac{1}{5}.$$

Exercises for section 2.28

Calculate the following limits:

1. $\lim_{x \to \infty} \frac{e^x}{\ln x}$.

2. $\lim_{x \to \infty} \frac{x}{\nu e^{\nu x}}$, where ν is an arbitrary constant.

3. $\lim_{x \to \infty} \frac{\ln(9x)}{\ln(2x)}$.

4. $\lim_{x \to \infty} \frac{x^3 + 3x + 1}{e^x}$.

5. $\lim_{x \to 0^+} x \ln x^2$.

6. $\lim_{x \to 0} \frac{\sin(x^2)}{x}$.

7. $\lim_{x \to 0} \frac{\cos x - 1}{x^2}$.

8. $\lim_{x \to \infty} \frac{x}{\ln(\ln x)}$.

9. $\lim_{x \to 0} \frac{x}{\tan^{-1} x}$.

10. $\lim_{x \to -\infty} x^2 e^{-x}$.

11. $\lim_{x \to 0} \frac{x}{\tan x - x}$.

12. $\lim_{x \to \infty} \frac{\sqrt{x}}{\ln(x^2 + x + 5)}$.

13. $\lim_{x \to \infty} x^{1/x}$ (Hint: Draw a graph of the function and note that $x^{1/x} = e^{\frac{1}{x} \ln x}$)

14. $\lim_{x \to \infty} \frac{\ln(x^2 + x + 1)}{\ln(x^3 + 1)}$.

15. $\lim_{x \to 0+} x^{1/x}$ (Hint: None of L'Hôpital's rules are needed.)

16. $\lim_{x \to 0+} (1 + \frac{1}{x})^x$.

17. $\lim_{x \to \infty} (1 + x)^{\frac{1}{\ln x}}$.

2.29   Improper Integrals

Given $f(x) = e^{-x}$, $x \geq 0$, is it possible to determine a value for the area of the region between the function and the x-axis? We start by noting that, when $b > 0$,

$$\int_0^b e^{-x} dx = [-e^{-x}]_0^b = -e^{-b} + 1.$$

Since $\lim_{b \to \infty} e^{-b} = 0$ we can conclude that $\lim_{b \to \infty} \int_0^b e^{-x} dx = 1$, a fact that is highlighted by writing $\int_0^\infty e^{-x} dx = 1$. An integral of the type $\int_a^\infty f(x) dx$, f continuous on $[a, \infty]$, is called an 'improper integral.' If $\lim_{b \to \infty} \int_a^b f(x) dx$ exists, we will say that the improper integral converges; otherwise it is said to be divergent. Beware that the same symbol is used to denote $\lim_{b \to \infty} \int_a^b f(x) dx$ and the value of this limit (when it exists).

Next let us consider the function $f(x) = 1/x$, $x \geq 1$. Since

$$\int_1^b \frac{1}{x} dx = [\ln x]_1^b = \ln b$$

and $\lim_{b \to \infty} \ln b = \infty$, it follows that $\int_1^\infty \frac{1}{x} dx$ is divergent. No value can be assigned to the area under f and above the x-axis. For instance $\int(1/x, x, 1, 100) \approx 4.6051702$, $\int(1/x, x, 1, 1000) \approx 6.9077553$, $\int(1/x, x, 1, 100000) \approx 11.512925$, and so on. The corresponding integrals do not tend to any limit, they just keep increasing. However, if $f(x) = 1/x^2$, $x \geq 1$, then

$$\int_1^b \tfrac{1}{x^2}dx = \int_1^b x^{-2}dx = [-x^{-1}]_1^b = -\tfrac{1}{b}+1.$$

Since $\lim_{b\to\infty} 1/b = 0$ we can conclude that $\int_1^\infty \tfrac{1}{x^2}dx$ is convergent, moreover $\int_1^\infty \tfrac{1}{x^2}dx = 1$. It should be expected by now that $\int_1^\infty \tfrac{1}{x^p}dx$ is convergent whenever $p > 1$ and divergent for $p \le 1$. Indeed

$$\int_1^b \tfrac{1}{x^p}dx = \int_1^b x^{-p}dx = \tfrac{1}{1-p}[x^{1-p}]_1^b = \tfrac{1}{1-p}(b^{1-p}-1) = \tfrac{1}{1-p}(\tfrac{1}{b^{p-1}}-1).$$

But $\lim_{b\to\infty} \tfrac{1}{b^{p-1}} = 0$ whenever $p - 1 > 0$. Therefore, if $p > 1$ then

$$\int_1^\infty \frac{1}{x^p}dx = \frac{1}{p-1}. \tag{2.51}$$

What happens if $p < 1$? Under these circumstances

$$\lim_{b\to\infty} \tfrac{1}{b^{p-1}} = \lim_{b\to\infty} b^{1-p} = \infty.$$

Hence $\int_1^\infty \tfrac{1}{x^p}dx$ diverges when $p \le 1$ (recall that divergence for $p = 1$ was proven before).

Four more examples

1. What could be said about $\int_0^\infty \tfrac{dx}{x^2+1}$? We have

$$\int_0^b \tfrac{dx}{x^2+1} = [\tan^{-1}x]_0^b = \tan^{-1}b - \tan^{-1}0 = \tan^{-1}b.$$

But $\lim_{b\to\infty} \tan^{-1}(b) = \pi/2$. Hence $\int_0^\infty \tfrac{dx}{x^2+1} = \tfrac{\pi}{2}$.

2. Right away it can be seen that $\int_1^\infty \tfrac{\ln x}{x}dx$ diverges because

$$\int_1^b \tfrac{\ln x}{x}dx = [\tfrac{1}{2}\ln^2 x]_1^b = \tfrac{1}{2}\ln^2 b \to \infty \text{ as } b \to \infty.$$

No wonder that $\int(\ln x/x, x, 1, 100) \approx 10.603796$, $\int(\ln x/x, x, 1, 1000) \approx 23.858541$, $\int(\ln x/x, x, 1, 100000) \approx 66.273726$, and so on; the integrals do not converge to any limit.

3. A different scenario takes place when dealing with $\int_1^\infty \tfrac{\ln x}{x^2}dx$ (Figure 2.32). Using integration by parts (choosing $f(x) = \ln x$, $g'(x) = x^{-2}$) we get

$$\int_1^b \tfrac{\ln x}{x^2}dx = -[\tfrac{\ln x}{x}]_1^b + \int_1^b \tfrac{dx}{x^2} = -\tfrac{\ln b}{b} - [\tfrac{1}{x}]_1^b = -\tfrac{\ln b}{b} - \tfrac{1}{b} + 1.$$

L'Hôpital's first rule leads to $\lim_{b\to\infty} \tfrac{\ln b}{b} = \lim_{b\to\infty} \tfrac{1}{b} = 0$. Therefore

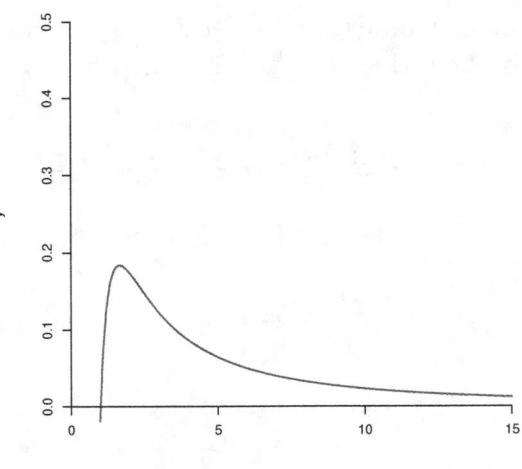

Figure 2.32: The curve $y = \frac{\ln x}{x^2}$

$$\int_1^\infty \frac{\ln x}{x^2} dx = 1.$$

4. Can we assign a value to the area between the curve $y = \arctan x$ $(0 \leq x < \infty)$ and the x−axis? Let us rephrase the question by asking whether $\int_0^\infty \arctan x \, dx$ is convergent (see Figure 1.41). For any $b > 0$, integration by parts (choosing $f(x) = \arctan x$, $g'(x) = 1$) leads to

$$\int_0^b \arctan x \, dx = [x \arctan x]_0^b - \frac{1}{2} \int_0^b \frac{2x dx}{x^2+1} = [x \arctan x]_0^b - \frac{1}{2}[\ln(x^2+1)]_0^b =$$
$$b \arctan b - \frac{1}{2} \ln(b^2+1).$$

What happens when $b \to \infty$? Using our intuition we might assert that $b \arctan b$ tends to infinity 'much faster' than $\frac{1}{2} \ln(b^2+1)$ so that

$$\lim_{b \to \infty}(b \arctan b - \frac{1}{2} \ln(b^2+1)) = \infty.$$

Thus, $\int_0^\infty \arctan x \, dx$ diverges.

If the reader prefers a mathematical argument to formalize our thoughts with regard to the above-mentioned limit, we have to realize that $\arctan b > \frac{\pi}{4}$ provided that $b > 1$ (Why? $\arctan 1 = \pi/4$ and $\arctan$ is strictly increasing). Hence

$$b \arctan b > \tfrac{\pi}{4} b, \quad b > 1.$$

Thanks to L'Hôpital's rule we have

$$\lim_{b \to \infty} \frac{\tfrac{\pi}{8} b}{\tfrac{1}{2} \ln(b^2+1)} = \tfrac{\pi}{8} \lim_{b \to \infty} \frac{b^2+1}{b} = \tfrac{\pi}{8} \lim_{b \to \infty} (b + \tfrac{1}{b}) = \infty.$$

Then there exists $p > 1$ such that, for $b > p$,

$$\frac{\tfrac{\pi}{8} b}{\tfrac{1}{2} \ln(b^2+1)} > 1.$$

Thus

$$\tfrac{1}{2} \ln(b^2 + 1) < \tfrac{\pi}{8} b, \quad b > p.$$

Finally, since $b \arctan b > \tfrac{\pi}{4} b$ and $-\tfrac{1}{2} \ln(b^2 + 1) > -\tfrac{\pi}{8} b$ (whenever $b > p$), we can conclude that

$$b \arctan b - \tfrac{1}{2} \ln(b^2 + 1) > \tfrac{\pi}{4} b - \tfrac{\pi}{8} b = \tfrac{\pi}{8} b$$

whenever $b > p$. Since $\lim_{b \to \infty} \tfrac{\pi}{8} b = \infty$, it follows that

$$\lim_{b \to \infty} b \arctan b - \tfrac{1}{2} \ln(b^2 + 1) = \infty,$$

thus confirming what we thought before starting a detailed argument.

Remark

Before we go any further let us point out that, given a continuous function on $[c, \infty]$ and $c < d$, $\int_c^\infty f(x)dx$ converges if and only if $\int_d^\infty f(x)dx$ converges. Moreover, if one of them converges, we can conclude that

$$\int_c^\infty f(x)dx = \int_c^d f(x)dx + \int_d^\infty f(x)dx.$$

For instance, $\int_3^\infty \frac{\ln x}{x^2} dx$ converges and $\int_1^\infty \frac{\ln x}{x^2} dx = \int_1^3 \frac{\ln x}{x^2} dx + \int_3^\infty \frac{\ln x}{x^2} dx$. Therefore $1 - [-\frac{\ln x}{x} - \frac{1}{x}]_1^3 = \int_3^\infty \frac{\ln x}{x^2}$, which in turn leads to

$$\int_3^\infty \frac{\ln x}{x^2} dx = \frac{\ln 3}{3} + \frac{1}{3} \approx 0.69953743.$$

A comparison test

In all the examples that we have discussed so far, it was possible to determine whether $\int_a^\infty f(x)dx$ converges or diverges because we were able to solve $\int_a^b f(x)dx$ in terms of well-known elementary functions. But what could we do with $\int_1^\infty e^{-x^2}dx$? The integral $\int_1^b e^{-x^2}dx$ cannot be expressed in terms of a finite collection of elementary functions. Luckily there is a powerful result that will allow us to determine whether an improper integral converges by comparing it with another improper integral whose convergence or divergence is known.

Theorem
Let f and g be continuous functions defined on $[a,\infty)$ such that $0 \le f(x) \le g(x)$, for every $x \ge a$. Actually, the inequalities have to be true 'eventually.' That is, there exists $b > a$ such that $0 \le f(x) \le g(x)$ for all $x \ge b$. The following implications are true:

(i) If $\int_a^\infty g(x)dx$ converges then $\int_a^\infty f(x)dx$ converges;

(ii) If $\int_a^\infty f(x)dx$ diverges then $\int_a^\infty g(x)dx$ diverges.

Moreover, in case (i) the inequality $\int_a^\infty f(x)dx \le \int_a^\infty g(x)dx$ holds.

Although at this stage we are not providing a proof of the comparison test[21], its geometrical interpretation makes it plausible: if the area under $g(x)$ is finite, then it is to be expected that the area under $f(x)$ will also be finite since $f(x) \le g(x)$ on $[a,\infty)$. Similarly, if the area under $f(x)$ is infinite, it is to be expected that the area under $g(x)$ will be infinite too.

Let us return to the study of $\int_1^\infty e^{-x^2}dx$ (Figure 2.33). For $x \ge 1$ we have $x^2 \ge x$, hence $-x^2 \le -x$. Consequently, $0 \le e^{-x^2} \le e^{-x}$. Since $\int_1^\infty e^{-x}dx$ converges, the comparison test implies that $\int_1^\infty e^{-x^2}dx$ converges and

$$\int_1^\infty e^{-x^2}dx \le \int_1^\infty e^{-x}dx = \tfrac{1}{e} \approx 0.36787944.$$

Using technology, we get $\int(e^{-x^2}, x, 1, 10) \approx 0.13940279$. As a matter of fact, $\int(e^{-x^2}, x, 1, \infty) \approx 0.13940279$, exactly the same value obtained when the upper limit was 10. What happens is that the function e^{-x^2} approaches zero very fast; beyond 10, the contribution to the total area is negligible. By the way, $\int_0^\infty e^{-x^2}dx = \int_0^1 e^{-x^2}dx + \int_1^\infty e^{-x^2}dx \approx 0.74682413 + 0.13940279 = 0.88622692$. It can be proven that $\int_0^\infty e^{-x^2}dx = \sqrt{\pi}/2$, but tools from advanced calculus are needed.

Let us discuss several other examples: $\int_1^\infty \frac{\sin^2 x}{x^3+1}dx$ converges because $0 \le \frac{\sin^2 x}{x^3+1} \le \frac{1}{x^3}$, $x \ge 1$, and $\int_1^\infty \frac{1}{x^3}dx$ is convergent. Similarly, $\int_1^\infty \frac{|\cos x|}{x^2}dx$ converges because $0 \le$

---

[21]A proof can be found in the appendix at the end of the book.

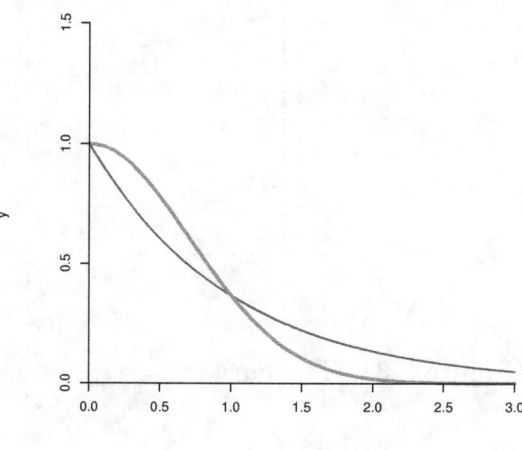

Figure 2.33: The curves $y = e^{-x^2}$ and $y = e^{-x}$

$\frac{|\cos x|}{x^2} \leq \frac{1}{x^2}$, $x \geq 1$, and $\int_1^\infty \frac{1}{x^2} dx$ is convergent. On the other hand, $\int_2^\infty \frac{1}{\ln x} dx$ diverges due to the fact[22] that $0 < \frac{1}{x} < \frac{1}{\ln x}$, $x \geq 2$, and $\int_2^\infty \frac{1}{x} dx$ diverges. However, $\int_0^\infty \frac{1}{e^x + 7} dx$ converges because $0 < \frac{1}{e^x + 7} < \frac{1}{e^x}$, $x \geq 0$, and $\int_0^\infty \frac{1}{e^x} dx$ converges.

Improper integrals over the whole real line

It is advisable to say something about integrals of the type $\int_{-\infty}^a f(x) dx$, where $f(x)$ is a continuous function defined on $(-\infty, a]$. We wish to determine whether $\lim_{b \to -\infty} \int_b^a f(x) dx$ exists. For instance, let us analyze $\int_{-\infty}^0 \frac{1}{x^2 + 1} dx$. We have

$$\int_b^0 \frac{1}{x^2 + 1} dx = [\tan^{-1} x]_b^0 = -\tan^{-1} b.$$

But $\lim_{b \to -\infty} \tan^{-1} b = -\frac{\pi}{2}$, consequently $\int_{-\infty}^0 \frac{1}{x^2 + 1} dx = \frac{\pi}{2}$. By definition,

$$\int_{-\infty}^\infty \frac{1}{x^2 + 1} dx = \int_{-\infty}^0 \frac{dx}{x^2 + 1} + \int_0^\infty \frac{dx}{x^2 + 1}.$$

Thus $\int_{-\infty}^\infty \frac{dx}{x^2 + 1} = \pi$. It is the area under the curve $y = \frac{1}{x^2 + 1}$ and the x-axis (Figure 2.34).

In general, given a continuous function $f(x)$, defined on $(-\infty, \infty)$, if $\int_{-\infty}^c f(x) dx$ and $\int_c^\infty f(x) dx$ are convergent then we will define

$$\int_{-\infty}^\infty f(x) dx = \int_{-\infty}^c f(x) dx + \int_c^\infty f(x) dx.$$

---

[22]In exercise 7, section 2.8, it is stated that $\ln x \leq x - 1$ for any $x > 0$.

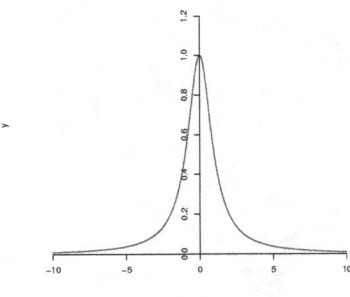

Figure 2.34: The curve $y = \frac{1}{x^2+1}$

The choice of the point c is immaterial in the sense that, for any d, if $\int_{-\infty}^{c} f(x)dx$ and $\int_{c}^{\infty} f(x)dx$ are convergent, then we can replace c by d and conclude that

$$\int_{-\infty}^{\infty} f(x)dx = \int_{-\infty}^{d} f(x)dx + \int_{d}^{\infty} f(x)dx.$$

Applications

A sensible question to ask is whether improper integrals appear in the applications of mathematics to the natural sciences. Indeed they do, as the following example from physics shows. In the first place, we pose the question about the work to be done to separate two opposite electrical charges, q_1 and q_2, from a distance d_1 to a bigger distance d_2. According to Coulomb's law we will have

$$W = k \int_{d_1}^{d_2} \frac{q_1 q_2}{x^2} dx.$$

If we wish to move q_2 and put it far away from the other charge, it is necessary to compute $k \int_{d_1}^{\infty} \frac{q_1 q_2}{x^2} dx$. Indeed,

$$W = \lim_{b \to \infty} k q_1 q_2 \int_{d_1}^{b} x^{-2} dx = k q_1 q_2 \lim_{b \to \infty} [-x^{-1}]_{d_1}^{b} = \frac{k q_1 q_2}{d_1}.$$

Improper integrals also appear in a biological context (McCarthy, 1997). Suppose that a solitary bird of a certain species leaves the nest where it was born and moves in a straight line until it finds a territorial vacancy. The probability density function (p.d.f.) $f(x) = \nu e^{-\nu x}$, $x > 0$, is considered a suitable model to represent the distance X traveled by the bird, if the vacancies are distributed in space at random ($\nu > 0$ is the parameter of the p.d.f., and is related to the abundance of vacancies). What is the mean or expected value $E(X)$? In statistics (Miller and Miller, 2004), the expected value of a random variable X, $0 < X < \infty$, is defined as

$$E(X) = \int_0^\infty x f(x) dx.$$

That is to say,

$$E(X) = \lim_{r \to \infty} \int_o^r \nu x e^{-\nu x} dx.$$

In the first place we will calculate $\int_0^r x e^{-\nu x} dx$. Let $f(x) = x$ and $g'(x) = e^{-\nu x}$; thus, $f'(x) = 1$ and $g(x) = -\frac{1}{\nu} e^{-\nu x}$. Therefore

$$\int_0^r x e^{-\nu x} dx = [-\tfrac{x}{\nu} e^{-\nu x}]_0^r + \tfrac{1}{\nu} \int_0^r e^{-\nu x} dx = -\tfrac{r}{\nu} e^{-\nu r} - \tfrac{1}{\nu^2}[e^{-\nu x}]_0^r$$

$$= -\tfrac{r}{\nu e^{\nu r}} - \tfrac{1}{\nu^2} e^{-\nu r} + \tfrac{1}{\nu^2}.$$

Using L'Hôpital's first rule we get $\lim_{r \to \infty} \frac{r}{e^{\nu r}} = 0$. Consequently

$$\lim_{r \to \infty} \int_0^r x e^{-\nu x} dx = \tfrac{1}{\nu^2}$$

which, in turn, implies $E(X) = 1/\nu$. This is an appealing result because it asserts that the mean distance that birds travel, from their place of birth, to make their first nest, is in inverse relation to the abundance of vacancies. It also suggests a way of estimating the value of the parameter ν, based on observations of several birds: $\nu = 1/(\text{average distance})$. Can we calculate the median M of the random variable X? Mathematically speaking this is a simpler task because all we have to do is solve

$$\int_0^M \nu e^{-\nu x} dx = \tfrac{1}{2}.$$

Indeed $\nu(\frac{-1}{\nu}[e^{-\nu x}]_0^M) = \tfrac{1}{2}$, which leads to $-(e^{-\nu M} - 1) = \tfrac{1}{2}$. That is to say, $e^{-\nu M} = \tfrac{1}{2}$. Taking natural logarithms to both sides we reach the answer, namely $M = \frac{\ln 2}{\nu}$. Since $2 < e$, we can conclude that $\ln 2 < 1$, thus the median is smaller than the mean. This makes sense if we look at the shape of the p.d.f. $f(x) = \nu e^{-\nu x}$. For example, let $\nu = 0.005$. Then the p.d.f. is $f(x) = 0.005 e^{-0.005x}$ (Figure 2.35), $E(X) = \frac{1}{0.005} = 200$ and $M = \frac{\ln 2}{0.005} = 138.63$. Thus, the mean distance traveled by birds, of the species being studied, is 200 meters while its median is 138.63 meters. The mean is larger than the median because the shape of the distribution is asymmetric, with a large tail to the right. The birds that travel very long distances will influence more the mean than the median.

Exercises for section 2.29

1. Determine whether $\int_0^\infty \frac{dx}{(2x+1)^2}$ converges or diverges. If it converges, find its value.

2. Show that $\int_{-\infty}^\infty x e^{-x^2} dx$ is convergent. Find its value and interpret your answer in terms of areas.

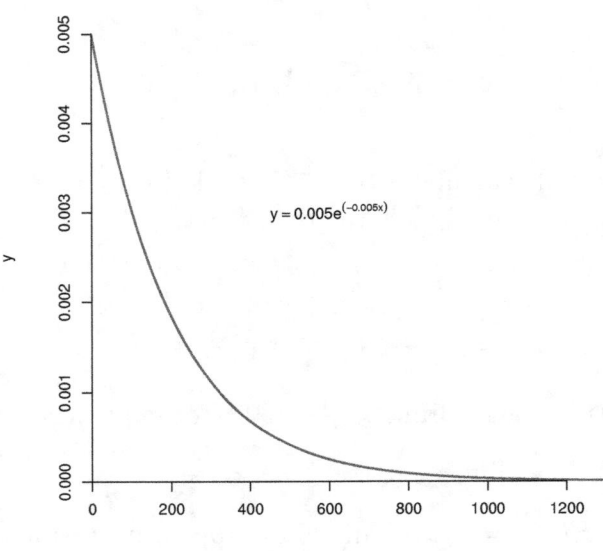

Figure 2.35: The probability density function $f(x) = 0.005e^{-0.005x}$

3. Is the improper integral $\int_0^\infty \frac{\tan^{-1} x}{x^2+1} dx$ convergent? If it converges, find its value.

4. Is $\int_{1.01}^\infty \frac{dx}{\ln x}$ convergent or divergent? (Hint: Keep in mind that $0 \le \ln x < x$ whenever $x \ge 1$.)

5. Show that $\int_1^\infty \frac{dx}{x^3+2}$ is convergent. Thereafter, use your graphing calculator to approximate its value to two decimals.

6. Is $\int_2^\infty \frac{dx}{x^3-1}$ convergent? (Hint: Observe that $x^3 - 1 > x^2$ eventually.)

7. What can you say about the convergence of $\int_0^\infty \frac{dx}{e^x+e^{-x}}$? If it is convergent, use your calculator to approximate its value up to four decimals. Can you calculate the exact value of the given improper integral?

8. We cannot calculate $\int_0^1 \ln x \, dx$ using the Evaluation Theorem because ln is not continuous on $[0, 1]$ (it is not even defined at $x = 0$). However, we can ask whether

$$\lim_{a \to 0^+} \int_a^1 \ln x \, dx$$

exists as a real number. This is a Type II improper integral. Calculate the above-mentioned limit and interpret it as an area.

9. Using a graphics calculator, we note that $\int_1^3 \frac{e^x}{x}dx = 8.0387148$, $\int_1^5 \frac{e^x}{x}dx = 38.290158$, $\int_1^{10} \frac{e^x}{x}dx = 2490.3339$. So, we might conjecture that $\int_1^\infty \frac{e^x}{x}dx$ is divergent. Test the conjecture using the comparison test.

10. Show that $\int_0^\infty \nu e^{-\nu x}dx = 1$, where $\nu > 0$. This is a property that $f(x) = \nu e^{-\nu x}$ has to fulfill in order to become a probability density function.

11. We can rotate the curve $y = e^{-x}$, $x \geq 0$, around the x-axis to obtain a solid of revolution with volume $V = \pi \int_0^\infty (e^{-x})^2 dx$. Calculate V.

12. Despite the fact that the area under the curve $y = \frac{1}{x}$, $x \geq 1$, is infinite, the solid of revolution obtained by rotating the curve around the x-axis happens to have a finite volume V. Calculate V.

2.30   Non-elementary Functions

Suppose we wish to calculate $\int \frac{e^x}{x}dx$. Integration by parts does not lead to an answer, so we might try using a graphing calculator. Surprisingly, if we write $\int(e^x/x, x)$ and press Enter , the symbol $\int(e^x/x, x)$ appears on the screen; as if the calculator refuses to provide an answer. What has happened? The function

$$F(x) = \int_a^x \frac{e^t}{t}dt$$

is well-defined for any positive x ($a > 0$ is fixed), thus FTC implies $F'(x) = \frac{e^x}{x}$. Therefore

$$\int \frac{e^x}{x}dx = F(x) + C.$$

However, $F(x)$ is not an elementary function. That is to say, $F(x)$ cannot be expressed in terms of a finite number of the usual functions (polynomials, n^{th} roots, exponential functions, logarithmic functions, power functions, trigonometric functions, and inverse trigonometric functions) through the operations of addition, subtraction, division, multiplication, and composition. In an advanced mathematics course, it can be proven that, indeed, $F(x)$ is non-elementary. Notwithstanding this fact, we can approximate the value of any definite integral $\int_a^b \frac{e^x}{x}dx$ either by using the TMA program or a graphing calculator; for instance, $\int_1^5 \frac{e^x}{x}dx = 38.2902$.

Very many indefinite integrals are non-solvable in the above sense: $\int e^{-x^2}dx$, $\int \frac{\cos x}{x}dx$, etc. In particular,

$$F(x) = \int_0^x e^{-t^2}dt$$

is an important non-elementary function linked to the error function in probability theory. In section 2.7, we noted that, thanks to FTC, $F'(x) = e^{-x^2}$ and $F''(x) = -2xe^{-x^2}$; thus, F is increasing everywhere, concave up on $(-\infty, 0)$, and concave down on $(0, \infty)$. Let us recall that one of the virtues of the fundamental theorem of calculus is its ability to define new non-elementary functions and draw consequences by taking derivatives. The function $F(x) = \int_1^x \frac{1}{t} dt$, studied in great detail in section 2.8, is a dramatic example with regard to the usefulness and central role played by FTC in calculus.

Exercises for section 2.30

1. The Fresnel Sine Function is defined by $S(x) = \sqrt{\frac{2}{\pi}} \int_0^x \sin(t^2) dt$. Calculate $S'(x)$ and $S''(x)$. Use your graphing calculator to conjecture whether $S(x)$ is an elementary function.

2. Find the solution of the initial value problem $y'(x) = \frac{2}{\sqrt{\pi}} e^{-x^2}$, $y(0) = 0$. Is the solution an elementary function? Where is the solution increasing or decreasing? Find its point of inflection and analyze its concavity. Use your calculator to make a table of values $(x, y(x))$ and guess who are going to be the two horizontal asymptotes.

3. Solve the initial value problem $y'(x) = xe^{x^3}$, $y(0) = 1$, and approximate the value of $y(0.1)$ using the TMA program.

4. Sketch the solution of the IVP $y'(x) = cos(x^2)$, $y(0) = \pi$. Use, for this purpose, a combination of 'paper and pencil' techniques and your graphing calculator.

5. The integral $\int \frac{\sin x}{x} dx$ is non-solvable. No wonder that a graphing calculator does not provide an answer when we type $\int (\sin x/x, x)$. However, an approximation to the definite integral $\int_1^2 \frac{\sin x}{x} dx$ can be found either by writing $\int (\sin x/x, x, 1, 2)$ (make sure that your calculator is set in the 'approximate' mode) or using the TMA program from section 2.3 ($n = 10$ should work fine). Compare both approximations.

2.31   Average Value of a Function

We all know that the average of n numbers $a_1, ..., a_n$ is $(a_1 + ... + a_n)/n$. Is it possible to define the average of a continuous function f defined on an interval $[a, b]$? Of course, we would like to have a definition that coincides with our intuition of what average means! Let us divide $[a, b]$ in n subintervals of equal length, namely $(b-a)/n$. Then we select a point c_i in each subinterval ($1 \le i \le n$). If n is really large, the number $(1/n)\Sigma_{i=1}^{n}f(c_i)$ will somehow represent an approximation to what we might think as the average of f. But

$$\frac{1}{n}\Sigma_{i=1}^{n}f(c_i) = \frac{1}{b-a}\Sigma_{i=1}^{n}f(c_i)\frac{b-a}{n}.$$

We realize that this is a Riemann sum corresponding to f. Therefore,

$$\lim_{n\to\infty}\frac{1}{b-a}\Sigma_{i=1}^{n}f(c_i)\frac{b-a}{n} = \frac{1}{b-a}\int_a^b f(x)dx$$

will be the 'exact' value of the average of f on the interval $[a,b]$, a number that is usually denoted by the symbol $av(f)$. Thus

$$av(f) = \frac{1}{b-a}\int_a^b f(x)dx.$$

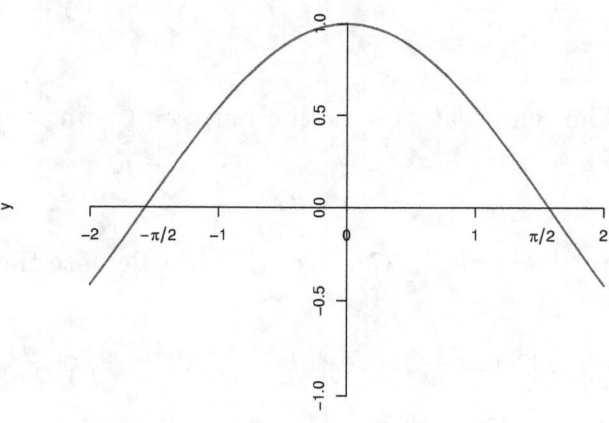

Figure 2.36: The cosine function on $[-\pi/2, \pi/2]$.

For instance, if $f(x) = x^2$, its average value on $[0,1]$ is $\int_0^1 x^2 dx = 1/3$, while its average value on $[0,2]$ is $\frac{1}{2}\int_0^2 x^2 dx = 4/3$. What is the average value of $f(x) = \cos x$ on $[-\pi/2, \pi/2]$? (Figure 2.36). We observe that

$$av(f) = \frac{1}{\frac{\pi}{2}-(-\frac{\pi}{2})}\int_{-\pi/2}^{\pi/2}\cos x dx = \frac{1}{\pi}[\sin x]_{-\pi/2}^{\pi/2} = \frac{2}{\pi}.$$

Suppose that the temperature (in degrees Celsius) in a certain day and place is given by

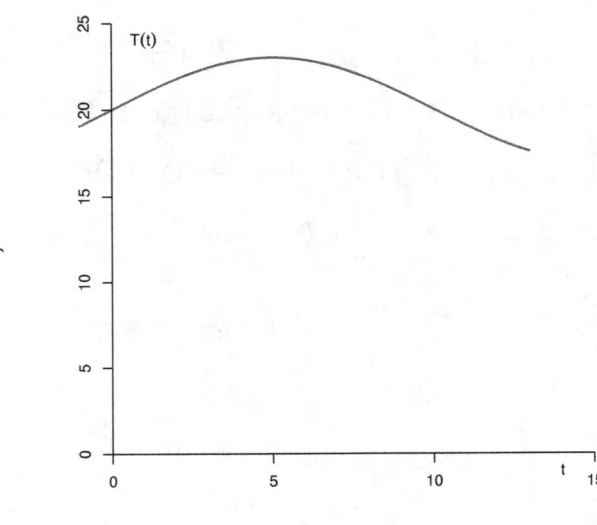

Figure 2.37: The temperature function between 6 a.m. and 7 p.m.

$$T(t) = 20 + 3\sin\tfrac{\pi t}{10}$$

between 6 a.m. and 7 p.m. (Figure 2.37). We wish to calculate the average temperature. Indeed,

$$av(T) = \tfrac{1}{13}\int_0^{13}(20 + 3\sin\tfrac{\pi t}{10})dt = \tfrac{1}{13}[20t - \tfrac{30}{\pi}\cos\tfrac{\pi t}{10}]_0^{13}$$

$$= \tfrac{1}{13}(260 - \tfrac{30}{\pi}\cos\tfrac{13\pi}{10} + \tfrac{30}{\pi}) \approx 21.17°.$$

Analysis of plant growth

As we will see next, the idea of average of a function plays a crucial role in the analysis of plant growth (Causton, 1977). Let $W(t)$ denote the dry mass of a plant at any instant t. While $\tfrac{dW}{dt}$ is the growth rate, we are particularly interested in

$$R(t) = \tfrac{1}{W(t)}\tfrac{dW}{dt},$$

which is the **relative growth rate**. Since neither $W(t)$ nor $\tfrac{dW}{dt}$ are usually known explicitly, there is no way to determine $R(t)$. Thus, we have to settle with something less ambitious, namely the average of $R(t)$ between t_1 and t_2 ($t_1 < t_2$):

$$av(R) = \tfrac{1}{t_2-t_1}\int_{t_1}^{t_2} R(t)dt.$$

Then

$$av(R) = \frac{1}{t_2-t_1} \int_{t_1}^{t_2} \frac{W'(t)}{W(t)} dt = \frac{1}{t_2-t_1}[\ln W(t)]_{t_1}^{t_2} = \frac{1}{t_2-t_1}(\ln W_2 - \ln W_1)$$

where $W_2 = W(t_2)$ and $W_1 = W(t_1)$. Although one has to destroy a plant in order to measure its dry mass, the use of several plants from the same species allows us to determine W_1 and W_2.

Through photosynthesis, plants increase their dry mass by converting carbon dioxide and water into sugar and oxygen. Sugar is, in turn, stored in the form of ATP, the main fuel of a cell. Minerals are incorporated into the plant through the root, but their contribution to the total dry mass of the plant is considerably less than the mass created by photosynthesis.

The process of photosynthesis takes place through the leaves and, as expected, the relative growth rate will depend on $L(t)$ (leaf area of foliage[23]), $E(t)$ (the net assimilation rate), and $W(t)$. As a matter of fact, $E(t)$ is *defined* by the equality

$$E(t) = \frac{W'(t)}{L(t)}.$$

The net assimilation rate of a plant is a measure of how efficient its leaves are in the process of photosynthesis.

Although $L(t)$ can be found, $E(t)$ cannot because it involves $W'(t)$. However, we can try to calculate the average of $E(t)$ between t_1 and t_2 ($t_1 < t_2$):

$$av(E) = \frac{1}{t_2-t_1} \int_{t_1}^{t_2} E(t)dt = \frac{1}{t_2-t_1} \int_{t_1}^{t_2} \frac{W'(t)}{L(t)} dt.$$

This integral cannot be evaluated as it stands unless we assume some relationship between $W(t)$ and $L(t)$. The first thing that comes to mind is a linear relationship. That is to say, $W(t) = a + bL(t)$ for some constants a, b. Then $W'(t) = bL'(t)$, which in turn leads to

$$av(E) = \frac{1}{t_2-t_1} \int_{t_1}^{t_2} \frac{bL'(t)}{L(t)} dt = \frac{b}{t_2-t_1}[\ln L(t)]_{t_1}^{t_2} = \frac{b}{t_2-t_1}(\ln L_2 - \ln L_1)$$

where $L_2 = L(t_2)$, $L_1 = L(t_1)$. Still, we need to calculate b, an easy task since $W_2 = a + bL_2$, $W_1 = a + bL_1$. Thus $b = \frac{W_2-W_1}{L_2-L_1}$, and the average net assimilation rate can then be found from two measurements of W and L.

---

[23]Sometimes plant physiologists deal with both the concept of leaf mass and leaf area; under these circumstances, one would have to use different symbols for each of them.

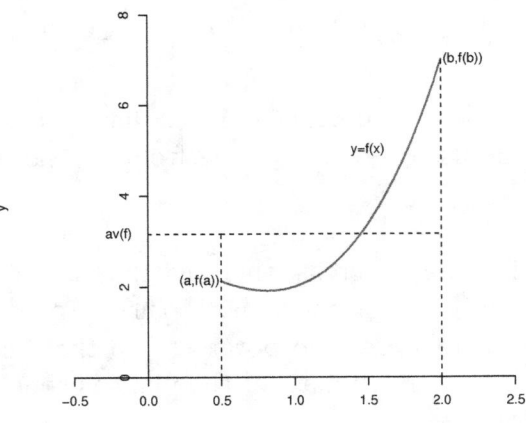

Figure 2.38: The geometrical interpretation of the average of a function

Remark

Suppose f is continuous on $[a, b]$. Then there exists θ in (a, b) such that $av(f) = f(\theta)$. Indeed[24], FTC asserts that there exists an antiderivative F of f on $[a, b]$. Then the Evaluation Theorem implies that

$$av(f) = \frac{1}{b-a} \int_a^b f(x)dx = \frac{1}{b-a}[F(x)]_a^b = \frac{1}{b-a}(F(b) - F(a)).$$

But the Mean Value Theorem implies that there exists θ in (a, b) such that $F(b) - F(a) = F'(\theta)(b - a) = f(\theta)(b - a)$. Consequently, $av(f) = f(\theta)$.

If f is not only continuous on $[a, b]$ but also positive, we may note (see Figure 2.38) that the area under f between a and b is equal to the area of the rectangle with sides of length $b - a$ and $av(f)$.

Exercises for section 2.31

1. Suppose that the temperature (in degrees Celsius) on a certain day and place is given by $T(t) = 22 + 2\sin\frac{\pi t}{9}$, between 10 a.m. and 8 p.m. Calculate the average temperature.

2. Calculate the average relative growth rate of a plant if two measurements, made four days apart, provide the values 2.7 kg. and 3.2 kg.

---

[24]This result was to be expected (see remark at the end of section 2.7).

3. Let us assume that $f(t) = 10 + 15e^{-0.3t}\sin(1.7t)$ describes the concentration of a human hormone after a certain drug is administered through a vein. The concentration of the hormone exhibits a damped oscillatory behavior. Using technology, find the average concentration of the hormone between $t = 0$ and $t = 6$ hours.

4. Calculate the average net assimilation rate of a plant if two measurements, made four days apart, provide the following values for mass and leaf area of foliage: 2.7 kg. and 89 cm^2, 3.2 kg. and 95 cm^2.

5. The Lotka-Volterra equations, namely

$$\frac{dx}{dt} = \alpha x - \beta xy$$

$$\frac{dy}{dt} = \gamma xy - \delta y$$

are used to analyze **prey-predator interactions** (Burghes and Borrie, 1981); x denotes the prey and y the predator ($\alpha, \beta, \gamma, \delta$ are the parameters of the Lotka-Volterra model). There is no easy way to solve the given system of differential equations and obtain an expression for $x(t)$ and an expression for $y(t)$. However, using tools from the qualitative theory of differential equations, it can be proven that both $x(t)$ and $y(t)$ exhibit a cyclic behavior (with the predator population 'lagging behind' the prey population). Let p be the common period of the cycle of prey and predator, and let

$$av(x) = \frac{1}{p}\int_0^p x(t)dt, \quad av(y) = \frac{1}{p}\int_0^p y(t)dt,$$

that is, $av(x)$ and $av(y)$ are the average values of the prey and the predator across a cycle, respectively. Show that $av(x) = \frac{\delta}{\gamma}$ and $av(y) = \frac{\alpha}{\beta}$ (Hint: Note that $x(p) = x(0)$ and $y(p) = y(0)$ because p is the period of the cycle).

Chapter 3

Series

3.1 Revisiting Sequences

Let us revisit the idea of limit of a sequence. We would like to point out that, strictly speaking, a sequence (a_n) of real numbers is also a function. It is a correspondence between each natural number n and a real number a_n. In other words, a sequence (a_n) happens to be a function $a : \mathbf{N} \to \Re$, where $\mathbf{N}$ denotes the set of natural numbers. Nonetheless, we use the word 'function' exclusively when the domain under consideration is an interval of the real line. Sequences are important on their own right and thus deserve to be singled out for study.

We say that a sequence (a_n) converges to a limit L if and only if given any $\epsilon > 0$ there exists N such that $|a_n - L| < \epsilon$ whenever $n \geq N$. One must stress that this statement resembles the definition of limit of a function $f : [a, \infty) \to \Re$ when $x \to \infty$. Actually, $\lim_{x\to\infty} f(x) = L$ if and only if for every $\epsilon > 0$ there exists M such that $|f(x) - L| < \epsilon$ whenever $x \geq M$. No wonder that if $\lim_{x\to\infty} f(x) = L$ then $\lim_{n\to\infty} f(n) = L$. This simple fact allows us to use L'Hôpital's rule to show, for example, that $\lim_{n\to\infty} \frac{\ln n}{n} = 0$. Indeed,

$$\lim_{x\to\infty} \tfrac{\ln x}{x} = \lim_{x\to\infty} \tfrac{1}{x} = 0.$$

Hence $\lim_{n\to\infty} \frac{\ln n}{n} = 0$. On the other hand, if $x > 0$ then $\lim_{n\to\infty} x^{\frac{1}{n}} = 1$ because $x^{\frac{1}{n}} = e^{\frac{1}{n}\ln x}$ and $\lim_{n\to\infty} \frac{1}{n}\ln x = 0$. Then the continuity of the exponential function implies that

$$\lim_{n\to\infty} e^{\frac{1}{n}\ln x} = e^{\lim_{n\to\infty}\frac{1}{n}\ln x} = e^0 = 1.$$

In a similar fashion, $n^{\frac{1}{n}} = e^{\frac{1}{n}\ln n}$. Since $\lim_{n\to\infty} \frac{1}{n}\ln n = 0$ we can conclude that

$$\lim_{n\to\infty} e^{\frac{1}{n}\ln n} = e^0 = 1.$$

How about $\lim_{n\to\infty} x^n$ whenever $|x| < 1$? Our intuition tells us that this limit will be zero, a fact that was used in section 2.1 when discussing geometric series. A proof can be provided: given any $\epsilon > 0$, since $\epsilon^{1/n} \to 1$ and $|x| < 1$, thanks to exercise 16, section 2.1, there exists N such that $|x| < \epsilon^{1/n}$ if $n \geq N$. Thus, $|x^n - 0| = |x|^n < \epsilon$ whenever $n \geq N$.

Let us put together the four limits of sequences that we have just obtained:

1. $\lim_{n\to\infty} \frac{\ln n}{n} = 0$.

2. $\lim_{n\to\infty} x^{1/n} = 1$, $x > 0$.

3. $\lim_{n\to\infty} n^{1/n} = 1$.

4. $\lim_{n\to\infty} x^n = 0$ whenever $|x| < 1$.

We should add to this list a fifth entry, one that was proven in section 2.9:

5. $\lim_{n\to\infty} (1 + \frac{x}{n})^n = e^x$.

Examples of convergent sequences

1. Obviously, $\lim_{n\to\infty} 3^{1/n} = 1$ while $\lim_{n\to\infty} (\frac{1}{5})^n = 0$.

2. Suppose that we are asked to calculate $\lim_{n\to\infty} \frac{\ln(n+7)}{\sqrt{n}}$. Applying L'Hôpital's rule twice we obtain:

$$\lim_{x\to\infty} \frac{\ln(x+7)}{\sqrt{x}} = \lim_{x\to\infty} \frac{\frac{1}{x+7}}{\frac{1}{2\sqrt{x}}} = \lim_{x\to\infty} \frac{2\sqrt{x}}{x+7} = 2\lim_{x\to\infty} \frac{1}{2\sqrt{x}} = 0.$$

   Hence $\lim_{n\to\infty} \frac{\ln(n+7)}{\sqrt{n}} = 0$.

3. A third example does not require the use of L'Hôpital's rule:

$$\lim_{n\to\infty} (\frac{2}{n})^{1/n} = \lim_{n\to\infty} \frac{2^{1/n}}{n^{1/n}} = \frac{\lim_{n\to\infty} 2^{1/n}}{\lim_{n\to\infty} n^{1/n}} = \frac{1}{1} = 1.$$

Divergent sequences

A sequence (a_n) is said to be **divergent** if it is not convergent. A particularly important class of divergent sequences are those that diverge to infinity in the sense that given any $M > 0$ there exists a natural number N such that if $n > N$ then $a_n > M$. Under these circumstances we will write $\lim_{n\to\infty} a_n = \infty$. Recall that in section 1.5 we introduced the symbol $\lim_{x\to\infty} f(x) = \infty$ to mean that for every $M > 0$ there exists p such that if $x > p$ then $f(x) > M$. Evidently, $\lim_{x\to\infty} f(x) = \infty$ implies $\lim_{n\to\infty} f(n) = \infty$. Thus, $\lim_{n\to\infty} 2^n = \infty$, $\lim_{n\to\infty} \ln n = \infty$, etc. It should be noted that the sequence $((-1)^n)$ neither converges nor diverges to infinity. It just diverges due to the fact that the sequence under discussion simply oscillates between -1 and 1.

Examples of divergent sequences

1. If $q > 0$ then we can conclude that $\lim_{n \to \infty} n^q = \infty$ because $n^q = e^{q \ln n}$ and $\lim_{n \to \infty} q \ln n = \infty$.

2. What can be said about $\lim_{n \to \infty} \frac{2^n}{n^3}$? Applying L'Hôpital's rule three times we get:

$$\lim_{x \to \infty} \frac{2^x}{x^3} = \lim_{x \to \infty} \frac{2^x \ln 2}{3x^2} = \lim_{x \to \infty} \frac{2^x (\ln 2)^2}{6x} = \lim_{x \to \infty} \frac{2^x (\ln 2)^3}{6}.$$

But $\lim_{x \to \infty} \frac{2^x (\ln 2)^3}{6} = \infty$, consequently $\lim_{n \to \infty} \frac{2^n}{n^3} = \infty$.

Ten properties of convergent sequences: a summary

Let us summarize the ten main properties of sequences, most of which were discussed in section 2.1. Given convergent sequences (a_n), (b_n), the following properties are true:

1. $\lim_{n \to \infty} (a_n + b_n) = \lim_{n \to \infty} a_n + \lim_{n \to \infty} b_n$.

2. $\lim_{n \to \infty} (a_n - b_n) = \lim_{n \to \infty} a_n - \lim_{n \to \infty} b_n$.

3. $\lim_{n \to \infty} (a_n b_n) = \lim_{n \to \infty} a_n * \lim_{n \to \infty} b_n$.

4. $\lim_{n \to \infty} \frac{a_n}{b_n} = \frac{\lim_{n \to \infty} a_n}{\lim_{n \to \infty} b_n}$, provided $b_n \neq 0$ for all n and $\lim_{n \to \infty} b_n \neq 0$.

5. If $a_n \leq b_n$ for all n (or eventually[1]) then $\lim_{n \to \infty} a_n \leq \lim_{n \to \infty} b_n$.

6. If $a_n \leq c_n \leq b_n$ for all n (or eventually) and $\lim_{n \to \infty} a_n = \lim_{n \to \infty} b_n$, then (c_n) converges. Moreover, $\lim_{n \to \infty} c_n = \lim_{n \to \infty} a_n = \lim_{n \to \infty} b_n$.

7. Every convergent sequence (a_n) is bounded, that is, there exists H such that $|a_n| \leq H$ for all n.

8. If $a_n \to L$ and $L < M$ then $a_n < M$ eventually. Similarly, if $a_n \to L$ and $M < L$ then $a_n > M$ eventually.

9. $\lim_{n \to \infty} f(a_n) = f(\lim_{n \to \infty} a_n)$ whenever f is continuous at the number $\lim_{n \to \infty} a_n$ and the terms of the sequence lie on the domain of the function f.

10. If $a_n \to \infty$ then $\frac{1}{a_n} \to 0$. Similarly, if $a_n \to 0$ then $\frac{1}{a_n} \to \infty$.

---

[1]In the context of the study of sequences, the word 'eventually' means that there exists N such that the given property is valid for all $n \geq N$.

Proofs of these ten properties can be found either in section 2.1 or in an appendix at the end of the book. It should be noted that this list does not exhaust all the properties that might be needed in a particular situation. For instance, if asked to calculate $\lim_{n\to\infty} \frac{e^n}{2^{1/n}}$, we might realize that since $\lim_{n\to\infty} e^n = \infty$ and $\lim_{n\to\infty} 2^{1/n} = 1$, the limit of the quotient probably is infinity. Could we justify such a claim? Well, our task is to show that if $a_n \to \infty$ and $b_n \to L > 0$ (both are positive sequences) then $\frac{a_n}{b_n} \to \infty$. Indeed, since (b_n) is convergent, it has to be bounded; thus, there exists $H > 0$ such that $b_n < H$ for all n. Given any $M > 0$, since $a_n \to \infty$ we can assert that $a_n > MH$ eventually. But $\frac{1}{b_n} > \frac{1}{H}$ for all n. Hence $\frac{a_n}{b_n} > MH\frac{1}{H} = M$ eventually. Therefore $a_n/b_n \to \infty$. Thus, we can assert with full confidence that $\lim_{n\to\infty} e^n/2^{1/n} = \infty$.

Digging deeper into the structure of $\Re$

We need to go deeper into the structure of the real number system in order to study several important tests in the theory of series. The Completeness Axiom for the real numbers states that any non-empty set of real numbers that is bounded above does have a least upper bound. In symbols, if A is a non-empty set among the reals and $a \le M$ for every a in A (for some number M), there exists c in $\Re$ such that:

(i) $a \le c$ for every a in A.
(ii) Given any $\epsilon > 0$ there exists a in A such that $a > c - \epsilon$.

If a number c satisfies (i) and (ii) then it is certainly unique, and we will write $c = l.u.b.\ A$. The reader may note that (ii) is equivalent to saying that no number less than c can be an upper bound of A. It should be noted that the rational numbers do not have the l.u.b. property despite the fact that they are dense in $\Re$, in the sense that between any two real numbers there is a rational number.

Proposition 1

Let (s_n) be an increasing sequence, bounded above. Then (s_n) converges, indeed $s_n \to L$ where L is the least upper bound of the set of values of the sequence. That is, $L = $ l.u.b $\{s_n : n \ge 1\}$.

The proof is rather straightforward: Let $\epsilon > 0$. Since $L - \epsilon$ is not an upper bound of (s_n) we can assert that there exists N such that $s_N > L - \epsilon$. But (s_n) is increasing, thus $s_n > s_N$ whenever $n > N$. Therefore $L - \epsilon < s_n < L + \epsilon$ for $n > N$, that is, $s_n \to L$. QED

Interestingly, as we will see next, the preceding proposition has its counterpart for functions defined on an interval of the type (b, ∞). Proposition 1a will be needed to prove the integral test in section 3.9.

Proposition 1a

Let $F : (b, \infty) \to \Re$ be increasing and bounded above[2]. Then $\lim_{t\to\infty} F(t) = L$, where $L=$ l.u.b.$\{F(t) : t > b\}$.

We have to 'mimic' the proof of proposition 1. Given any $\epsilon > 0$, since $L - \epsilon$ cannot be an upper bound of $\{F(t) : t > b\}$, there exists $\hat{t} > b$ such that $F(\hat{t}) > L - \epsilon$. But F is increasing, hence $F(t) > F(\hat{t}) > L - \epsilon$ whenever $t > \hat{t}$. Therefore,

$$L - \epsilon < F(t) \leq L < L + \epsilon$$

if $t > \hat{t}$. Thus, $\lim_{t\to\infty} F(t) = L$. QED

We can prove that if a set of real numbers is bounded below then it has a greatest lower bound (see exercise 21 at the end of the section). That is to say, if A is a non-empty set of real numbers and $L \leq a$ for every a in A (for some number L), there exists a unique real number d (denoted g.l.b.A, the 'greatest lower bound' of A) such that:

(i) $d \leq a$ for every a in A.
(ii) Given any $\epsilon > 0$ there exists a in A such that $a < d + \epsilon$.

Property (ii) means that no number bigger than d can be a lower bound of A. Similar to proposition 1 we have the following result, which will be used when discussing alternating series. Its proof is left to the reader.

Proposition 2

. Every bounded below decreasing sequence (s_n) converges and does so to g.l.b. $\{s_n : n \geq 1\}$, that is, if there exists L such that $L \leq s_n$ for every n and (s_n) is decreasing, then (s_n) converges to g.l.b. $\{s_n : n \geq 1\}$.

The factorial of a number

Given any positive integer n we define $n! = n*(n-1)*(n-2)...*3*2*1$. The symbol $n!$ is known as the **factorial** of n and, as we will see later on, plays an important role in the theory of series. Rather unexpectedly, it is intimately linked to the exponential function and Taylor series in general (section 3.12).

The factorial of a number increases dramatically as the number increases. For instance, $5! = 5*4*3*2 = 120$ while $8! = 8*7*6*5! = 40,320$. We might wonder whether the factorial increases more rapidly than the power of any number. Actually, we will see that, given any number x,

---

[2]The function could well be defined on an interval of the type $[b, \infty)$.

$$\lim_{n \to \infty} \frac{x^n}{n!} = 0.$$

With this purpose in mind let us establish a result that is interesting in its own right.

Proposition 3

Let (a_n) be a sequence of non-zero terms such that $\lim_{n \to \infty} \left| \frac{a_{n+1}}{a_n} \right| = L$, where $0 \leq L < 1$. Then $a_n \to 0$.

A proof goes as follows: choose c such that $L < c < 1$. Since $\left| \frac{a_{n+1}}{a_n} \right| \to L$, the eighth property of sequences asserts that there exists N such that

$$\left| \frac{a_{n+1}}{a_n} \right| < c \text{ whenever } n \geq N.$$

Hence $|a_{N+1}| < c|a_N|$. In a similar fashion, we also get $\left| \frac{a_{N+2}}{a_{N+1}} \right| < c$; thus, $|a_{N+2}| < c|a_{N+1}| < c^2|a_N|$ and, in general,

$$|a_{N+n}| < c^n |a_N|.$$

Since $0 < c < 1$ it follows that $c^n \to 0$. Hence $a_{N+n} \to 0$, which in turn implies that $a_n \to 0$ as we wished to prove. Interestingly enough, the method of proof employed foreshadows what will be learned in the context of the ratio test, one of the most important topics in the theory of series. QED

Let us apply the above-mentioned proposition to the sequence $\left(\frac{x^n}{n!} \right)$:

$$\frac{\left| \frac{x^{n+1}}{(n+1)!} \right|}{\left| \frac{x^n}{n!} \right|} = \frac{n!}{(n+1)!} |x| = \frac{1}{n+1} |x| \to 0.$$

Then $\frac{x^n}{n!} \to 0$. This result allows us to conclude, for instance, that

$$\lim_{n \to \infty} \frac{2^n * 5^n}{n!} = \lim_{n \to \infty} \frac{(2*5)^n}{n!} = \lim_{n \to \infty} \frac{10^n}{n!} = 0$$

and

$$\lim_{n \to \infty} \frac{5^{2n}}{n!} = \lim_{n \to \infty} \frac{(5^2)^n}{n!} = 0.$$

The reader might meet the challenge of proving that if $\left| \frac{a_{n+1}}{a_n} \right| \to L$, where $L > 1$ or $L = \infty$, then $a_n \to \infty$.

Remark

When working with power series (section 3.11) we will need to define $0! = 1$. Similarly, the symbol 0^0 will denote the number 1.

Exercises for section 3.1

Determine whether each of the following eighteen sequences is convergent or divergent. If it converges, find the limit.

1. $a_n = (-\frac{1}{4})^n$.

2. $a_n = \frac{\sin n}{n^2}$.

3. $a_n = (1 - \frac{2}{n})^n$.

4. $a_n = \ln(1 + \frac{3}{n})^n$.

5. $a_n = \frac{\arctan n}{n}$.

6. $a_n = \frac{\ln n}{\ln(5n)}$.

7. $a_n = \ln n - \ln(n + 6)$.

8. $a_n = \frac{n+1}{n^2+3n+1}$.

9. $a_n = (n^3)^{1/n}$.

10. $a_n = (2n)^{1/n}$.

11. $a_n = \frac{\ln(4n)}{n^{1/n}}$.

12. $a_n = \frac{\ln n}{\sqrt{n}}$.

13. $a_n = \frac{e^{1/n}}{n}$.

14. $a_n = \sqrt{n + 3} - \sqrt{n}$.

15. $a_n = \frac{3^n}{e^n + 7}$.

16. $\frac{n!}{5^n}$.

17. $\frac{n^3}{n!}$.

18. $\frac{(\ln n)^2}{n!}$.

19. At the beginning of the section we saw that $\lim_{x \to \infty} f(x) = L$ implies $\lim_{r \to \infty} f(n) = L$. Is the converse true?

20. Show that if (a_n) is a sequence of non-zero terms and $|\frac{a_{n+1}}{a_n}| \to L$, where $L > 1$ or $L = \infty$, then $a_n \to \infty$.

21. Show that if a set A of real numbers is bounded below then g.l.b.A exists (Hint: Use the l.u.b. property on the set $B = \{-a : a \text{ in } A\}$.)

3.2   Basic Characteristics of Series

In section 2.1 the idea of a geometric series was introduced: given any number r, the sequence (s_n) is built, where

$$s_n = 1 + r + \ldots + r^n.$$

That is, $s_n = \Sigma_{i=0}^n r^i$. The sequence (s_n), called the 'series' determined by the sequence (r^n), is customarily denoted $\Sigma_{n,0} r^n$. As we know (recall (2.5)), if $|r| < 1$ the geometric series converges. Moreover, $\Sigma_{n=0}^\infty r^n = \frac{1}{1-r}$ if $|r| < 1$, where $\Sigma_{n=0}^\infty r^n$ denotes the limit of the series $\Sigma_{n,0} r^n$. Since

$$1 + r + r^2 + \ldots + r^n = \frac{1-r^{n+1}}{1-r} = \frac{1}{1-r} - \frac{r^{n+1}}{1-r} = \frac{1}{1-r} - \frac{r^{n+1}}{1-r}$$

we can see that the geometric series does not converge whenever $|r| > 1$ because, under these circumstances, $r^{n+1} \to \infty$. When $r = 1$, obviously $s_n = 1 + \ldots + 1 = n$, which tends to infinity, while if $r = -1$ the corresponding series $\Sigma_{n,0}(-1)^n$ oscillates between 1 and 0. Thus, it does not converge either. It should be mentioned that $\Sigma_{n,1} r^n$, $|r| < 1$, converges too. Moreover,

$$\Sigma_{n=1}^\infty r^n = \frac{1}{1-r} - 1 = \frac{r}{1-r}.$$

Although geometric series are quite important, there are many series that are not geometric, for instance $\Sigma_{n,1}\frac{1}{n}$, $\Sigma_{n,1}\frac{1}{n(n+1)}$, $\Sigma_{n,1}\frac{1}{n^2}$, $\Sigma_{n,1}\frac{\log n}{n^2}$, etc. In general, given any sequence (a_n), called the 'sequence of terms', we build the sequence (s_n) where $s_n = a_1 + \ldots + a_n$. The latter is called the 'sequence of partial sums' or the series determined by the sequence (a_n), and is denoted by the symbol $\Sigma_{n,1} a_n$. If it converges, its limit is written $\Sigma_{n=1}^\infty a_n$. Sometimes it is convenient to denote the series $\Sigma_{n,1} a_n$ by the symbol $a_1 + \ldots + a_n + \ldots$. For instance,

$$1 + \tfrac{1}{2} + \tfrac{1}{3} + \ldots + \tfrac{1}{n} + \ldots$$

is another way to denote the series $\Sigma_{n,1}\frac{1}{n}$.

The bouncing ball

Suppose that a rubber ball starts from rest at a height h, reaches the bottom and bounces back. The ball will reach a height smaller than h and then it will go down and bounce again. This process is repeated until the ball comes to a stop. Disregarding the resistance of air, we would like to find the total time of the whole process.

In the first place, we need to calculate the time, t_1, it takes to reach the bottom from the original height h. From (1.25) we get

$$0 = -\tfrac{1}{2}gt^2 + h.$$

Hence $t_1 = \sqrt{2h/g}$. When the ball touches the bottom, its velocity is obtained using (1.24):

$$v\left(\sqrt{\tfrac{2h}{g}}\right) = -g\sqrt{\tfrac{2h}{g}} = -\sqrt{2gh}.$$

Then the ball bounces back with a positive velocity $\alpha\sqrt{2gh}$, where $0 < \alpha < 1$ (the parameter α depends on the material used to manufacture the rubber ball; we are assuming that α is independent of the velocity at which the ball hits the bottom). The time, t_2, that the ball needs to reach its highest point will be obtained from (1.24). Indeed

$$0 = -gt_2 + \alpha\sqrt{2gh}.$$

Hence

$$t_2 = \frac{\alpha\sqrt{2gh}}{g} = \alpha\sqrt{\tfrac{2h}{g}}.$$

The time, t_3, it takes the ball from the top to the bottom is the same as t_2. Thus, the total time since the ball was launched from a height h until it reaches the bottom for a second time, happens to be

$$t_1 + t_2 + t_3 = \sqrt{\tfrac{2h}{g}} + 2\alpha\sqrt{\tfrac{2h}{g}}.$$

The rubber ball touches the bottom, for a second time, with a velocity obtained from (1.24) using t_2, namely $-g\alpha\sqrt{2h/g}$, that is, $-\alpha\sqrt{2gh}$. Then it bounces back with a velocity $\alpha(\alpha\sqrt{2gh})$; that is to say, $\alpha^2\sqrt{2gh}$. By now, a pattern has been found: the total time will be

$$\sqrt{\tfrac{2h}{g}} + 2\alpha\sqrt{\tfrac{2h}{g}} + 2\alpha^2\sqrt{\tfrac{2h}{g}} + \dots.$$

That is to say,

$$\sqrt{\tfrac{2h}{g}}\left(1 + 2(\alpha + \alpha^2 + \dots)\right).$$

We are dealing with the geometric series $\alpha + \alpha^2 + \dots$, whose sum is $\frac{\alpha}{1-\alpha}$. Therefore, the total time it takes the ball to finally be at rest is

$$\sqrt{\tfrac{2h}{g}}\left(1 + \tfrac{2\alpha}{1-\alpha}\right)$$

or, what is the same,

$$\sqrt{\tfrac{2h}{g}}\,\tfrac{1+\alpha}{1-\alpha}.$$

Remark

In many calculus textbooks the symbol $\Sigma_{n=1}^{\infty} a_n$ is used to denote both a series and its limit (when $\lim_{n\to\infty} s_n$ exists). This double meaning causes some confusion to beginners; thus, we prefer to keep the symbols $\Sigma_{n,1} a_n$ and $\Sigma_{n=1}^{\infty} a_n$. Once students feel comfortable with series, it is not necessary to use the symbol $\Sigma_{n,1} a_n$ anymore.

Examples of series

1. Let us analyze the series $\Sigma_{n,1} \frac{1}{n(n+1)}$. We have to pay close attention to the corresponding sequence of partial sums (s_n), where

$$s_n = \tfrac{1}{1(1+1)} + \tfrac{1}{2(2+1)} + ... + \tfrac{1}{n(n+1)}.$$

But $\frac{1}{n(n+1)} = \frac{1}{n} - \frac{1}{n+1}$ for any n$\geq$ 1. Thus,

$$s_n = \left(1 - \tfrac{1}{2}\right) + \left(\tfrac{1}{2} - \tfrac{1}{3}\right) + ... + \left(\tfrac{1}{n} - \tfrac{1}{n+1}\right) = 1 - \tfrac{1}{n+1}.$$

Therefore $\lim_{n\to\infty} s_n = 1$. That is, $\Sigma_{n=1}^{\infty} \frac{1}{n(n+1)} = 1$.

2. To what value does the series $\Sigma_{n,1}[\tan^{-1}(n+1) - \tan^{-1} n]$ converge? We observe that

$$s_n = \tan^{-1} 2 - \tan^{-1} 1 + \tan^{-1} 3 - \tan^{-1} 2 + ... + \tan^{-1}(n+1) - \tan^{-1} n = \tan^{-1}(n+1) - \tan^{-1} 1.$$

But $\lim_{n\to\infty} \tan^{-1}(n+1) = \pi/2$ and $\tan^{-1} 1 = \pi/4$. Hence $\lim_{n\to\infty} s_n = \pi/2 - \pi/4$, that is,

$$\Sigma_{n=1}^{\infty}[\tan^{-1}(n+1) - \tan^{-1}(n)] = \tfrac{\pi}{4}.$$

The last two series are quite particular, so much so that they belong to the class of 'telescoping series' whose very structure (all the terms of the series cancel out, except the last and first) permits a simple analysis of convergence.

Not every telescoping series converges. Indeed, let us analyze the series $\Sigma_{n,1}[\ln(n+1) - \ln n]$. We have

$$s_n = \ln(2) - \ln(1) + \ln(3) - \ln(2) + ... + \ln(n+1) - \ln(n) = \ln(n+1).$$

But $\ln(n+1) \to \infty$ as $n \to \infty$. So, the series diverges.

Table 3.1: The sequence of partial sums corresponding to $\Sigma_{n,1}\frac{1}{n}$

m	$\sum_{n=1}^{m}\frac{1}{n}$
10	2.9289683
100	5.1873775
1000	7.4854709
10,000	9.787606

3. For what values of x is the geometric series $\Sigma_{n,0}\cos^n x$ convergent? This will happen whenever $|\cos x| < 1$. From basic trigonometry we know that $|\cos x| \leq 1$ and $|\cos x| = 1$, that is, $\cos x = 1$ or $\cos x = -1$, whenever $x = n\pi$ (n is any integer). Thus, $\Sigma_{n,0}\cos^n x$ converges for any real number x that is *not* of the form $n\pi$.

4. Repeated decimals are a well-known example of geometric series. For instance, $0.99999...$ represents the series $\frac{9}{10} + \frac{9}{10^2} + \frac{9}{10^3} + ...$, that is, the geometric series $\frac{9}{10}(1 + \frac{1}{10} + \frac{1}{10^2} + \frac{1}{10^3} + ...)$, whose value is

$$\frac{9}{10}\frac{1}{1-\frac{1}{10}} = \frac{9}{10}\frac{10}{9} = 1.$$

Similarly, $0.4767676...$ represents the geometric series $0.4 + \frac{76}{10^3} + \frac{76}{10^5} + \frac{76}{10^7} + ... = \frac{4}{10} + \frac{76}{10^3}(1 + \frac{1}{10^2} + \frac{1}{10^4} + ...)$ whose value is

$$\frac{4}{10} + \frac{76}{10^3}\left(\frac{1}{1-\frac{1}{10^2}}\right) = \frac{4}{10} + \frac{76}{10^3}\frac{10^2}{99} = \frac{4}{10} + \frac{76}{990} = \frac{396+76}{990} = \frac{236}{495}.$$

5. One of the simplest imaginable series is $\Sigma_{n,1}\frac{1}{n}$, called the **harmonic series**. We notice that

$$s_n = 1 + \frac{1}{2} + ... + \frac{1}{n}$$

and since $\frac{1}{n} \to 0$ our intuition might lead us to believe that the series converges, but a graphing calculator provides some disturbing news (see table 3.1). As expected, the values of the sums increase; however, they do not seem to converge to any number. Actually, if we try to find the value of $\Sigma(\frac{1}{n}, n, 1, \infty)$, our calculator will not provide an answer either in the approximate mode or the exact mode. These numerical explorations suggest that the series under study is not convergent. Indeed (see Figure 3.1)

$$1 > \int_1^2 \frac{1}{x}dx,$$

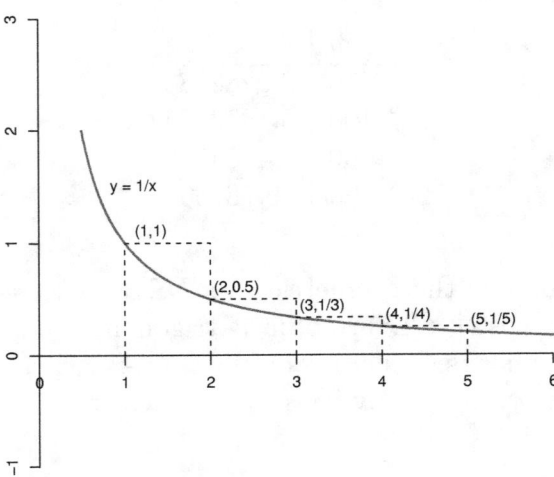

Figure 3.1: The divergence of the series $\Sigma_{n,1} \frac{1}{n}$.

$$1 + \tfrac{1}{2} > \int_1^3 \tfrac{1}{x} dx,$$

$$1 + \tfrac{1}{2} + \tfrac{1}{3} > \int_1^4 \tfrac{1}{x} dx$$

and, in general,

$$s_n = 1 + \tfrac{1}{2} + \tfrac{1}{3} + \ldots + \tfrac{1}{n} > \int_1^{n+1} \tfrac{1}{x} dx = [\ln x]_1^{n+1} = \ln(n+1).$$

But $\ln(n+1) \to \infty$, hence $s_n \to \infty$. Despite the fact that $\frac{1}{n} \to 0$, we have shown that $\Sigma_{n,1} \frac{1}{n}$ does not converge! The idea of using integral calculus, in order to solve a problem about series, foreshadows a systematic approach to be learned in section 3.9.

Exercises for Section 3.2

1. Is the series $\Sigma_{n,0} \frac{3^{n+1}}{7^n}$ convergent? If it were convergent, what is its limit?

2. Calculate 0.176176176... and 0.32131313....

3. For what values of x does the series $\Sigma_{n,0} \frac{1}{x^n}$ converge? To what number does it converge?

4. What can you say about the convergence of $\Sigma_{n,0}(\ln x)^n$?

5. Is the series $\Sigma_{n,0}\frac{\cos(n\pi)}{2^n}$ convergent? If it were convergent, to what number would it converge?

6. Does the telescoping series $\Sigma_{n,1}\ln(\frac{n}{n+1})$ converge?

7. Show that $\Sigma_{n=k}^{\infty}x^n = \frac{x^k}{1-x}$, $|x| < 1$.

3.3 Main Properties of Convergent Series

Two are the main properties of convergent series:

1. Suppose that $\Sigma_{n,1}a_n$ and $\Sigma_{n,1}b_n$ converge. Then $\Sigma_{n,1}(a_n + b_n)$ converges and

$$\Sigma_{n=1}^{\infty}(a_n + b_n) = \Sigma_{n=1}^{\infty}a_n + \Sigma_{n=1}^{\infty}b_n.$$

2. Suppose that $\Sigma_{n,1}a_n$ converges and let λ be any real number. Then $\Sigma_{n,1}(\lambda a_n)$ converges and

$$\Sigma_{n=1}^{\infty}(\lambda a_n) = \lambda\Sigma_{n=1}^{\infty}a_n.$$

Let us prove both properties. Define

$$s_n = a_1 + ... + a_n,$$

$$t_n = b_1 + ... + b_n.$$

By hypothesis $s_n \to \Sigma_{n=1}^{\infty}a_n$ and $t_n \to \Sigma_{n=1}^{\infty}b_n$. Hence

$$s_n + t_n \to \Sigma_{n=1}^{\infty}a_n + \Sigma_{n=1}^{\infty}b_n.$$

But $s_n + t_n = (a_1 + b_1) + ... + (a_n + b_n)$. We have then shown that

$$(a_1 + b_1) + ... + (a_n + b_n) \to \Sigma_{n=1}^{\infty}a_n + \Sigma_{n=1}^{\infty}b_n.$$

Therefore $\Sigma_{n,1}(a_n + b_n)$ converges and

$$\Sigma_{n=1}^{\infty}(a_n + b_n) = \Sigma_{n=1}^{\infty}a_n + \Sigma_{n=1}^{\infty}b_n.$$

Analogously, $\lambda s_n = \lambda(a_1 + ... + a_n) = \lambda a_1 + ... + \lambda a_n$. Since $\lambda s_n \to \lambda\Sigma_{n=1}^{\infty}a_n$ we can conclude that $\Sigma_{n,1}(\lambda a_n)$ converges and $\Sigma_{n=1}^{\infty}(\lambda a_n) = \lambda\Sigma_{n=1}^{\infty}a_n$. QED

For instance, $\Sigma_{n,1}[(\frac{1}{2})^n + \frac{1}{n(n+1)}]$ converges since $\Sigma_{n,1}(\frac{1}{2})^n$ and $\Sigma_{n,1}\frac{1}{n(n+1)}$ converge. Moreover,

$$\Sigma_{n=1}^{\infty}[(\tfrac{1}{2})^n + \tfrac{1}{n(n+1)}] = \Sigma_{n=1}^{\infty}(\tfrac{1}{2})^n + \Sigma_{n=1}^{\infty}\tfrac{1}{n(n+1)} = \tfrac{\frac{1}{2}}{1-\frac{1}{2}} + 1 = 2.$$

In a similar vein, $\Sigma_{n,1} 4(\tfrac{1}{3})^n$ converges since $\Sigma_{n,1}(\tfrac{1}{3})^n$ converges. Moreover,

$$\Sigma_{n,1} 4(\tfrac{1}{3})^n = 4\Sigma_{n=1}^{\infty}(\tfrac{1}{3})^n = 4\tfrac{\frac{1}{3}}{1-\frac{1}{3}} = 4\tfrac{\frac{1}{3}}{\frac{2}{3}} = 2.$$

It should be noted that the convergence of $\Sigma_{n,1} a_n$ and $\Sigma_{n,1} b_n$ implies the convergence of $\Sigma_{n,1}(a_n - b_n)$, and

$$\Sigma_{n=1}^{\infty}(a_n - b_n) = \Sigma_{n=1}^{\infty} a_n - \Sigma_{n=1}^{\infty} b_n.$$

Why is this so? Since $\Sigma_{n,1} b_n$ is convergent we can conclude that $\Sigma_{n,1}(-b_n)$ is convergent and $\Sigma_{n=1}^{\infty}(-b_n) = -\Sigma_{n=1}^{\infty} b_n$. Then $\Sigma_{n,1}(a_n + (-b_n))$ is convergent and

$$\Sigma_{n=1}^{\infty}(a_n + (-b_n)) = \Sigma_{n=1}^{\infty} a_n + \Sigma_{n=1}^{\infty}(-b_n) = \Sigma_{n=1}^{\infty} a_n - \Sigma_{n=1}^{\infty} b_n.$$

Thus, $\Sigma_{n,1}(a_n - b_n)$ is convergent and $\Sigma_{n=1}^{\infty}(a_n - b_n) = \Sigma_{n=1}^{\infty} a_n - \Sigma_{n=1}^{\infty} b_n$.

Remark

The preceding properties of convergent series are valid irrespective of whether we are working with a series like $\Sigma_{n,1} a_n$ or $\Sigma_{n,0} a_n$. If one is convergent, the other is too and

$$\Sigma_{n=0}^{\infty} a_n = a_0 + \Sigma_{n=1}^{\infty} a_n.$$

Indeed, let $s_n = a_0 + a_1 + \ldots + a_n$, $t_n = a_1 + \ldots + a_n$. Suppose $\Sigma_{n,0} a_n$ is convergent. Then $s_n \to \Sigma_{n=0}^{\infty} a_n$. Hence $s_n - a_0 \to \Sigma_{n=0}^{\infty} a_n - a_0$, that is, $t_n \to \Sigma_{n=0}^{\infty} a_n - a_0$. Thus $\Sigma_{n,1} a_n$ converges and $\Sigma_{n=1}^{\infty} a_n = \Sigma_{n=0}^{\infty} a_n - a_0$. Therefore $\Sigma_{n=0}^{\infty} a_n = a_0 + \Sigma_{n=1}^{\infty} a_n$. In a similar fashion we can show that the convergence of $\Sigma_{n,1} a_n$ implies the convergence of $\Sigma_{n,0} a_n$ and $\Sigma_{n=0}^{\infty} a_n = a_0 + \Sigma_{n=1}^{\infty} a_n$.

As a matter of fact, given any sequence (a_n) and arbitrary positive integers p, q (say, $p < q$), $\Sigma_{n,p} a_n$ converges if and only if $\Sigma_{n,q} a_n$ converges. Moreover,

$$\Sigma_{n=p}^{\infty} a_n = a_p + a_{p+1} + \ldots + a_{q-1} + \Sigma_{n=q}^{\infty} a_n. \tag{3.1}$$

In summary, the convergence of a series is not affected if we change the initial value of the index. However, we have to keep in mind the relationship (3.1). For instance, $\Sigma_{n,2}\tfrac{1}{n(n+1)}$ is convergent since $\Sigma_{n,1}\tfrac{1}{n(n+1)}$ is convergent. Moreover,

$$1 = \Sigma_{n=1}^{\infty}\tfrac{1}{n(n+1)} = \tfrac{1}{1(1+1)} + \Sigma_{n=2}^{\infty}\tfrac{1}{n(n+1)}.$$

Hence $\Sigma_{n=2}^{\infty}\tfrac{1}{n(n+1)} = 1 - \tfrac{1}{2} = \tfrac{1}{2}$.

The remainder of a series

Given a convergent series $\Sigma_{n,1} a_n$, define $R_n = \Sigma_{i=n+1}^{\infty} a_i$. The sequence (R_n) is called the 'remainder' of the given series. Note that (R_n) is a well-defined sequence because $\Sigma_{i=n+1}^{\infty} a_i$ is a real number for each n. Undoubtedly,

$$\Sigma_{i=1}^{\infty} a_i = a_1 + \dots + a_n + R_n,$$

that is, $\Sigma_{i=1}^{\infty} a_i = \Sigma_{i=1}^{n} a_i + R_n$. Hence

$$R_n = \Sigma_{i=1}^{\infty} a_i - \Sigma_{i=1}^{n} a_i.$$

Since $\lim_{n\to\infty} \Sigma_{i=1}^{n} a_i = \Sigma_{i=1}^{\infty} a_i$ we can conclude that $\lim_{n\to\infty} R_n = 0$. The idea of remainder will play an important role in section 3.9.

Exercises for Section 3.3

Calculate the value of the following convergent series.

1. $\Sigma_{n=1}^{\infty} \left(\frac{3}{4^n} + \frac{1}{2^n} \right)$.

2. $\Sigma_{n=3}^{\infty} \left(\frac{1}{4} \right)^n$.

3. $\Sigma_{n=2}^{\infty} \left[\frac{1}{n^2+n} - \left(\frac{1}{2} \right)^n \right]$.

4. $\Sigma_{n=3}^{\infty} \left[\sin^{-1} \left(\frac{1}{n+1} \right) - \sin^{-1} \left(\frac{1}{n} \right) \right]$.

5. $\Sigma_{n=4}^{\infty} \frac{3}{e^n}$.

3.4 A Criterion for Non-Convergence

So far we only know which geometric series are convergent and that some telescoping series like $\Sigma_{n,1} \frac{1}{n(n+1)}$ are convergent. Soon we will considerably enlarge the collection of convergent series at our disposal. In the meantime let us discuss a simple criterion to determine the non-convergence of a series. *Customarily we will use the word 'divergence' to denote non-convergence.*

The Divergence Test

If $\Sigma_{n,1} a_n$ is convergent then $a_n \to 0$.

The proof is quite short: Let $s_n = a_1 + \dots + a_n$. By hypothesis $s_n \to L$, where $L = \Sigma_{n=1}^{\infty} a_n$. But (s_{n-1}) converges to the same limit (see exercise 15, section 2.1). Hence $s_{n-1} \to L$, which in turn leads to $a_n = s_n - s_{n-1} \to L - L = 0$. QED

We are especially interested in the logical contrapositive of the preceding result, namely:

If (a_n) does not converge to zero then $\Sigma_{n,1} a_n$ diverges.

Thus, $\Sigma_{n,1} n$ and $\Sigma_{n,1} \log n$ diverge, as well as $\Sigma_{n,0} n!$ and $\Sigma_{n,2} \frac{n}{\log n}$ because the sequences (n), $(\log n)$, $(n!)$ and $(\frac{n}{\log n})$ do not converge to zero; actually, they tend to infinity. Also, $\Sigma_{n,1} \cos(n\pi)$ diverges because $(\cos(n\pi))$ does not tend to zero (it oscillates between -1 and 1) while $\Sigma_{n,2} \ln(\frac{1}{n^2})$ diverges because $(\ln(\frac{1}{n^2}))$ does not converge to zero either (it tends to $-\infty$). **We should be careful in the sense that the converse of the preceding result is not true.** That is to say, $a_n \to 0$ does not imply that $\Sigma_{n,1} a_n$ is convergent. For instance, $\frac{1}{n} \to 0$ but $\Sigma_{n,1} \frac{1}{n}$ is divergent, as we proved in section 3.2.

Exercises for Section 3.4

In each of the following questions provide a detailed argument to justify your answer.

1. Does the series $\Sigma_{n,0} \cos(\frac{1}{n+3})$ converge?

2. Is $\Sigma_{n,0} \frac{n}{n+1}$ convergent or divergent?

3. Why can we assert that $\Sigma_{n,2} (1 + \frac{4}{n})^n$ is divergent?

4. Does $\Sigma_{n,1} \cos(\frac{\pi}{n})$ converge or diverge?

5. Is $\Sigma_{n,1} \sin(n + \frac{1}{2})\pi$ divergent?

3.5   The Comparison Test

Let us discuss a test that plays a central role in the theory of series. This new test has not only intrinsic value but it will allow us to justify two tests in the next section, namely the ratio test and the root test.

The Comparison Test (CT)

Suppose (a_n) and (b_n) are two sequences such that $0 \le a_n \le b_n$ for every n. The following two implications are true:

1. If $\Sigma_{n,1} b_n$ converges then $\Sigma_{n,1} a_n$ converges.

2. If $\Sigma_{n,1} a_n$ diverges then $\Sigma_{n,1} b_n$ diverges.

Table 3.2: The sequence of partial sums corresponding to $\Sigma_{n,1}\frac{1}{n^2}$

m	$\Sigma_{n=1}^{m}\frac{1}{n^2}$
10	1.5497677
100	1.6349839
1000	1.6439346
10,000	1.6448341

We need to prove only the first implication because the second is just the logical contrapositive of the preceding implication. With this purpose in mind, let

$$s_n = a_1 + \ldots + a_n,$$

$$t_n = b_1 + \ldots + b_n.$$

Right away we observe that $s_n \leq t_n$ for all n. The fact that $\Sigma_{n,1}b_n$ converges means that (t_n) is convergent, hence (t_n) is bounded; thus, there exists a number M such that $t_n \leq M$ for all n. Therefore $s_n \leq M$ for all n. Since (s_n) is increasing, proposition 1 (section 3.1) allows us to conclude that (s_n) is convergent, that is, $\Sigma_{n,1}a_n$ is convergent. QED

Remark

Under the hypothesis of the comparison test, if $\Sigma_{n,1}b_n$ converges then $\Sigma_{n=1}^{\infty}a_n \leq \Sigma_{n=1}^{\infty}b_n$. Why is this so? The answer is very simple: since $s_n \leq t_n$ for all n and $s_n \to \Sigma_{n=1}^{\infty}a_n$ while $t_n \to \Sigma_{n=1}^{\infty}b_n$, the fifth property of convergent sequences (section 3.1) implies that $\Sigma_{n=1}^{\infty}a_n \leq \Sigma_{n=1}^{\infty}b_n$.

It should also be noted that, when applying the comparison test, the inequality $a_n \leq b_n$ does not have to hold for every n but eventually, say $0 \leq a_n \leq b_n$ for all $n \geq N$. Indeed, the fact that $\Sigma_{n,1}b_n$ converges implies that $\Sigma_{n,N}b_n$ converges. Hence $\Sigma_{n,N}a_n$ converges, which in turn implies that $\Sigma_{n,1}a_n$ converges. By the way,

$$0 \leq \Sigma_{n=N}^{\infty}a_n \leq \Sigma_{n=N}^{\infty}b_n$$

if $\Sigma_{n,1}b_n$ converges.

Examples

1. We already know that $\Sigma_{n,1}\frac{1}{n}$ is divergent. The next natural question is to ask whether $\Sigma_{n,1}\frac{1}{n^2}$ is convergent; table 3.2 suggests that it is indeed convergent! Let us try to provide a proof of convergence. Since

$$n^2 + n^2 \geq n^2 + n = n(n+1)$$

it follows that

$$0 < \tfrac{1}{n^2} \leq 2\tfrac{1}{n(n+1)}.$$

But the telescoping series $\Sigma_{n,1}\tfrac{1}{n(n+1)}$ was proven to be convergent. Then the comparison test implies that $\Sigma\tfrac{1}{n^2}$ is convergent. Where does it converge to? As of now, all we know is that

$$\Sigma_{n=1}^{\infty}\tfrac{1}{n^2} \leq 2\Sigma_{n=1}^{\infty}\tfrac{1}{n(n+1)} = 2 \times 1 = 2.$$

Evidently $\Sigma_{n=1}^{\infty}\tfrac{1}{n^2} > 1$, thus

$$1 < \Sigma_{n=1}^{\infty}\tfrac{1}{n^2} \leq 2.$$

Later, after learning about the integral test, we will be able to assert that, up to the first three decimals, $\Sigma_{n=1}^{\infty}\tfrac{1}{n^2} = 1.644$.

2. For what values of p does the series $\Sigma_{n,1}\tfrac{1}{n^p}$ converge? This is a very important series, which is called a p-**series**. So far we know that a p-series converges when $p = 2$ (the series diverges when $p = 1$). If $p > 2$ we have

$$0 < \tfrac{1}{n^p} < \tfrac{1}{n^2} \quad \text{for all } n,$$

hence the comparison test implies that $\Sigma_{n,1}\tfrac{1}{n^p}$ converges. If $p \leq 0$ the sequence (n^{-p}) does not converge to zero, consequently $\Sigma_{n,1}\tfrac{1}{n^p}$ diverges. Next let us suppose that $0 < p < 1$. Under these circumstances

$$0 < \tfrac{1}{n} < \tfrac{1}{n^p} \quad \text{for all } n,$$

hence the comparison test implies that $\Sigma_{n,1}\tfrac{1}{n^p}$ is divergent.Thus, we have shown that a p-series converges when $p \geq 2$ and diverges when $p \leq 1$. What happens when $1 < p < 2$? For the time being let us accept that **a p-series converges for $p > 1$**. In section 3.9 it will be proven that this is indeed the case.

3. Let us consider the series $\Sigma_{n,1}\tfrac{\sin^2 n}{n^3}$. Does it converge? We note that

$$0 \leq \tfrac{\sin^2 n}{n^3} \leq \tfrac{1}{n^3}.$$

Since $\Sigma_{n,1}\frac{1}{n^3}$ converges, the comparison test implies that $\Sigma_{n,1}\frac{\sin^2 n}{n^3}$ converges. It is interesting to note that so far we do not know whether $\Sigma_{n,1}\frac{\sin n}{n^3}$ converges because $\sin n$ adopts positive and negative values as n increases. Due to this fact, the comparison test cannot be applied.

4. An important series is $\Sigma_{n,1}\frac{1}{n!}$. We observe that

$$0 < \tfrac{1}{n!} < \tfrac{1}{2^n}$$

for all $n \geq 4$ since $2^n < n!$ whenever $n \geq 4$. The latter inequality follows from the observation that $2 \times 2 \dots \times 2 < n(n-1)\dots 2$ whenever $n \geq 4$ or, if more rigor is needed, a proof by induction can be provided. The series $\Sigma_{n,1}\left(\frac{1}{2}\right)^n$ is a geometric series with $r = 1/2$, thus it converges. The comparison test implies that $\Sigma_{n,1}\frac{1}{n!}$ converges, hence $\Sigma_{n,0}\frac{1}{n!}$ converges. In a later section we will show that $\Sigma_{n=0}^{\infty}\frac{1}{n!} = e$. Right now, all we know is that

$$\Sigma_{n=4}^{\infty}\tfrac{1}{n!} \leq \Sigma_{n=4}^{\infty}\left(\tfrac{1}{2}\right)^n = 2 - \left(1 + \tfrac{1}{2} + \tfrac{1}{4} + \tfrac{1}{8}\right) = \tfrac{1}{8}.$$

Thus

$$\Sigma_{n=0}^{\infty}\tfrac{1}{n!} = 1 + \tfrac{1}{1} + \tfrac{1}{2} + \tfrac{1}{6} + \Sigma_{n=4}^{\infty}\tfrac{1}{n!} \leq 2 + \tfrac{4}{6} + \tfrac{1}{8} = 2 + \tfrac{19}{24} \approx 2.791666.$$

5. What can be said about the convergence of $\Sigma_{n,2}\frac{1}{\ln n}$? We know that $\ln n < n$, hence

$$0 < \tfrac{1}{n} < \tfrac{1}{\ln n}$$

whenever $n \geq 2$. So, the given series diverges since $\Sigma_{n,2}\frac{1}{n}$ diverges. Despite the fact that $\frac{1}{\ln n} \to 0$, the series $\Sigma_{n,2}\frac{1}{\ln n}$ diverges!

6. Let us analyze the series $\Sigma_{n,0}\frac{3^n}{1+5^n}$. We can see that

$$\tfrac{3^n}{1+5^n} < \tfrac{3^n}{5^n} = \left(\tfrac{3}{5}\right)^n.$$

Since $\Sigma_{n,0}\left(\frac{3}{5}\right)^n$ converges, the given series has to converge too.

7. Does $\Sigma_{n,1}\frac{1}{n2^n}$ converge? It certainly does, thanks to the comparison test, because

$$0 < \tfrac{1}{n2^n} \leq \tfrac{1}{2^n},$$

whenever $n \geq 1$, and the geometric series $\Sigma_{n,1}\left(\frac{1}{2}\right)^n$ converges. Moreover,

$$0 < \Sigma_{n=1}^{\infty} \frac{1}{n2^n} \leq \Sigma_{n=1}^{\infty} (\tfrac{1}{2})^n = \frac{\frac{1}{2}}{1 - \frac{1}{2}} = 1.$$

Using a graphing calculator we can check that $\Sigma_{n=1}^{50} \frac{1}{n2^n} = 0.69314718$ and $\Sigma_{n=1}^{500} \frac{1}{n2^n} = 0.69314718$. Thus, we can feel confident that the value of the series is 0.6931471 (up to seven decimals).

Next we will study the limit comparison test, which happens to be an almost direct consequence of the comparison test.

The Limit Comparison Test (LCT)

Let (a_n) and (b_n) be sequences of positive terms. The following three implications are true:

1. If $\lim_{n \to \infty} \frac{a_n}{b_n} = L > 0$, then $\Sigma_{n,1} a_n$ converges if and only if $\Sigma_{n,1} b_n$ converges.

2. If $\lim_{n \to \infty} \frac{a_n}{b_n} = 0$ and $\Sigma_{n,1} b_n$ converges then $\Sigma_{n,1} a_n$ converges.

3. If $\lim_{n \to \infty} \frac{a_n}{b_n} = \infty$ and $\Sigma_{n,1} b_n$ diverges then $\Sigma_{n,1} a_n$ diverges.

Let us prove each of the three implications separately.

1. Since $\frac{a_n}{b_n} \to L$ it follows that $|\frac{a_n}{b_n} - L| < \frac{L}{2}$ eventually. Hence, for sufficiently large n we will have

$$-\frac{L}{2} < \frac{a_n}{b_n} - L < \frac{L}{2},$$

that is, there exists N such that if $n \geq N$, then

$$\frac{L}{2} < \frac{a_n}{b_n} < \frac{3}{2}L.$$

Thus, provided that $n \geq N$ we will have

$$\frac{L}{2}b_n < a_n < \frac{3}{2}Lb_n.$$

Assume that $\Sigma_{n,1} a_n$ converges. Since $0 < b_n < \frac{2}{L}a_n$ for $n \geq N$, the comparison test implies that $\Sigma_{n,1} b_n$ converges. On the other hand, suppose that $\Sigma_{n,1} b_n$ converges. Since $0 < a_n < \frac{3}{2}Lb_n$ eventually, the comparison test leads to the conclusion that $\Sigma_{n,1} a_n$ converges.

2. By hypothesis $\frac{a_n}{b_n} \to 0$, so there exists N such that $\frac{a_n}{b_n} < 1$ whenever $n \geq N$, that is, $0 < a_n < b_n$ eventually. Since Σb_n converges, the comparison test implies that $\Sigma_{n,1} a_n$ converges.

3. Since $\frac{a_n}{b_n} \to \infty$ we can assert that $\frac{a_n}{b_n} > 1$ eventually, that is, $0 < b_n < a_n$ eventually. Since $\Sigma_{n,1} b_n$ diverges, the comparison test allows us to assert that $\Sigma_{n,1} a_n$ diverges. QED

Examples

1. Does $\Sigma_{n,1} \frac{n+1}{n^3+n+1}$ converge? For large n the series behaves like $\Sigma_{n,1} \frac{1}{n^2}$. Hence it seems to be a good idea to compare the given series with $\Sigma_{n,1} \frac{1}{n^2}$. Indeed

$$\frac{\frac{n+1}{n^3+n+1}}{\frac{1}{n^2}} = \frac{n^3+n^2}{n^3+n+1} = \frac{1+\frac{1}{n}}{1+\frac{1}{n^2}+\frac{1}{n^3}} \to 1.$$

Since $\Sigma_{n,1} \frac{1}{n^2}$ converges, the limit comparison test (part 1) implies that the given series does converge.

2. $\Sigma_{n,0} \frac{1}{2n+1}$ diverges since

$$\frac{\frac{1}{2n+1}}{\frac{1}{n}} \to \frac{1}{2} > 0$$

and $\Sigma_{n,1} \frac{1}{n}$ diverges. We have applied LCT (part 1).

3. Does $\Sigma_{n,1} \frac{n}{n^2+6n+5}$ converge? We observe that, for large n, the given series behaves like $\Sigma_{n,1} \frac{1}{n}$. Thus, let us compare the series with $\Sigma_{n,1} \frac{1}{n}$:

$$\frac{\frac{n}{n^2+6n+5}}{\frac{1}{n}} = \frac{n^2}{n^2+6n+5} = \frac{1}{1+\frac{6}{n}+\frac{5}{n^2}} \to 1.$$

The limit comparison test (part 1) implies that $\Sigma_{n,1} \frac{n}{n^2+6n+5}$ diverges.

4. Given the series $\Sigma_{n,1} \frac{\ln n}{n}$ we might be tempted to guess that it converges due to the fact that $\frac{\ln n}{n} \to 0$. However,

$$\frac{\frac{\ln n}{n}}{\frac{1}{n}} = \ln n \to \infty.$$

Since $\Sigma_{n,1} \frac{1}{n}$ diverges, part 3 of the limit comparison test implies that $\Sigma_{n,1} \frac{\ln n}{n}$ diverges.

5. Next let us consider the series $\Sigma_{n,1} \frac{\ln n}{n^2}$. We cannot compare it with either $\Sigma_{n,1} \frac{1}{n}$ or $\Sigma_{n,1} \frac{1}{n^2}$ because

$$\frac{\frac{\ln n}{n^2}}{\frac{1}{n}} = \frac{\ln n}{n} \to 0,$$

but $\Sigma_{n,1} \frac{1}{n}$ diverges. Similarly,

$$\frac{\frac{\ln n}{n^2}}{\frac{1}{n^2}} = \ln n \to \infty,$$

but $\Sigma \frac{1}{n^2}$ converges. Since we haven't been successful comparing with a p-series (either $p = 1$ or $p = 2$) we might try to compare the given series with $\Sigma_{n,1} \frac{1}{n^{3/2}}$:

$$\frac{\frac{\ln n}{n^2}}{\frac{1}{n^{3/2}}} = \frac{\ln n}{\sqrt{n}} \to 0.$$

Since $\Sigma_{n,1} \frac{1}{n^{3/2}}$ converges (it is a p-series with $p > 1$), the series $\Sigma_{n,1} \frac{\ln n}{n^2}$ converges. LCT (part 2) has been the tool of choice.

6. The series $\Sigma_{n,1} \frac{1}{\sqrt{n}+4}$ looks a little bit puzzling. We might start by realizing that

$$0 < \frac{1}{\sqrt{n}+4} < \frac{1}{\sqrt{n}},$$

but $\Sigma_{n,1} \frac{1}{n^{1/2}}$ diverges. However,

$$\frac{\frac{1}{\sqrt{n}+4}}{\frac{1}{\sqrt{n}}} = \frac{1}{1+\frac{4}{\sqrt{n}}} \to 1.$$

Hence, the given series diverges thanks to LCT (part 1).

Exercises for Section 3.5

Use either the comparison test or the limit comparison test to determine whether the following series are convergent or divergent.

1. $\Sigma_{n,1} \frac{1}{n4^n}$.

2. $\Sigma_{n,1} \frac{1}{n+6}$.

3. $\Sigma_{n,1} \frac{\sin^2 n}{n^2}$.

4. $\Sigma_{n,1} \frac{1}{n^4+11}$.

5. $\Sigma_{n,1} \ln\left(1 + \frac{1}{n}\right)$.

6. $\Sigma_{n,1} \ln\left(1 + \frac{1}{n^p}\right)$, $p \geq 2$.

7. $\Sigma_{n,1} \frac{3+\sin n}{n^3}$.

8. $\Sigma_{n,1} \frac{(\ln n)^2}{n^2}$.

9. $\Sigma_{n,2} \frac{1}{(\ln n)^2}$.

10. $\Sigma_{n,2} \frac{1}{\ln(\ln n)}$.

11. $\Sigma_{n,0} \frac{\sqrt{n}+8}{\sqrt{n^2+1}}$.

12. $\Sigma_{n,2} \frac{1}{n^2 \ln n}$.

Table 3.3: The sequence of partial sums corresponding to $\Sigma_{n,1}\frac{(-1)^n}{2n+1}$

m	$\Sigma_{n=1}^{m}\frac{(-1)^n}{2n+1}$
500	-0.214103
600	-0.214186
5000	-0.214552
10,000	-0.214577

3.6   The Alternating Series Test

So far we have only studied series of non-negative terms. As we will discover in section 3.11, series whose terms alternate between positive and negative values are quite important. Suppose that we have to deal with the series $\Sigma_{n,1}\frac{(-1)^n}{2n+1}$. Table 3.3 suggests that it converges. Actually, it seems that

$$\Sigma_{n=1}^{\infty}\frac{(-1)^n}{2n+1} \approx -0.2145.$$

The series under consideration has the properties that its terms oscillate in sign, $(\frac{1}{2n+1})$ is decreasing and converges to zero. Exploring different series with a graphing calculator we might suspect that $\Sigma_{n,1}(-1)^{n+1}a_n$ is convergent provided that $a_n > 0$, (a_n) is decreasing and $a_n \to 0$. Luckily there is a test that can be readily applied.

The Alternating Series Test

Let us consider a series $\Sigma_{n,1}(-1)^{n+1}a_n$ such that:

1. $a_n > 0$.

2. $a_1 > a_2 > a_3 > ...$

3. $a_n \to 0$.

Then the given series converges[3].

Before considering several examples let us discuss a proof of the alternating series test. With this purpose in mind define

$$s_n = a_1 - a_2 + ... + (-1)^{n+1}a_n.$$

Then

---

[3]Observe that the even terms of the series are preceded by a negative sign while the odd terms are preceded by a positive sign.

$$s_{2n} = (a_1 - a_2) + (a_3 - a_4) + \ldots + (a_{2n-1} - a_{2n})$$
$$= a_1 - (a_2 - a_3) - (a_4 - a_5) - \ldots - (a_{2n-2} - a_{2n-1}) - a_{2n}.$$

We observe that $s_{2n} < a_1$ because each term in parenthesis is positive and $a_{2n} > 0$. Next we will see that (s_{2n}) is an increasing sequence. Indeed,

$$s_{2(n+1)} = s_{2n+2} = s_{2n} + (a_{2n+1} - a_{2n+2}).$$

But $a_{2n+1} - a_{2n+2} > 0$ since (a_n) is, by hypothesis, a decreasing sequence. Hence $s_{2(n+1)} > s_{2n}$. Having proven that (s_{2n}) is increasing and bounded above (recall that a_1 is an upper bound), proposition 1 from section 3.1 implies that (s_{2n}) converges; actually, $s_{2n} \to L$ where $L = l.u.b.\{s_{2n}\}$. What happens to the sequence (s_{2n-1})? Evidently, $s_{2n} = s_{2n-1} - a_{2n}$. Since $a_n \to 0$ we can conclude that $a_{2n} \to 0$ and consequently $s_{2n-1} \to L + 0 = L$. Finally, having proven that $s_{2n} \to L$ and $s_{2n-1} \to L$ we will have that $s_n \to L$, that is, $\Sigma_{n,1}(-1)^{n+1} a_n$ converges. QED

Remark

In the preceding proof we have used the three conditions under which the test works. Thus, we might suspect that dropping any of the conditions does not guarantee convergence of the alternating series. Indeed, defining (a_n) in such a way that $a_n = \frac{1}{n}$ (n odd), $a_n = \frac{1}{n^2}$ (n even) we can see that $a_n \to 0$ but the sequence is not decreasing. It can be shown that $\Sigma_{n,1}(-1)^{n+1} a_n$ actually diverges!

Examples

1. The simplest imaginable example of an alternating series is $\Sigma_{n,1}(-1)^{n+1}\frac{1}{n}$. Obviously $\frac{1}{n} > 0$, $\frac{1}{n} > \frac{1}{n+1}$, and $\frac{1}{n} \to 0$. Thus, the alternating series test implies that the given series converges. Pretty soon we will show that, surprisingly, this series converges to $\ln 2$.

2. Both the series $\Sigma_{n,1}(-1)^{n+1}\frac{1}{n!}$ and $\Sigma_{n,1}(-1)^{n+1}\frac{1}{\sqrt{n}}$ are convergent because the three conditions required by the alternating series test are met.

3. Is $\Sigma_{n,2}(-1)^{n+1}\frac{\ln n}{n}$ convergent? We do know that $\frac{\ln n}{n} > 0$ and $\frac{\ln n}{n} \to 0$, but it is necessary to check whether the sequence $(\frac{\ln n}{n})$ is decreasing. With this purpose in mind let us define the function $f(x) = \frac{\ln x}{x}$. Then

$$f'(x) = \frac{\frac{1}{x}x - \ln x}{x^2} = \frac{1 - \ln x}{x^2} < 0 \text{ whenever } x > e.$$

Thus, $f(x)$ is decreasing eventually, which in turn implies that $(\frac{\ln n}{n})$ is decreasing eventually. Therefore $\Sigma_{n,2}(-1)^{n+1}\frac{\ln n}{n}$ converges[4].

---

[4]When applying the alternating series test we may relax the second condition and demand only that the terms are decreasing eventually.

Approximating the value of an alternating series

Under the hypothesis of the alternating series test, a remarkable inequality is true:

$$|s_n - \Sigma_{n=1}^{\infty}(-1)^{n+1}a_n| \leq a_{n+1},$$

where $s_n = \Sigma_{i=1}^{n}(-1)^{i+1}a_i$. A proof can be found as an enrichment note at the end of the section. Let us use it in some particular cases. For instance, letting $s = \Sigma_{n=1}^{\infty}(-1)^{n+1}\frac{1}{n}$,

$$|(1 - \tfrac{1}{2} + \tfrac{1}{3}) - s| \leq \tfrac{1}{4},$$

$$|(1 - \tfrac{1}{2} + \tfrac{1}{3} - \tfrac{1}{4}) - s| \leq \tfrac{1}{5},$$

and, in general,

$$|(1 - \tfrac{1}{2} + \tfrac{1}{3} - \tfrac{1}{4} + \ldots + (-1)^{n+1}\tfrac{1}{n} - s| \leq \tfrac{1}{n+1}.$$

Thus, if we wish that

$$|(1 - \tfrac{1}{2} + \tfrac{1}{3} - \tfrac{1}{4} + \ldots + (-1)^{n+1}\tfrac{1}{n} - s| \leq \tfrac{1}{100}$$

it is enough to choose n such that $\frac{1}{n+1} < \frac{1}{100}$. Obviously, $n = 100$ will do the job. So $\Sigma_{n=1}^{100}(-1)^{n+1}\frac{1}{n}$ is an approximation of $\Sigma_{n=1}^{\infty}(-1)^{n+1}\frac{1}{n}$ with an error of less than 0.01.

In a similar fashion, if $s = \Sigma_{n=1}^{\infty}(-1)^{n+1}\frac{1}{n!}$ then

$$|(1 - \tfrac{1}{2} + \tfrac{1}{6} - \tfrac{1}{24} + \ldots + (-1)^{n+1}\tfrac{1}{n!}) - s| \leq \tfrac{1}{(n+1)!}.$$

From this inequality we can get as close to the exact value of the series as we wish.

The exact value of a series

Let us start observing that

$$\tfrac{1-q^n}{1-q} = 1 + q + q^2 + \ldots + q^{n-1}, \; q \neq 1.$$

Thus

$$\tfrac{1}{1-q} = 1 + q + q^2 + \ldots + q^{n-1} + \tfrac{q^n}{1-q}.$$

Making $q = -x$ we can then assert that, whenever $x \neq -1$,

$$\tfrac{1}{1+x} = 1 - x + x^2 + \ldots + (-x)^{n-1} + \tfrac{(-x)^n}{1+x}.$$

Therefore

$$\textstyle\int_0^1 \tfrac{dx}{1+x} = \int_0^1 dx - \int_0^1 x dx + \int_0^1 x^2 dx + \ldots + \int_0^1 (-x)^{n-1} dx + \int_0^1 \tfrac{(-x)^n}{1+x} dx.$$

But $\ln 2 = \int_1^2 \frac{du}{u} = \int_0^1 \frac{dx}{x+1}$. Hence

$$\ln 2 = 1 - \tfrac{1}{2} + \tfrac{1}{3} + \dots + \frac{(-1)^{n-1}}{n} + \int_0^1 \frac{(-x)^n}{1+x} dx.$$

However,

$$\left| (-1)^n \int_0^1 \frac{x^n}{1+x} dx \right| \leq \int_0^1 x^n dx = [\tfrac{x^{n+1}}{n+1}]_0^1 = \tfrac{1}{n+1} \to 0 \text{ as } n \to \infty.$$

Consequently

$$\ln 2 = 1 - \tfrac{1}{2} + \tfrac{1}{3} - \tfrac{1}{4} + \dots$$

Exercises for Section 3.6

Use the alternating series test to determine whether the first four series are convergent. Do not forget to check the hypotheses needed to apply the above-mentioned test.

1. $\Sigma_{n,2}(-1)^{n+1} \frac{1}{\ln n}$.

2. $\Sigma_{n,2}(-1)^{n+1} \frac{1}{n \ln n}$.

3. $\Sigma_{n,1}(-1)^{n+1} \frac{\sqrt{n}}{n+3}$.

4. $\Sigma_{n,1}(-1)^{n+1} \frac{1}{n^2}$.

5. Why you cannot apply the alternating series test to $\Sigma_{n,2}(-1)^{n+1} \ln(\frac{1}{n})$? What alternative would you use?

6. Does the series $\Sigma_{n,1}(-1)^{n+1} \sin(\frac{1}{n})$ converge? If it does, use your graphing calculator to find the value of the series up to the first two decimals. What is the least number of terms needed to reach this goal?

Enrichment Note: Proving an Inequality

Linked to the alternating series test we have portrayed the inequality

$$\left| s_n - \Sigma_{n=1}^{\infty}(-1)^{n+1} a_n \right| \leq a_{n+1},$$

where $s_n = \Sigma_{i=1}^{n}(-1)^{i+1} a_i$. Our task is to provide a proof of it. Let us start by recalling that (s_{2n}) is increasing and bounded above, and $s_{2n} \to L$ where $L = l.u.b.\{s_{2n}\}$. In particular, $s_{2n} \leq L$ for every n.

Is the sequence (s_{2n-1}) decreasing? We would have to show that $s_{2n-1} > s_{2(n+1)-1}$, that is, $s_{2n-1} > s_{2n+1}$. Indeed,

$$s_{2n+1} = s_{2n-1} - a_{2n} + a_{2n+1} = s_{2n-1} - (a_{2n} - a_{2n+1}).$$

But $a_{2n} - a_{2n+1} > 0$, hence $s_{2n+1} < s_{2n-1}$. On the other hand, recall that $s_{2n-1} \to L$. Since (s_{2n-1}) is decreasing and convergent, it has to converge to its greatest lower bound; that is, $L = g.l.b.\{s_{2n-1}\}$. In particular $L \leq s_{2n-1}$ for all n. So far we have shown that

$$s_{2n} \leq L \leq s_{2n-1} \qquad (3.2)$$

for all n. The proof would come to an end if we can show that $|s_{2n} - L| \leq a_{2n+1}$ and $|s_{2n-1} - L| \leq a_{2n}$. Let us prove these two inequalities separately:

1. From (3.2) we get $0 \leq L - s_{2n}$ and $L \leq s_{2(n+1)-1} = s_{2n+1}$. Thus $0 \leq L - s_{2n} \leq s_{2n+1} - s_{2n} = a_{2n+1}$, which in turn leads to $|L - s_{2n}| \leq a_{2n+1}$; that is, $|s_{2n} - L| \leq a_{2n+1}$.

2. Again, from (3.2) we have $0 \leq s_{2n-1} - L$ and $-L \leq -s_{2n}$. Hence $0 \leq s_{2n-1} - L \leq s_{2n-1} - s_{2n}$. But $s_{2n} = s_{2n-1} - a_{2n}$, consequently $s_{2n-1} - s_{2n} = a_{2n}$. We may then conclude that $0 \leq s_{2n-1} - L \leq a_{2n}$, thus $|s_{2n-1} - L| \leq a_{2n}$. QED

3.7 Absolute and Conditional Convergence

It is to be noted that a series $\Sigma_{n,1} a_n$ may be convergent but $\Sigma_{n,1} |a_n|$ may be divergent. For instance, $\Sigma_{n,1} (-1)^{n+1} \frac{1}{n}$ is convergent but $\Sigma_{n,1} \frac{1}{n}$ is divergent. In this case we say that the given series **converges conditionally**. A natural question arises: Does the convergence of $\Sigma_{n,1} |a_n|$ imply the convergence of $\Sigma_{n,1} a_n$? The answer will come in the next proposition. If $\Sigma_{n,1} |a_n|$ converges we say that $\Sigma_{n,1} a_n$ **converges absolutely**.

Absolute convergence implies convergence

Assume that $\Sigma_{n,1} |a_n|$ converges. Then $\Sigma_{n,1} a_n$ converges.

The proof is quite straightforward. Since

$$0 \leq |a_n| - a_n \leq |a_n| + |a_n| = 2|a_n|,$$

the comparison test implies that $\Sigma_{n,1} (|a_n| - a_n)$ converges. Therefore $\Sigma_{n,1} (a_n - |a_n|)$ converges. But

$$a_n = a_n - |a_n| + |a_n|.$$

This last equality and the convergence of $\Sigma_{n,1} (a_n - |a_n|)$ and $\Sigma_{n,1} |a_n|$ imply the convergence of $\Sigma_{n,1} a_n$. QED

Examples

1. Is the series $\Sigma_{n,1}\frac{\cos n}{n^2}$ convergent? The series $\Sigma_{n,1}\frac{|\cos n|}{n^2}$ converges since

$$0 \leq \frac{|\cos n|}{n^2} \leq \frac{1}{n^2}$$

   and $\Sigma\frac{1}{n^2}$ converges. Then the given series converges converges because absolute convergence implies convergence.

2. What can be said about the convergence of $\Sigma_{n,1}\frac{\sin n}{n^3}$? Since

$$0 \leq \frac{|\sin n|}{n^3} \leq \frac{1}{n^3}$$

   and $\Sigma_{n,1}\frac{1}{n^3}$ converges, the comparison test implies that $\Sigma_{n,1}\frac{|\sin n|}{n^3}$ converges. Then the given series converges since absolute convergence implies convergence.

Remarks

1. If $\Sigma_{n,1}|a_n|$ converges, then $|\Sigma_{n=1}^{\infty}a_n| \leq \Sigma_{n=1}^{\infty}|a_n|$. Why is this so? Let $s_n = a_1 + ... + a_n$, $t_n = |a_1| + ... + |a_n|$. From $s_n \to \Sigma_{n=1}^{\infty}a_n$ it follows[5] that $|s_n| \to |\Sigma_{n=1}^{\infty}a_n|$. Since $t_n \to \Sigma_{n=1}^{\infty}|a_n|$, and $|s_n| \leq t_n$, we can then conclude that $|\Sigma_{n=1}^{\infty}a_n| \leq \Sigma_{n=1}^{\infty}|a_n|$. As expected, the inequality is also valid if instead of start adding from $n = 1$ we need to start adding from $n = k$ onwards.

2. The sum and the difference of two convergent series is a convergent series. However, the product of two convergent series may be a divergent series. Indeed, $\Sigma_{n,1}(-1)^{n+1}\frac{1}{\sqrt{n}}$ is convergent but

$$\Sigma_{n,1}[(-1)^{n+1}\tfrac{1}{\sqrt{n}}][(-1)^{n+1}\tfrac{1}{\sqrt{n}}] = \Sigma_{n,1}(-1)^{2n+2}\tfrac{1}{n} = \Sigma_{n,1}\tfrac{1}{n}$$

   is divergent. Interestingly, if $\Sigma_{n,1}a_n^2$ and $\Sigma_{n,1}b_n^2$ are convergent then $\Sigma_{n,1}a_nb_n$ is convergent! Why is this so? Let us start by noting that, for any two real numbers u, v, the inequality $2|uv| \leq u^2 + v^2$ is true (just use the fact that $0 \leq (u - v)^2$ and $0 \leq (u + v)^2$). Therefore

$$2|a_nb_n| \leq a_n^2 + b_n^2.$$

   Since $\Sigma_{n,1}(a_n^2 + b_n^2)$ is convergent, the comparison test implies that $\Sigma_{n,1}|a_nb_n|$ is convergent. Then $\Sigma_{n,1}a_nb_n$ is convergent.

---

[5]Recall exercise 17, section 2.1.

Exercises for Section 3.7

1. Does the series $\Sigma_{n,1}(-1)^{n+1}\sin(\frac{1}{n})$ converge conditionally?

2. Does the series $\Sigma_{n,2}(-1)^{n+1}\ln(\frac{1}{n})$ converge conditionally?

3. Does the series $\Sigma_{n,1}(-1)^{n+1}\frac{\sqrt{n}}{n+5}$ converge absolutely?

4. If asked whether $\Sigma_{n,1}(-1)^{n+1}\frac{\arctan(n)}{n^2+1}$ is convergent we could try to use the alternating series test. But it seems easier to check the convergence of $\Sigma_{n,1}\frac{\arctan(n)}{n^2+1}$ first, and then conclude that the original series is convergent.

5. Does the series $\Sigma_{n,1}\frac{\cos(n\pi)}{n}$ converge conditionally?

6. Does the series $\Sigma_{n,2}(-1)^{n+1}\frac{1}{\ln n}$ converge conditionally?

7. Analyze whether the series $\Sigma_{n,1}(-1)^{n+1}\frac{1}{n!}$ converges absolutely or conditionally.

8. Is $\Sigma_{n,1}(-1)^{n+1}\frac{\ln n}{n^2}$ absolutely or conditionally convergent?

9. Is $\Sigma_{n,1}(-1)^{n+1}n\sin(\frac{1}{n})$ divergent, conditionally convergent or absolutely convergent?

10. Is $\Sigma_{n,1}(-1)^{n+1}\frac{n+1}{3^n}$ conditionally convergent or absolutely convergent?

3.8   The Ratio Test and the Root Test

The comparison test and the limit comparison test have a drawback: they require us to guess with what to compare the given series. To avoid this difficulty, mathematicians created two tests that do not require finding candidates for comparison.

The Ratio Test

Let $\Sigma_{n,1}a_n$ be a series with $a_n \neq 0$ for all n. Assume that $|\frac{a_{n+1}}{a_n}| \to L$. The following two implications are true:

1. If $0 \leq L < 1$, then $\Sigma_{n,1}|a_n|$ converges.

2. If $L > 1$ or $L = \infty$, then $\Sigma_{n,1}a_n$ diverges.

Let us provide a proof of each of them.

1. Choose M such that $L < M < 1$. Since $|\frac{a_{n+1}}{a_n}| \to L$ we can assert, thanks to property 8 (section 3.1), that there exists k such that

$$\left|\tfrac{a_{n+1}}{a_n}\right| < M \text{ whenever } n \geq k.$$

Thus

$$\left|\tfrac{a_{k+1}}{a_k}\right| < M, \text{ that is, } |a_{k+1}| < M|a_k|.$$

Similarly,

$$\left|\tfrac{a_{k+2}}{a_{k+1}}\right| < M, \text{ that is, } |a_{k+2}| < M|a_{k+1}|.$$

Therefore $|a_{k+2}| < M^2|a_k|$. A pattern emerges:

$$0 < |a_{k+n}| < M^n|a_k| \text{ for } n = 1, 2, 3, \dots$$

However, $\Sigma_{n,1}M^n$ converges because $0 < M < 1$. Then the comparison test implies that $\Sigma_{n,1}|a_{k+n}|$ converges, which in turn implies that $\Sigma_{n,1}|a_n|$ converges.

2. Assume that $L > 1$. Since $\left|\tfrac{a_{n+1}}{a_n}\right| \to L$, property 8 of sequences (section 3.1) allows us to conclude that there exists r such that $\left|\tfrac{a_{n+1}}{a_n}\right| > 1$ whenever $n \geq r$. So $|a_{r+1}| > |a_r|$, $|a_{r+2}| > |a_{r+1}|$ and, in general, $|a_r| < |a_{r+1}| < |a_{r+2}| < |a_{r+3}| < \dots$. Therefore (a_n) does not converge to zero, which in turn implies that $\Sigma_{n,1}a_n$ diverges. In a similar fashion we can show that $L = \infty$ implies divergence. QED

Remarks

1. A reader might raise the following question: What happens if $L = 1$? Let us consider the well-known series $\Sigma_{n,1}\tfrac{1}{n}$ and $\Sigma_{n,1}\tfrac{1}{n^2}$. Applying the ratio test to the first series we get

$$\frac{\tfrac{1}{n+1}}{\tfrac{1}{n}} = \frac{n}{n+1} \to 1$$

while, for the second series, the ratio test leads to

$$\frac{\tfrac{1}{(n+1)^2}}{\tfrac{1}{n^2}} = \frac{n^2}{n^2+2n+1} \to 1.$$

But $\Sigma_{n,1}\tfrac{1}{n}$ diverges while $\Sigma_{n,1}\tfrac{1}{n^2}$ converges. Thus, the ratio test does not provide an answer when $L = 1$. It is inconclusive.

2. If $0 \leq L < 1$, the ratio test implies that $\Sigma_{n,1}|a_n|$ converges. Therefore $\Sigma_{n,1}a_n$ converges since absolute convergence leads to convergence. On the other hand, if $L > 1$ or $L = \infty$, the series $\Sigma_{n,1}a_n$ diverges; hence $\Sigma_{n,1}|a_n|$ diverges too.

Examples

1. The series $\Sigma_{n,1} \frac{3^n+5}{7^n}$ converges because

$$\frac{\frac{3^{n+1}+5}{7^{n+1}}}{\frac{3^n+5}{7^n}} = \frac{1}{7}\frac{3^{n+1}+5}{3^n+5} \rightarrow \frac{3}{7} < 1.$$

2. The series $\Sigma_{n,1} \frac{3^n+5}{2^n}$ diverges because

$$\frac{\frac{3^{n+1}+5}{2^{n+1}}}{\frac{3^n+5}{2^n}} = \frac{1}{2}\frac{3^{n+1}+5}{3^n+5} \rightarrow \frac{3}{2} > 1.$$

3. In section 3.5 we were able to show that $\Sigma_{n,1} \frac{1}{n!}$ is convergent after realizing that $2^n < n!$ for $n \geq 4$. The ratio test provides a shorter route:

$$\frac{\frac{1}{(n+1)!}}{\frac{1}{n!}} = \frac{1}{n+1} \rightarrow 0.$$

4. The series $\Sigma_{n,1} \frac{6^n}{n!}$ converges because

$$\frac{\frac{6^{n+1}}{(n+1)!}}{\frac{6^n}{n!}} = \frac{6}{n+1} \rightarrow 0.$$

The reader must have noticed that the ratio test is the tool of choice whenever a power or a factorial are involved.

5. Does the series $\Sigma_{n,1} \frac{(-1)^n n}{3^n}$ converge?. We have

$$\left| \frac{\frac{(-1)^{n+1}(n+1)}{3^{n+1}}}{\frac{(-1)^n n}{3^n}} \right| = \frac{1}{3}\frac{n+1}{n} \rightarrow \frac{1}{3}.$$

Hence, the series converges absolutely. In particular, it converges.

6. Is $\Sigma_{n,1}(-1)^{n+1}\frac{5^n}{n!}$ convergent? Applying the ratio test, right away we can assert that $\Sigma_{n,1}\frac{5^n}{n!}$ is convergent. Therefore the original series is convergent.

The Root Test

Let $\Sigma_{n,1} a_n$ be a series. Assume that $|a_n|^{1/n} \to L$. The following two implications are true:

1. If $0 \le L < 1$, then $\Sigma_{n,1} |a_n|$ converges.

2. If $L > 1$ or $L = \infty$, then $\Sigma_{n,1} a_n$ diverges.

The proof is even simpler than the one we discussed for the ratio test. Indeed, assume that $0 \le L < 1$ and choose M such that $L < M < 1$. Then there exists a positive integer p such that $|a_n|^{1/n} < M$ whenever $n \ge p$. Hence $0 \le |a_n| < M^n$ provided that $n \ge p$. The series $\Sigma_{n,p} M^n$ is convergent because it is a geometric series with $0 < M < 1$. The comparison test then implies that $\Sigma_{n,p} |a_n|$ converges, which in turn leads to the conclusion that $\Sigma_{n,1} |a_n|$ converges.

On the other hand, assume that $L > 1$. Since $|a_n|^{1/n} \to L$ we can assert that $|a_n|^{1/n} > 1$ eventually; thus, $|a_n| > 1$ eventually. Then (a_n) does not converge to zero, consequently $\Sigma_{n,1} a_n$ diverges. A similar reasoning can be used if $L = \infty$. QED

Remark

What happens if we apply the root test and $a_n^{1/n} \to 1$? Under these circumstances the test is inconclusive. Indeed, $\Sigma_{n,1} \frac{1}{n}$ is divergent and $\frac{1}{n^{1/n}} \to 1$, while $\Sigma_{n,1} \frac{1}{n^2}$ is convergent and

$$\frac{1}{(n^2)^{1/n}} = \frac{1}{n^{1/n} n^{1/n}} = \frac{1}{n^{1/n}} \frac{1}{n^{1/n}} \to 1 \times 1 = 1.$$

Examples

1. The series $\Sigma \frac{(-1)^n}{n^n}$ converges absolutely because $|\frac{(-1)^n}{n^n}|^{1/n} = \frac{1}{n} \to 0$. In particular, it converges.

2. Does the series $\Sigma_{n,1} (1 - \frac{1}{n})^n$ converge or diverge? The root test is inconclusive since $((1 - \frac{1}{n})^n)^{1/n} = 1 - \frac{1}{n} \to 1$. However,

$$(1 - \tfrac{1}{n})^n = (1 + \tfrac{-1}{n})^n \to e^{-1} \ne 0.$$

Hence, the given series diverges.

3. The scenario changes if we have to deal with the series $\Sigma_{n,1} (1 - \frac{1}{n})^{n^2}$. Applying the root test we observe that

$$((1 - \tfrac{1}{n})^{n^2})^{1/n} = (1 - \tfrac{1}{n})^n \to e^{-1} < 1.$$

Thus, the given series converges.

4. It seems natural to ask whether the series $\Sigma_{n,1}(1 + \frac{1}{n})^{n^2}$ converges. Applying the root test we note that

$$((1 + \tfrac{1}{n})^{n^2})^{1/n} = (1 + \tfrac{1}{n})^n \to e > 1.$$

Thus, the given series diverges.

5. We could apply the ratio test to the series $\Sigma_{n,1}\frac{n^2}{3^n}$. But let us see whether the root test provides an answer too:

$$\frac{(n^2)^{1/n}}{3} = \tfrac{1}{3}n^{1/n}n^{1/n} \to \tfrac{1}{3}.$$

Consequently, the given series converges.

Exercises for Section 3.8

Use either the ratio test or the root test to determine whether the following series are convergent or divergent

1. $\Sigma_{n,1}\frac{n^3}{2^n}$.

2. $\Sigma_{n,1}\frac{n^5}{n!}$.

3. $\Sigma_{n,1}(-1)^n \sin^n(\frac{1}{n})$.

4. $\Sigma_{n,1}(-1)^n \frac{n!}{2^n}$.

5. $\Sigma_{n,1}\frac{(\ln n)^n}{n}$.

6. $\Sigma_{n,1}\frac{n^n}{n!}$.

7. $\Sigma_{n,1}\frac{n!}{x^n}, x \neq 0$.

8. $\Sigma_{n,2}\left(\frac{\ln n}{n+2}\right)^n$.

9. $\Sigma_{n,0}e^{-n^2}$.

10. Can we apply the ratio test to an arbitrary p-series $\Sigma_{n,1}\frac{1}{n^p}$ and reach an answer on whether it converges or diverges?

3.9   The Integral Test

The idea of using integrals to analyze series should come as no surprise since it was employed at the end of section 3.2 when we found that the series $\Sigma_{n,1}\frac{1}{n}$ diverges. Actually, the following test is extremely powerful and useful. It allows us to utilize the machinery of integral calculus with the purpose of determining whether a series converges or diverges.

The test

Assume that f is a positive, continuous, and decreasing function defined on $[1,\infty)$. Then

$$\Sigma_{n,1}f(n) \text{ converges if and only if } \int_1^\infty f(x)dx \text{ converges.}$$

Using the integral test

1. We already know that the series $\Sigma_{n,1}\frac{1}{n^p}$ converges for $p \geq 2$ and diverges for $p \leq 1$. What happens when $1 < p < 2$? Assume that $p > 1$ and let $f(x) = \frac{1}{x^p} = x^{-p}$, for every $x > 0$. Then

$$f'(x) = -px^{-p-1} = -p\frac{1}{x^{p+1}} < 0.$$

Thus, f is decreasing, continuous and positive. Moreover[6],

$$\int_1^r \frac{1}{x^p}dx = \int_1^r (x^{-p})dx = \left[\frac{x^{-p+1}}{-p+1}\right]_1^r = \frac{r^{1-p}}{1-p} - \frac{1}{1-p} = \frac{1}{1-p}\frac{1}{r^{p-1}} + \frac{1}{p-1}.$$

Since $p - 1 > 0$ we can conclude that $\frac{1}{r^{p-1}} \to 0$ as $r \to \infty$. Therefore $\int_1^\infty \frac{1}{x^p}dx$ converges whenever $p > 1$. The integral test implies that $\Sigma_{n,1}\frac{1}{n^p}$ converges if $p > 1$. In a similar fashion we can confirm that the series under consideration diverges for $p < 1$. Of course, it is a well-known fact by now that the series diverges also for $p = 1$.

2. Does the series $\Sigma_{n,2}\frac{1}{n\ln n}$ converge? We could try to compare the given series with a p-series. Indeed,

$$0 < \frac{1}{n\ln n} < \frac{1}{n}, n \geq 3 \text{ and}$$

$$0 < \frac{1}{n^2} < \frac{1}{n\ln n}, \ n \geq 2.$$

---

[6]The reader might recall that this type of calculations were done in section 2.29.

Table 3.4: The sequence of partial sums corresponding to $\Sigma_{n,1}\frac{1}{n\ln n}$

m	$\sum_{n=1}^{m}\frac{1}{n\ln n}$
50	2.16128
100	2.32294
200	2.46254
1000	2.7274
5000	2.93678

But $\Sigma_{n,3}\frac{1}{n}$ diverges while $\Sigma_{n,2}\frac{1}{n^2}$ converges. No other p-series seems to work. A numerical exploration (see table 3.4) might suggest that the series under consideration is divergent. However, since the value of the series increases very slowly as m increases, some lingering doubts remain. Let us apply the integral test. We define $f(x) = \frac{1}{x\ln x} = (x\ln x)^{-1}$, $x \geq 2$. Then

$$f'(x) = -(x\ln x)^{-2}(\ln x + 1) = -\frac{\ln x + 1}{(x\ln x)^2} < 0,\ x \geq 2.$$

Thus, f is decreasing, continuous and positive on $[2,\infty)$. Defining $u(x) = \ln x$ we get $u'(x) = 1/x$; hence, as $r \to \infty$,

$$\int_2^r \frac{1}{x\ln x}dx = \int_2^r \frac{u'(x)}{u(x)}dx = [\ln(\ln x)]_2^r = \ln(\ln r) - \ln(\ln 2)) \to \infty.$$

Consequently $\Sigma_{n,2}\frac{1}{n\ln n}$ diverges.

3. Next let us try to investigate whether $\Sigma_{n,2}\frac{1}{n(\ln n)^2}$ converges. Defining $f(x) = \frac{1}{x(\ln x)^2}$ one can see that it is positive, continuous and decreasing on $[2,\infty)$. It is necessary to check whether $\int_2^\infty f(x)dx$ converges. Letting $u(x) = \ln x$ as in the previous example, we get

$$\int_2^r \frac{dx}{x(\ln x)^2} = \int_2^r u^{-2}(x)u'(x)dx = -[u^{-1}(x)]_2^r = -[\frac{1}{\ln x}]_2^r = -(\frac{1}{\ln r} - \frac{1}{\ln 2}) \to \frac{1}{\ln 2}\ \text{as}$$
$$r \to \infty.$$

Since the improper integral converges, the given series will converge too.

Justifying the test

Before going any further we might as well provide a justification of the test. We observe (Figure 3.2) that the area of the rectangle with height $f(2)$ and width 1 is $f(2)$, an area smaller than the area under $y = f(x)$ between 1 and 2. Thus $f(2) < \int_1^2 f(x)dx$. A similar argument leads to the inequalities $f(3) < \int_2^3 f(x)dx$, ..., $f(n) < \int_{n-1}^n f(x)dx$. Therefore

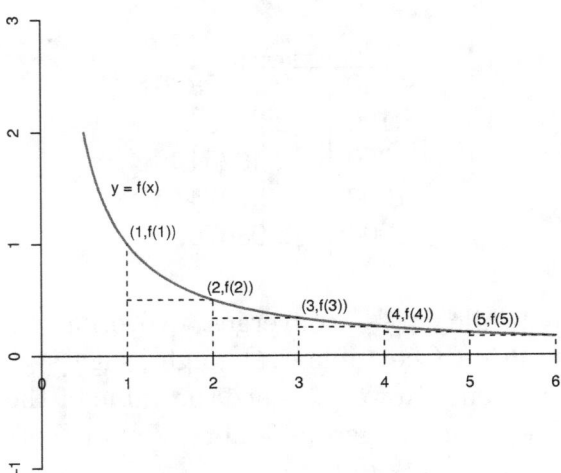

Figure 3.2: A positive, continuous, and decreasing function on $(0, \infty)$.

$$f(2) + f(3) + \ldots + f(n) < \int_1^2 f + \int_2^3 f + \ldots + \int_{n-1}^n f = \int_1^n f.$$

Consequently,

$$s_n = f(1) + f(2) + f(3) + \ldots + f(n) < f(1) + \int_1^n f. \tag{3.3}$$

Next we will show that $\int_1^{n+1} f < s_n$ for every n ($n = 1, 2, \ldots$). Indeed, the area of the rectangle of height $f(1)$ and width 1 is bigger than the area under the curve $y = f(x)$ between 1 and 2. Thus, $\int_1^2 f < f(1)$. Similarly, $\int_2^3 f < f(2)$, $\ldots$, $\int_n^{n+1} f < f(n)$. Therefore,

$$\int_1^{n+1} f = \int_1^2 f + \int_2^3 f + \ldots + \int_n^{n+1} f < f(1) + f(2) + \ldots + f(n) = s_n. \tag{3.4}$$

From (3.3) and (3.4) it follows that

$$\int_1^{n+1} f < s_n < f(1) + \int_1^n f. \tag{3.5}$$

Assume that $\int_1^\infty f$ converges. Then $\lim_{n \to \infty} \int_1^n f$ exists and consequently the sequence $(\int_1^n f)$ is bounded[7]. Thus, there exists $M > 0$ such that $\int_1^n f < M$ for every n. From (3.5) it follows that $s_n < f(1) + M$ for every n. But (s_n) is an increasing

---

[7]Every convergent sequence is bounded; see appendix B1.

sequence because, as n increases, we are adding positive numbers. Thanks to proposition 1 (section 3.1), we can then conclude that (s_n) is convergent. That is to say, $\Sigma_{n,1} f(n)$ converges.

On the other hand, suppose that (s_n) is convergent. Hence, this sequence is bounded, which in turn implies that there exists $H > 0$ such that $s_n < H$ for all n. Then from (3.5) we can conclude that $\int_1^{n+1} f < H$ for all n. Given any $t > 1$ there exists n such that $t < n+1$, consequently $\int_1^t f < \int_1^{n+1} f < H$. Let $F(t) = \int_1^t f$, $t > 1$. Evidently, $F(t)$ is an increasing function since f is positive. It is bounded above (H is an upper bound), hence proposition 1a (section 3.1) leads to the conclusion that $\lim_{t \to \infty} F(t)$ exists. That is, $\int_1^\infty f$ is convergent. QED

Remark

It should be noted that, under the hypothesis of the integral test, and assuming the convergence of the corresponding improper integral, from (3.5) it follows that

$$\int_1^\infty f(x)dx \leq \Sigma_{n=1}^\infty f(n) \leq f(1) + \int_1^\infty f(x)dx. \qquad (3.6)$$

In the particular case of the p-series, when $p > 1$ the preceding inequalities become

$$\frac{1}{p-1} \leq \Sigma_{n=1}^\infty \frac{1}{n^p} \leq 1 + \frac{1}{p-1} = \frac{p}{p-1}.$$

Beware that, when applying the integral test, $\Sigma_{n=1}^\infty f(n)$ does not have to be equal to $\int_1^\infty f(x)dx$. All we can say is that the inequalities in (3.6) are valid.

Moreover, the reader will surely realize that if f is positive, continuous, and decreasing on $[2, \infty)$ then

$$\Sigma_{n,2} f(n) \text{ converges if and only if } \int_2^\infty f(x)dx \text{ converges.}$$

The test can be applied if f is positive, continuous, and decreasing on any interval of the form $[m, \infty)$, where $m \geq 1$.

Approximating the value of a series through the integral test

Assume that f, defined on $[1, \infty)$, is positive, continuous, and decreasing, and $\int_1^\infty f(x)dx$ converges. Let n be an arbitrary (fixed) natural number. The same type of analysis, which we used when trying to justify the test, allows us to assert that, for $i > 1$,

$$f(n+1) < \int_n^{n+1} f, \ f(n+2) < \int_{n+1}^{n+2} f, \ ..., \ f(n+i) < \int_{n+i-1}^{n+i} f.$$

Therefore

$$f(n+1) + f(n+2) + ... + f(n+i) < \int_n^{n+i} f.$$

Letting $i \to \infty$ we reach the inequality $\Sigma_{i=n+1}^{\infty} f(i) \le \int_n^{\infty} f$, that is,

$$R_n \le \int_n^{\infty} f \tag{3.7}$$

where $R_n = \Sigma_{i=n+1}^{\infty} f(i)$ is the remainder, a concept that was introduced at the end of section 3.3. Obviously $R_n > 0$.

Let us consider two examples in detail.

1. The series $\Sigma_{n,1} \frac{1}{n^2}$ is a well-known convergent series. According to (3.7) we will have $R_n \le \int_n^{\infty} \frac{1}{x^2} dx$. But, as $r \to \infty$,

$$\int_n^r \frac{1}{x^2} dx = -\left[\frac{1}{x}\right]_n^r = -\frac{1}{r} + \frac{1}{n} \to \frac{1}{n}.$$

Therefore $R_n \le \frac{1}{n}$. For instance, how many terms do we have to add in order to make sure that $R_n < 10^{-2}$? We would need to choose a natural number n so that $\frac{1}{n} < 10^{-2}$. Thus, $n > 100$ will do the job, say $n = 101$. Then

$$s_{101} = 1 + \frac{1}{2^2} + ... + \frac{1}{100^2} + \frac{1}{101^2} \approx 1.635081$$

is an approximation of $\Sigma_{n=1}^{\infty} \frac{1}{n^2}$ such that

$$\Sigma_{n=1}^{\infty} \frac{1}{n^2} - s_{101} = \Sigma_{n=102}^{\infty} \frac{1}{n^2} < 0.01.$$

Actually, one can prove that if f, defined on $[1, \infty)$, is positive, continuous, and decreasing, and $\int_1^{\infty} f$ converges, then

$$\int_{n+1}^{\infty} f \le R_n. \tag{3.8}$$

Putting together (3.7) and (3.8) we get

$$\int_{n+1}^{\infty} f \le R_n \le \int_n^{\infty} f.$$

That is to say, $\int_{n+1}^{\infty} f \le \Sigma_{i=1}^{\infty} f(i) - s_n \le \int_n^{\infty} f$, which in turn leads to

$$s_n + \int_{n+1}^{\infty} f \le \Sigma_{i=1}^{\infty} f(i) \le s_n + \int_n^{\infty} f. \tag{3.9}$$

In the particular case when $f(x) = \frac{1}{x^2}$, the preceding inequalities become

$$s_n + \int_{n+1}^{\infty} \frac{1}{x^2} \le \Sigma_{i=1}^{\infty} \frac{1}{i^2} \le s_n + \int_n^{\infty} \frac{1}{x^2} dx,$$

where $s_n = 1 + \frac{1}{2^2} + \ldots + \frac{1}{n^2}$. But

$$\int_n^\infty \tfrac{1}{x^2}dx = \lim_{b\to\infty} \int_n^b x^{-2}dx = \lim_{b\to\infty}[-x^{-1}]_n^b = -\lim_{b\to\infty}(\tfrac{1}{b} - \tfrac{1}{n}) = \tfrac{1}{n}.$$

In a similar fashion one can show that $\int_{n+1}^\infty \frac{1}{x^2}dx = \frac{1}{n+1}$. Therefore

$$s_n + \frac{1}{n+1} \leq \Sigma_{i=1}^\infty \frac{1}{i^2} \leq s_n + \frac{1}{n}. \tag{3.10}$$

For example, let us choose $n = 20$. Using a calculator we get $\Sigma(\frac{1}{i^2}, i, 1, 20) = 1.5961632$. Thus

$$1.5961632 + \tfrac{1}{21} \leq \Sigma_{i=1}^\infty \tfrac{1}{i^2} \leq 1.5961632 + \tfrac{1}{20}.$$

Then

$$1.6437822 \leq \Sigma_{i=1}^\infty \tfrac{1}{i^2} \leq 1.6461632.$$

We have found the first two 'true' decimals of $\Sigma_{i=1}^\infty \frac{1}{i^2}$. It is to be expected that the approximation will improve as n gets bigger. Indeed, if $n = 100$ we get $\Sigma(\frac{1}{i^2}, i, 1, 100) = 1.6349839$. Therefore

$$1.6349839 + \tfrac{1}{101} \leq \Sigma_{i=1}^\infty \tfrac{1}{i^2} \leq 1.6349839 + \tfrac{1}{100},$$

which in turn leads to

$$1.6448849 \leq \Sigma_{i=1}^\infty \tfrac{1}{i^2} \leq 1.6449839.$$

Thus, we can be confident that, up to the first three decimals, $\Sigma_{i=1}^\infty \frac{1}{i^2} = 1.644$.

Does the series $\Sigma_{i,1} \frac{1}{i^2}$ have an exact value? This was an open problem for several decades, the famous Basel Problem, until Leonhard Euler (1707-1783) solved it in 1734. He found that $\Sigma_{i=1}^\infty \frac{1}{i^2} = \frac{\pi^2}{6}$. We will say more about this amazing equality in an enrichment note at the end of section 3.14.

2. Let us approximate the value of $\Sigma_{i=1}^\infty \frac{\arctan(i)}{i^2+1}$. It is certainly convergent because $\Sigma_{i,1} \frac{1}{i^2}$ is convergent and

$$\frac{\arctan(i)}{i^2+1} < \frac{\pi}{2}\frac{1}{i^2+1} < \frac{\pi}{2}\frac{1}{i^2}.$$

Hence

$$\Sigma_{i=1}^{\infty} \frac{\arctan(i)}{i^2+1} \leq \frac{\pi}{2}\Sigma_{i=1}^{\infty}\frac{1}{i^2} = \frac{\pi}{2}\frac{\pi^2}{6} \approx 2.583856.$$

This is a rather 'crude' approximation, so let us use the integral test instead. Defining $f(x) = \frac{\arctan x}{x^2+1}$ we can check that $f(x)$ is defined on $[1, \infty)$, and is positive, continuous, and decreasing (actually, $f'(x) < 0$ eventually). Moreover, $\int_1^{\infty} f(x)dx$ is convergent because, as $r \to \infty$,

$$\int_1^r f(x)dx = \frac{1}{2}[\arctan^2 x]_1^r = \frac{1}{2}\arctan^2 r - \frac{1}{2}\arctan 1 \to \frac{1}{2}\frac{\pi^2}{4} - \frac{1}{2}\frac{\pi}{4}.$$

In a similar fashion we can prove that $\int_n^{\infty} f(x)dx = \frac{1}{2}\frac{\pi^2}{4} - \frac{1}{2}\arctan^2(n)$ and $\int_{n+1}^{\infty} f(x)dx = \frac{1}{2}\frac{\pi^2}{4} - \frac{1}{2}\arctan^2(n+1)$. Then using (3.9) we get

$$s_n + \frac{1}{2}\frac{\pi^2}{4} - \frac{1}{2}\arctan^2(n+1) \leq \Sigma_{i=1}^{\infty}\frac{\arctan(i)}{i^2+1} \leq s_n + \frac{1}{2}\frac{\pi^2}{4} - \frac{1}{2}\arctan^2(n).$$

For instance, for $n = 100$ we get

$$1.1355038 \leq \Sigma_{i=1}^{\infty}\frac{\arctan(i)}{i^2+1} \leq 1.1356584.$$

Thus, we can feel confident that the value of the series is 1.135 (up to the third decimal).

Exercises for Section 3.9

1. In section 3.5 we showed that $\Sigma_{n,2}\frac{\ln n}{n}$ is divergent. Reach the same answer using the integral test.

2. In section 3.5 we showed, after some false starts, that $\Sigma_{i,1}\frac{\ln i}{i^2}$ is convergent. Use the integral test to show that the series does converge indeed. Apply the inequalities from (3.9) to find the first two 'true' decimals of $\Sigma_{i=1}^{\infty}\frac{\ln i}{i^2}$.

3. Does the series $\Sigma_{n,2}\frac{1}{n(\ln n)^p}$, $p > 1$, converge?

4. Does the series $\Sigma_{n,2}\frac{1}{n\sqrt{\ln n}}$ converge?

5. Find n such that $\Sigma_{n=1}^{\infty}\frac{1}{n^3} - s_n < 10^{-2}$.

6. Find n such that $\Sigma_{n=1}^{\infty}\frac{1}{n^4} - s_n < 10^{-3}$.

3.10 Different Strategies

Given a series $\Sigma_{n,1} a_n$, we may not realize right away what test to use. There is a bewildering number of different tests! However, there are some strategies that can be followed:

1. If we identify the series as geometric there is not much of a problem because $\Sigma_{n,0} r^n$ converges if and only if $|r| < 1$ and its sum is $1/(1-r)$.

2. Telescoping series are usually easy to handle, but one has to distinguish them as such from the start.

3. The test of divergence can be applied readily to check whether the series is divergent: If (a_n) does not converge to zero then $\Sigma_{n,1} a_n$ diverges.

4. Alternating series can be identified easily. We might suspect that $\Sigma_{n,1}(-1)^{n+1} a_n$ converges but we have to check that $a_n > 0$, $a_n \to 0$, and (a_n) is decreasing. For the latter property we might have to work with the corresponding function of a real variable and verify that its derivative is negative (eventually).

5. Using the comparison test or the limit comparison test is a convenient strategy if we can employ either the p-series $\Sigma_{n,1} \frac{1}{n^p}$ or a geometric series for purposes of comparison. The drawback of these two tests is that sometimes it is not easy to figure out with what to compare the given series.

6. The ratio test and the root test have a distinctive advantage: they can be applied directly without having to worry about finding another series to compare with. A 'rule of thumb' is to try the ratio test when a power function or factorials are involved, and to try the root test if $a_n^{1/n}$ happens to be a more manageable and simpler expression than a_n.

7. Absolute convergence implies convergence, with the added advantage that proving absolute convergence is sometimes easier than proving convergence. Thus, if the series under scrutiny has positive and negative terms, dealing with the series of absolute values of the terms is an option that we have to keep in mind.

8. The integral test is quite powerful. However, we have to check that $f(x)$ is positive, continuous, and eventually decreasing, as well as to find out whether $\int_1^\infty f(x)dx$ is convergent or not. The latter task may be difficult if $\int f(x)dx$ is not expressible as an elementary function.

9. The integral test has a 'sharp' inequality accompanying it, namely $\int_{n+1}^\infty f \leq R_n \leq \int_n^\infty f$, where $R_n = \Sigma_{i=n+1}^\infty f(i)$. Thus, this test is to be preferred if one wishes to approximate the value of a convergent series.

10. If none of the tests seem to work and you have access to a graphing calculator, try to explore the convergence or divergence of the series using the Σ option with different values for the corresponding finite sum. If the numbers on the screen seem to get closer and closer to each other as n increases, it is highly probable that the series converges. Then we can go back and try to prove convergence using all the knowledge at our command.

Errors to avoid

Here is a sample of errors made by students on a test, when asked to determine whether a given series converges or not. They are included in this section with the purpose of warning the reader about common mistakes that can be done.

1. $\Sigma_{n,2} \frac{1}{\ln(2n)}$. Applying the ratio test, a student arrived at

$$\frac{\frac{1}{\ln(2n+2)}}{\frac{1}{\ln(2n)}} = \frac{\ln(2n)}{\ln(2n+2)} \to 0 \text{ as } n \to \infty.$$

Thus, he concluded that the series converges. Actually,

$$\lim_{n\to\infty} \frac{\ln(2n)}{\ln(2n+2)} = 1.$$

Hence, the ratio test is inconclusive when applied to the given series. Rather we should try the limit comparison test, comparing the given series with $\Sigma_{n,1} \frac{1}{n}$. Indeed,

$$\frac{\frac{1}{\ln(2n)}}{\frac{1}{n}} = \frac{n}{\ln(2n)} \to \infty \text{ as } n \to \infty.$$

Since $\Sigma_{n,1} \frac{1}{n}$ diverges, thanks to LCT we can conclude that $\Sigma_{n,2} \frac{1}{\ln(2n)}$ diverges!

2. $\Sigma_{n,2} \frac{1}{n(\ln(n))^3}$. Several students tried to use LCT, comparing with $\Sigma_{n,2} \frac{1}{n^2}$. It is not a bad idea, but they soon ran into trouble when writing

$$\frac{\frac{1}{n(\ln(n))^3}}{\frac{1}{n^2}} = \frac{n}{(\ln(n))^3} \to 0 \text{ as } n \to \infty,$$

and concluding that $\Sigma_{n,2} \frac{1}{n(\ln(n))^3}$ is convergent. The answer is correct but the justification is not. Why? Using L'Hôpital's rule, three times, we get

$$\lim_{n\to\infty} \frac{n}{(\ln(n))^3} = \infty.$$

Thus, LCT is inconclusive because $\Sigma_{n,2}\frac{1}{n^2}$ is convergent. We should try to use the integral test, as we did with a similar problem in section 3.9. Let

$$f(x) = \frac{1}{x(\ln x)^3}, \; x \geq 2.$$

This is a continuous, positive, and decreasing function. Using the substitution $u = \ln x$ we get

$$\int \frac{dx}{x(\ln x)^3} = \int u^{-3}du = -\tfrac{1}{2}u^{-2} + C = -\tfrac{1}{2}\frac{1}{(\ln x)^2} + C.$$

Then

$$\int_2^r \frac{dx}{x(\ln x)^3} = -\tfrac{1}{2}\big[\tfrac{1}{(\ln x)^2}\big]_2^r = -\tfrac{1}{2}\big(\tfrac{1}{(\ln r)^2} - \tfrac{1}{(\ln 2)^2}\big) \to \tfrac{1}{2}\frac{1}{(\ln(2))^2} \text{ as } r \to \infty.$$

Thus, $\int_2^\infty \frac{dx}{x(\ln x)^3}$ is convergent, which in turn implies the convergence of $\Sigma_{n,2}\frac{1}{n(\ln(n))^3}$.

3. $\Sigma_{n,1}\frac{\arcsin(1/n)}{n}$. A common mistake was to write

$$\frac{\frac{\arcsin(1/n)}{n}}{\frac{1}{n}} = \arcsin(1/n) \to \frac{\pi}{2} \text{ as } n \to \infty$$

and then assert that the given series is divergent. Of course, we should realize that $\arcsin(1/n) \to 0$ as $n \to \infty$. This fact makes it impossible to draw a valid conclusion when comparing with $\Sigma_{n,1}\frac{1}{n}$. Rather let us compare with $\Sigma\frac{1}{n^2}$. Indeed

$$\frac{\frac{\arcsin(1/n)}{n}}{\frac{1}{n^2}} = n\arcsin(1/n).$$

But, using L'Hôpital's rule we get

$$\lim_{x\to\infty} x\arcsin(\tfrac{1}{x}) = \lim_{x\to\infty} \frac{\arcsin(\frac{1}{x})}{\frac{1}{x}} = 1.$$

Therefore $\lim_{n\to\infty} n\arcsin(1/n) = 1$. Due to the convergence of $\Sigma_{n,1}\frac{1}{n^2}$, LCT allows us to conclude that $\Sigma_{n,1}\frac{\arcsin(1/n)}{n}$ is convergent.

4. $\Sigma_{n,1}\frac{2+\cos(n)}{n^2}$. Several students started by comparing the given series with $\Sigma\frac{1}{n^2}$ through LCT and writing

$$\frac{\frac{2+\cos(n)}{n^2}}{\frac{1}{n^2}} = 2 + \cos(n) \to L > 0 \text{ as } n \to \infty.$$

Then they claimed that LCT implies that the series converges. Well, something is wrong: $\lim_{n \to \infty} \cos(n)$ does not exist! Thus, although the answer is correct, the path taken is not. A simple application of the comparison test solves the problem at hand. Let us recall that $-1 \leq \cos(n) \leq 1$. Hence

$$0 < \frac{2+\cos n}{n^2} \leq \frac{3}{n^2}.$$

Since $\Sigma_{n,1} \frac{3}{n^2}$ converges, we can infer that $\Sigma_{n,1} \frac{2+\cos(n)}{n^2}$ in turn converges.

5. $\Sigma_{n,1} \frac{e^{2/n}}{n^2}$. More than one student tried the ratio test, which is not a surprise since there is a tendency of over-reliance on the above-mentioned test. Soon they got into trouble since the ratio test leads to

$$\frac{n^2}{(n+1)^2} e^{-\frac{2}{n^2+n}} \to 1, \text{ as } n \to \infty.$$

Thus, the test is inconclusive when applied to the series under study. Rather we should use LCT by comparing with $\Sigma_{n,1} \frac{1}{n^2}$. Indeed,

$$\frac{\frac{e^{2/n}}{n^2}}{\frac{1}{n^2}} = e^{2/n} \to 1 \text{ as } n \to \infty.$$

Since $\Sigma_{n,1} \frac{1}{n^2}$ converges, we can conclude that $\Sigma_{n,1} \frac{e^{2/n}}{n^2}$ also converges.

Exercises for Section 3.10

In the following problems determine whether the given series converges or diverges. Clearly state what test you are using to reach a conclusion and show all your steps.

1. $\Sigma_{n,1} \left(\frac{1}{n} - \frac{1}{n^2} \right)$.

2. $\Sigma_{n,2} \frac{1}{n^2 \ln n}$.

3. $\Sigma_{n,1} \left(1 - \frac{5}{n} \right)^n$.

4. $\Sigma_{n,1} (-1)^n \ln\left(1 + \frac{2}{n} \right)$.

5. $\Sigma_{n,1} \frac{e^{1/n}}{n^3}$.

6. $\Sigma_{n,1} \left(2 + \frac{1}{n} \right)^n$.

7. $\Sigma_{n,1} \frac{(-1)^n}{n+3^n}$.

8. $\Sigma_{n,1} \frac{n \ln n}{4^n}$.

9. $\Sigma_{n,1} \frac{\sin[(n+\frac{1}{2})\pi]}{n^{3/2}}$.

10. $\Sigma_{n,2} \frac{1}{\ln n}$.

11. $\Sigma_{n,1} \frac{1}{\cos^2(n)}$.

12. $\Sigma_{n,1} n e^{-n^2}$.

13. $\Sigma_{n,1} \frac{\arcsin(\frac{1}{n})}{n^3}$.

3.11   Power Series

Let us start this section by asking a question: For what values of x do the series $\Sigma_{n,0} \frac{x^n}{n!}$, $\Sigma_{n,1} \frac{x^n}{n}$, and $\Sigma_{n,0} n! x^n$ converge? Observe that x is a variable that ranges over the real numbers. So far we know that $\Sigma_{n,0} x^n$ converges only when $|x| < 1$ because it is a geometric series. Moreover,

$$\Sigma_{n=0}^{\infty} x^n = \frac{1}{1-x}, \; |x| < 1.$$

This fact was proven in section 2.1 and played a significant role in many examples across the book. Our first task is to analyze with care the series $\Sigma_{n,0} \frac{x^n}{n!}$. When $x = 0$ the series is obviously convergent, so let us assume that x is an arbitrary (fixed) non-zero real number. Applying the ratio test it follows that

$$\frac{|\frac{x^{n+1}}{(n+1)!}|}{|\frac{x^n}{n!}|} = \frac{1}{n+1} |x| \to 0 \text{ as } n \to \infty.$$

Hence $\Sigma_{n,0} \frac{x^n}{n!}$ converges absolutely for every x. For each x, to what number does the series converge? This is a question that will be answered later in this section. For the time being, all we know is that the series converges absolutely for all x; thus, it converges for every x. This assertion goes beyond what we knew with regard to the convergence of $\Sigma_{n,0} \frac{1}{n!}$. The latter's convergence was independently justified on the basis of the comparison test and also through the ratio test.

Next let us deal with $\Sigma_{n,1} \frac{x^n}{n}$. It is convergent for $x = 0$, so we might well consider that x is different from zero and is an arbitrary, fixed, real number. Applying the ratio test it follows that

$$\frac{|\frac{x^{n+1}}{n+1}|}{|\frac{x^n}{n}|} = \frac{n}{n+1} |x| \to |x| \text{ as } n \to \infty.$$

Hence $\Sigma_{n,1} \frac{x^n}{n}$ converges absolutely for $|x| < 1$ and diverges for $|x| > 1$. What happens if $x = 1$ or $x = -1$? We do know, from work done in previous sections, that $\Sigma_{n,1} \frac{1}{n}$ diverges while $\Sigma_{n,1} (-1)^n \frac{1}{n}$ converges. Indeed, the first one is a well-known divergent series while the second one is convergent thanks to the alternating series test. Finally, the series $\Sigma_{n,0} n! x^n$ converges only when $x = 0$ because the sequence $(n! x^n)$ does not converge to zero whenever $x \neq 0$.

The three examples that we have analyzed are series of the type $\Sigma_{n,0} a_n x^n$, where (a_n) is a given sequence of real numbers. They are called **power series**, a quite appropriate name because they are a generalization of polynomials $a_0 + a_1 x + ... + a_n x^n$. Interestingly, the behavior of any power series, with regard to convergence, is similar to that observed in the previous three examples.

Remark

Once we learn, later in the section, that power series can be differentiated and integrated term by term, it will be possible to show that $\Sigma_{n=0}^{\infty} \frac{x^n}{n!} = e^x$ for every x, and $\Sigma_{n=1}^{\infty} \frac{x^n}{n} = -\ln(1-x)$, $-1 \leq x < 1$ (see exercise 11 at the end of the section).

Finding the interval of convergence

Given any power series $\Sigma_{n,0} a_n x^n$, $a_n \neq 0$, suppose that

$$\lim_{n \to \infty} \left| \frac{a_{n+1}}{a_n} \right| = L.$$

If $L > 0$ then the series converges absolutely whenever $|x| < \frac{1}{L}$ and diverges if $|x| > \frac{1}{L}$. Furthermore, if $L = 0$ the series converges absolutely for all x while if $L = \infty$ the series converges only for $x = 0$.

The proof is not difficult. Indeed, the power series obviously converges for $x = 0$. For any $x \neq 0$ we have

$$\left| \frac{a_{n+1} x^{n+1}}{a_n x^n} \right| = \left| \frac{a_{n+1}}{a_n} \right| |x|.$$

If $L > 0$, it follows that $\left| \frac{a_{n+1} x^{n+1}}{a_n x^n} \right| \to L|x|$. By the ratio test we will have absolute convergence provided that $L|x| < 1$ and divergence if $L|x| > 1$. On the other hand, if $L = 0$ then $\left| \frac{a_{n+1} x^{n+1}}{a_n x^n} \right| \to 0$ for any $x \neq 0$. Thus, under these circumstances the given series converges absolutely for any x. Finally, suppose that $L = \infty$, that is, for any $x \neq 0$

$$\left| \frac{a_{n+1} x^{n+1}}{a_n x^n} \right| \to \infty.$$

Then, according to the ratio test, $\Sigma_{n,0} a_n x^n$ diverges for $x \neq 0$. Thus, the power series converges only for $x = 0$. QED

It should be noted that the previous analysis does not tell us what happens when $|x| = \frac{1}{L}$, that is, when $x = \frac{1}{L}$ or $x = -\frac{1}{L}$. At both points one has to make a separate analysis using any of the tests learned so far. The number $R = \frac{1}{L}$ is called the **radius of convergence** of the given power series, while the **interval of convergence** is the set of all values x for which the power series converges. This interval can be $(-R, R)$, $[-R, R)$, $[-R, R]$, or $(-R, R]$. It can also happen that the power series is absolutely convergent for every x, in which case we say that $\Re$, the whole real line, is the interval of convergence. Finally, the power series might converge only when $x = 0$. The latter type of series are not very interesting.

Examples

1. Let us consider the power series $\Sigma_{n,1} n x^n$. Since $\frac{n+1}{n} \to 1$, the series converges absolutely on $(-1, 1)$ and diverges on $(-\infty, -1)$ and $(1, \infty)$. But $\Sigma_{r,1} n$ and $\Sigma_{n,1} n(-1)^n$ diverge because neither the sequence (n) nor the sequence $(n(-1)^n)$ converge to zero.

2. The power series $\Sigma_{n,0} (-1)^n \frac{x^{2n+1}}{2n+1}$ appears to be more of a challenge. Right away we get

$$\frac{\left|\frac{(-1)^{n+1}}{2(n+1)+1}\right|}{\left|\frac{(-1)^n}{2n+1}\right|} = \frac{2n+1}{2n+3} \to 1.$$

Hence, the series converges on $(-1, 1)$ and diverges on $(-\infty, -1)$ and $(1, \infty)$. Does $\Sigma_{n,0} (-1)^n \frac{1}{2n+1}$ converge? The alternating series test assures us that it does converge. So, the given series converges at $x = 1$ too. On the other hand,

$$\Sigma_{n,0} (-1)^n \frac{(-1)^{2n+1}}{2n+1} = \Sigma_{n,0} \frac{(-1)^{3n+1}}{2n+1},$$

which is also convergent due to the alternating series test. Thus, the interval of convergence happens to be $[-1, 1]$. But neither at -1 nor 1 there is absolute convergence since $\Sigma_{n,0} \frac{1}{2n+1}$ is divergent (the limit comparison test, comparing with $\Sigma_{n,1} \frac{1}{n}$, leads to divergence).

3. What is the interval of convergence of the series $\Sigma_{n,0} (-1)^n \frac{x^{2n+1}}{(2n+1)!}$? We observe that

$$\frac{\frac{1}{(2n+3)!}}{\frac{1}{(2n+1)!}} = \frac{1}{(2n+3)(2n+2)} \to 0 \text{ as } n \to \infty.$$

Thus, the given power series converges absolutely for all x. Its interval of convergence happens to be the whole real line. In an identical fashion we can prove that $\Sigma_{n,0} (-1)^n \frac{x^{2n}}{(2n)!}$ converges absolutely for all x.

An alternative

Given any power series, suppose that $\lim_{n\to\infty} |a_n|^{1/n} = L$. If $L > 0$, the series converges absolutely whenever $|x| < \frac{1}{L}$ and diverges if $|x| > \frac{1}{L}$. Furthermore, if $L = 0$ the series converges absolutely for all x while if $L = \infty$ the series only converges for $x = 0$.

This result was to be expected and can be proven readily using the root test. It is enough to note that, for any x,

$$(|a_n x^n|)^{1/n} = (|a_n|)^{1/n}|x| \to L|x|.$$

Examples

1. What is the interval of convergence of $\Sigma_{n,2} \frac{1}{(\ln n)^n} x^n$? The answer is the whole real line because

$$\frac{1}{((\ln n)^n)^{1/n}} = \frac{1}{\ln n} \to 0 \text{ as } n \to \infty.$$

So, the given power series converges absolutely everywhere.

2. At first sight the series $\Sigma_{n,1}(1+\frac{1}{n})^n x^n$ might seem difficult to analyze. However,

$$((1 + \tfrac{1}{n})^n)^{1/n} = 1 + \tfrac{1}{n} \to 1.$$

Hence, the given power series converges absolutely on $(-1, 1)$ while it diverges on $(-\infty, 1)$ and $(1, \infty)$. At $x = 1$ we have to deal with $\Sigma_{n,1}(1+\frac{1}{n})^n$. It diverges since $(1+\frac{1}{n})^n \to e \neq 0$. At $x = -1$ the series becomes $\Sigma_{n,1}(1+\frac{1}{n})^n(-1)^n$, which diverges since the sequence $((1 + \frac{1}{n})^n(-1)^n)$ does not converge to zero.

Remark

Given a power series, it can be shown, independently from the ratio test and the root test, that $\Sigma_{n,0}a_n x^n$ either converges absolutely on the whole real line or there exists a number $R > 0$ such that the series converges absolutely on $(-R, R)$ and diverges for $|x| > R$, or the series converges only at $x = 0$. This result does not provide an answer about what happens at R or $-R$.

Differentiability and integrability of power series

Let $\Sigma_{n,0}a_n x^n$ be a power series with radius of convergence $R > 0$ or suppose that it converges on the whole real line. Define

$$s(x) = \Sigma_{n=0}^{\infty} a_n x^n, \; x \text{ on } (-R, R) \text{ or } \Re.$$

Then $s(x)$ (the 'sum function') is differentiable on $(-R, R)$ or $\Re$ and

$$s'(x) = \Sigma_{n=0}^{\infty} (a_n x^n)',$$

thus

$$s'(x) = \Sigma_{n=1}^{\infty} n a_n x^{n-1}.$$

That is to say, if $s(x) = a_0 + a_1 x + a_2 x^2 + a_3 x^3 + ... + a_n x^n + ...$ then $s'(x) = a_1 + 2a_2 x + 3a_3 x^2 + ... + n a_n x^{n-1} + ...$

Using Leibniz's notation we would write

$$\frac{d}{dx} s(x) = \Sigma_{n=0}^{\infty} \frac{d}{dx} (a_n x^n).$$

Moreover, for any x in $(-R, R)$ or $\Re$,

$$\int_0^x s(t) dt = \Sigma_{n=0}^{\infty} \int_0^x a_n t^n dt = \Sigma_{n=0}^{\infty} \frac{a_n}{n+1} x^{n+1},$$

and

$$\int_a^b s(t) dt = \Sigma_{n=0}^{\infty} \int_a^b a_n t^n dt, \text{ where } [a, b] \subseteq (-R, R).$$

The differentiability and integrability of a power series, as shown above, is a powerful theorem whose validity we will accept without proof[8]. Even though we are not providing a proof, the statement of the theorem under discussion seems plausible.

It is to be noted that $s'(x)$ is also a power series, hence it is differentiable and

$$s''(x) = \frac{d}{dx} \Sigma_{n=1}^{\infty} n a_n x^{n-1} = \Sigma_{n=1}^{\infty} \frac{d}{dx} (n a_n x^{n-1}) = \Sigma_{n=2}^{\infty} n(n-1) a_n x^{n-2}.$$

Actually, $s(x)$ is infinitely differentiable since a power series appears each time that we take the derivative.

Examples

1. We found that the interval of convergence of $\Sigma_{n,0} \frac{x^n}{n!}$ is the whole real line. Let $s(x) = \Sigma_{n=0}^{\infty} \frac{x^n}{n!}$, that is,

$$s(x) = 1 + x + \frac{x^2}{2!} + \frac{x^3}{3!} + \frac{x^4}{4!} + ...$$

Then

$$s'(x) = 1 + \frac{2}{2!} x + \frac{3}{3!} x^2 + \frac{4}{4!} x^3 + ... = 1 + x + \frac{x^2}{2!} + \frac{x^3}{3!} + ... = s(x).$$

---

[8]The interested reader can consult (Trench, 2002) for a proof.

Since $s'(x) = s(x)$ and $s(0) = 1$, using the techniques from section 1.15 we can conclude that $s(x) = e^x$. Indeed, $s'(x) - s(x) = 0$ for all x implies $e^{-x}(s'(x) - s(x)) = 0$. Thus $(e^{-x}s(x))' = 0$ for all x. Consequently, there exists C such that $e^{-x}s(x) = C$, so $s(x) = Ce^x$ for all x. In particular $1 = s(0) = C$, therefore $s(x) = e^x$ for all x. We have just shown that

$$\Sigma_{n=0}^{\infty} \frac{x^n}{n!} = e^x \quad \text{for all } x.$$

2. The power series $\Sigma_{n,1} nx^{n-1}$ has radius of convergence 1, as can be easily determined. To what value does it converge for each x in $(-1, 1)$? We observe that

$$\Sigma_{n=1}^{\infty} nx^{n-1} = \Sigma_{n=1}^{\infty} \frac{d}{dx} x^n = \frac{d}{dx} \Sigma_{n=1}^{\infty} x^n = \frac{d}{dx}\left(\frac{x}{1-x}\right) = \frac{1}{(1-x)^2}.$$

This question has considerable interest in statistics due to the fact that it is linked to the problem of finding the expected value of a geometric random variable.

3. Is it possible to express the function $\arctan x$ as a power series? Starting with the well-known equality

$$\frac{1}{1-x} = 1 + x + x^2 + x^3 + \ldots + x^n + \ldots, \quad -1 < x < 1,$$

we get

$$\frac{1}{1+x} = 1 - x + x^2 - x^3 + \ldots + (-1)^n x^n + \ldots, \quad -1 < x < 1.$$

Therefore

$$\frac{1}{1+x^2} = 1 - x^2 + x^4 - x^6 + \ldots + (-1)^n x^{2n} + \ldots, \quad -1 < x < 1.$$

Hence, for any x in $(-1, 1)$, we will have

$$\int_0^x \frac{dt}{1+t^2} = \int_0^x dt - \int_0^x t^2 dt + \int_0^x t^4 dt - \int_0^x t^6 dt + \ldots$$

Thus, for any x in $(-1, 1)$

$$\arctan x = x - \frac{x^3}{3} + \frac{x^5}{5} - \frac{x^7}{7} + \ldots = \Sigma_{n=0}^{\infty} \frac{(-1)^n}{2n+1} x^{2n+1}.$$

In the next section we will see that this equality is also valid for $x = 1$ and $x = -1$.

4. Can we find the power series expansion of $\ln(x+1)$? Starting from the fact that

$$\tfrac{1}{1+t} = 1 - t + t^2 - t^3 + \dots \quad -1 < t < 1,$$

we get

$$\textstyle\int_0^x \tfrac{dt}{1+t} = \int_0^x dt - \int_0^x t\,dt + \int_0^x t^2 dt - \int_0^x t^3 dt + \dots$$

But

$$\textstyle\int_0^x \tfrac{dt}{1+t} = \int_1^{1+x} \tfrac{du}{u} = \ln(1+x).$$

Hence

$$\ln(1+x) = x - \tfrac{x^2}{2} + \tfrac{x^3}{3} - \tfrac{x^4}{4} + \dots = \Sigma_{n=0}^{\infty}(-1)^n \tfrac{x^{n+1}}{n+1}, \quad -1 < x < 1.$$

At the end of section 3.6 we found that $\ln 2 = 1 - \tfrac{1}{2} + \tfrac{1}{3} - \tfrac{1}{4} + \dots$, so the previous expression for $\ln(1+x)$ is also valid for $x = 1$.

5. Sometimes we might need to derivate and then integrate. For instance, what is the sum of the power series $\Sigma_{n,0} \tfrac{x^{2n+1}}{2n+1}$? We can easily show that it converges on $(-1, 1)$ to a certain function $s(x)$. Then, for $|x| < 1$

$$s'(x) = \Sigma_{n=0}^{\infty}\big(\tfrac{x^{2n+1}}{2n+1}\big)' = \Sigma_{n=0}^{\infty}x^{2n} = \Sigma_{n=0}^{\infty}(x^2)^n = \tfrac{1}{1-x^2}.$$

Thus

$$s(x) = s(x) - s(0) = \textstyle\int_0^x s'(t)dt = \int_0^x \tfrac{dt}{1-t^2} = \tfrac{1}{2}\ln\tfrac{1+x}{1-x}.$$

In the last step we were lucky, in the sense that using partial fractions it was possible to calculate the integral. It may happen, in similar problems, that the integral obtained is not expressible in terms of elementary functions (see exercise 12 at the end of the section).

Remark

More general power series, of the type $\Sigma_{n,0} a_n(x - c)^n$ (c an arbitrary real number), can be studied. We have chosen to concentrate our attention on the case when $c = 0$ because it exemplifies quite well the main properties of general power series.

Exercises for Section 3.11

In each of the following ten problems find the largest interval around the origin where the given power series is convergent.

1. $\Sigma_{n,0}(\ln x)^n$.

2. $\Sigma_{n,0} 2^n x^n$.

3. $\Sigma_{n,0} \frac{2^n}{n!} x^n$.

4. $\Sigma_{n,0} 3^n x^n$.

5. $\Sigma_{n,0} \frac{x^n}{5^n}$.

6. $\Sigma_{n,0} \frac{5^n x^n}{n+1}$.

7. $\Sigma_{n,0} \frac{x^n}{\sqrt{n^2+1}}$.

8. $\Sigma_{n,1} \frac{1}{n^n} x^n$.

9. $\Sigma_{n,1} (1 - \frac{1}{n})^n x^n$.

10. $\Sigma_{n,2} \frac{1}{n \ln n} x^n$.

11. Show that $\Sigma_{n,1} \frac{x^n}{n} = -\ln(1-x)$, $-1 \le x < 1$ (Hint: Take the derivative of the given series and thereafter integrate).

12. Express $\Sigma_{n=1}^{\infty} \frac{x^n}{n^2}$ in terms of a definite integral (Hint: Take the derivative of the given series, keep in mind the previous exercise, and thereafter integrate).

3.12   Taylor Series

Let us suppose that a function $f(x)$ is expressible as a power series on an interval $(-R, R)$ or the whole real line. That is,

$$f(x) = a_0 + a_1 x + a_2 x^2 + a_3 x^3 + a_4 x^4 + \ldots$$

Then $a_n = \frac{f^n(0)}{n!}$, $n = 0, 1, 2, \ldots$. Why is this so? We have that

$$f^{(1)}(x) = a_1 + 2a_2 x + 3a_3 x^2 + 4a_4 x^3 + \ldots$$

$$f^{(2)}(x) = 2a_2 + 6a_3 x + 12a_4 x^2 + \ldots$$

$$f^{(3)}(x) = 6a_3 + 24a_4 x + \ldots$$

and so on. In particular, $f^{(1)}(0) = a_1$, $f^{(2)}(0) = 2!a_2$, $f^{(3)}(0) = 3!a_3$, $f^{(4)}(0) = 4!a_4,....$ Hence

$$f(x) = \Sigma_{n=0}^{\infty} \frac{f^{(n)}(0)}{n!} x^n, \; x \text{ in } (-R, R) \text{ or } \Re.$$

The reader may have noticed that what we have done is a generalization of the type of arguments developed in section 1.17.

Definition

Let f be defined on an open interval around $x = 0$ and infinitely differentiable on its domain. That is to say, $f^{(n)}(x)$ exists on an open interval around the origin (n is an arbitrary natural number). The power series $\Sigma_{n,0} \frac{f^{(n)}(0)}{n!} x^n$ is called the **Taylor series** of f at the origin (Brook Taylor, 1685-1731, was a disciple of Newton).

Is there an interval I that contains the number 0 in its interior, or the whole real line, such that $f(x) = \Sigma_{n=0}^{\infty} \frac{f^{(n)}(0)}{n!} x^n$ for every x in I or $\Re$? Two questions have to be answered:

1. What is the interval of convergence of $\Sigma_{n,0} \frac{f^{(n)}(0)}{n!} x^n$?

2. Is it true that $f(x) = \Sigma_{n=0}^{\infty} \frac{f^{(n)}(0)}{n!} x^n$ for every x in the interval of convergence of the Taylor series of f?

For instance, we already know that

$$\frac{1}{1-x} = \Sigma_{n=0}^{\infty} x^n, \quad |x| < 1.$$

Right away we can see that $\Sigma_{n,0} x^n$ is the Taylor series of $f(x) = \frac{1}{1-x}$ and that it converges to $f(x)$ for each x in $(-1, 1)$ and nowhere else. But it is not easy to provide an immediate answer when $f(x) = \sin x$ or $f(x) = \sqrt{1 + x}$.

Definition

A function is said to be **analytic** at the origin if its Taylor series is equal to the function on an interval around the number 0, that is, $f(x) = \Sigma_{n=0}^{\infty} \frac{f^{(n)}(0)}{n!} x^n$ on an interval that contains 0 in its interior. If the function is analytic at the origin, its **interval of analyticity** is the interval where the function adopts the same value as its Taylor series.

We have just shown that $f(x) = \frac{1}{1-x}$ is analytic at the origin with $(-1, 1)$ as its interval of analyticity. On the other hand, let $f(x) = e^x$. Then $f^{(n)}(0) = 1$ for every n. Thus, its Taylor series at the origin is $\Sigma_{n,0} \frac{1}{n!} x^n$. We know that this series is convergent for all x. Moreover, in the previous section we proved that

$$e^x = \Sigma_{n=0}^{\infty} \frac{1}{n!} x^n$$

for every real number x. Thus, e^x is analytic at the origin with $\Re$ as its interval of analyticity.

We will accept the fact that functions of common usage such as $\sin x$, $\cos x$, $\sqrt{1+x}$, $\ln(1+x)$, $\arctan x$, *etc. are analytic at the origin.* A rigorous approach to the theory of analytic functions is a topic of mathematics that goes beyond the realm of a textbook intended for first-year calculus students. However, in section 3.14 we will analyze Taylor's theorem, which will give us an inkling of the theory behind analytic functions.

Examples

1. Let $f(x) = \ln(1+x)$, $x > -1$. Then $f^{(1)}(x) = \frac{1}{1+x}$, $f^{(2)}(x) = -(1+x)^{-2}$, $f^{(3)}(x) = (-1)^2 2(1+x)^{-3}$, $f^{(4)}(x) = (-1)^3 3!(1+x)^{-5}$, ..., $f^{(n)}(x) = (-1)^{n-1}(n-1)!$ Consequently $f(0) = 0$ and

$$\frac{f^{(n)}(0)}{n!} = \frac{(-1)^{n-1}(n-1)!}{n!} = \frac{(-1)^{n-1}}{n}, \ n \geq 1.$$

Thus, the Taylor series of $\ln(1+x)$ around the origin is $x - \frac{1}{2}x^2 + \frac{1}{3}x^3 - \frac{1}{4}x^4 +$ ... or, in compact notation, $\Sigma_{n,1} \frac{(-1)^{n-1}}{n} x^n$; that is, $\Sigma_{n,0} \frac{(-1)^n}{n+1} x^{n+1}$. The series $\Sigma_{n,0} \frac{(-1)^n}{n+1} x^{n+1}$ converges on $(-1, 1]$, hence we expect that

$$\ln(1+x) = x - \tfrac{1}{2}x^2 + \tfrac{1}{3}x^3 - \tfrac{1}{4}x^4 + ..., \ -1 < x \leq 1.$$

This result confirms what we found in the previous section.

We cannot enlarge the interval of analyticity any further because $\ln(1+x)$ is not defined for $x \leq -1$ and the series $\Sigma_{n,0} \frac{(-1)^n}{n+1} x^{n+1}$ is divergent for $x > 1$.

2. Let $f(x) = \arctan x$, a function defined over the whole real line. We wish to find its Taylor series at the origin. With this purpose in mind we calculate successive derivatives of $f(x)$: $f^{(1)}(x) = (x^2 + 1)^{-1}$, $f^{(2)}(x) = -2x(x^2 + 1)^{-2}$, $f^{(3)}(x) = -2(x^2 + 1)^{-2} + 8x^2(x^2 + 1)^{-3}$, ... It gets increasingly more difficult to calculate the higher derivatives of f and hence the task of finding $f^{(n)}(0)$ seems out of reach. Luckily, in section 3.11 we found that

$$\arctan x = x - \tfrac{x^3}{3} + \tfrac{x^5}{5} - \tfrac{x^7}{7} + ..., \ -1 < x < 1,$$

or, in compact notation, for any x in $(-1, 1)$ we have

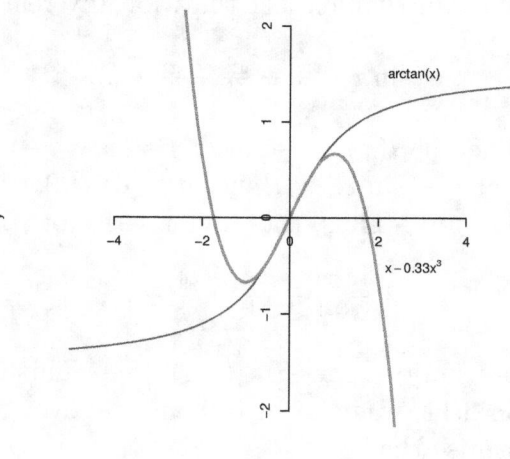

Figure 3.3: The approximation of the inverse tangent by a cubic polynomial

$$\arctan x = \Sigma_{n=0}^{\infty} \frac{(-1)^n}{2n+1} x^{2n+1}.$$

Since we found a power series representation of $\arctan x$, on an interval around the origin, it must be its Taylor series[9]. In an enrichment note, at the end of the section, it is shown that the previous equality does hold on $[-1, 1]$.

Interestingly, since

$$\frac{\pi}{6} = \arctan \frac{1}{\sqrt{3}} = \Sigma_{n=0}^{\infty} \frac{(-1)^n}{2n+1} \left(\frac{1}{\sqrt{3}}\right)^{2n+1}$$

it follows that

$$\pi = 6\Sigma_{n=0}^{\infty} \frac{(-1)^n}{2n+1} \left(\frac{1}{\sqrt{3}}\right)^{2n+1}.$$

For instance, $6\Sigma_{n=0}^{15} \frac{(-1)^n}{2n+1} \left(\frac{1}{\sqrt{3}}\right)^{2n+1} = 3.141592652$. After adding only 15 terms of the series we got the first eight decimals of π, a remarkable feat!

3. Let $f(x) = \sin x$. Then $f^{(1)}(x) = \cos x$, $f^{(2)}(x) = -\sin x$, $f^{(3)}(x) = -\cos x$, $f^{(4)}(x) = \sin x$, ... So $f^{(1)}(0) = 1$, $f^{(2)}(0) = 0$, $f^{(3)}(0) = -1$, $f^{(4)}(0) = 0$, $f^{(5)}(0) = 1$, ... Hence, the Taylor series of $\sin x$ is

$$x - \frac{x^3}{3!} + \frac{x^5}{5!} - \cdots,$$

---

[9]This result was proven at the beginning of the section.

that is, $\Sigma_{n,0}(-1)^n \frac{x^{2n+1}}{(2n+1)!}$. In section 3.11 we found that this power series converges everywhere. We can then expect that, for any real number x,

$$\sin x = x - \frac{x^3}{3!} + \frac{x^5}{5!} - \cdots$$

4. Let $f(x) = \cos x$. Then $f^{(1)}(x) = -\sin x$, $f^{(2)}(x) = -\cos x$, $f^{(3)}(x) = \sin x$, $f^{(4)}(x) = \cos x$, $f^{(5)}(x) = -\sin x$, ... Hence $f^{(1)}(0) = 0$, $f^{(2)}(0) = -1$, $f^{(3)}(0) = 0$, $f^{(4)}(0) = 1$, $f^{(5)}(0) = 0$,... Thus, the Taylor series of $\cos x$ is

$$1 - \frac{x^2}{2!} + \frac{x^4}{4!} - \frac{x^6}{6!} + \cdots,$$

which in compact notation becomes $\Sigma_{n,0}(-1)^n \frac{x^{2n}}{(2n)!}$. This is a well-known power series that, in section 3.11, was shown to converge everywhere. Then, for any real number x we surmise that

$$\cos x = 1 - \frac{x^2}{2!} + \frac{x^4}{4!} - \frac{x^6}{6!} + \cdots$$

5. Let $f(x) = \sqrt{1+x}$. Then $f^{(1)}(x) = \frac{1}{2}(1+x)^{-1/2}$, $f^{(2)}(x) = -\frac{1}{4}(1+x)^{-3/2}$, $f^{(3)}(x) = (-1)^2 \frac{1}{4}\frac{3}{2}(1+x)^{-5/2}$, $f^{(4)}(x) = (-1)^3 \frac{1}{4}\frac{3}{2}\frac{5}{2}(1+x)^{-7/2}$,... Hence

$$f^{(1)}(0) = \frac{1}{2}, \; f^{(n)}(0) = (-1)^{n-1}\frac{1}{2^n}3 \times 5 \times 7 \times \ldots(2n-3), \; n > 1.$$

Thus, the Taylor series of $\sqrt{1+x}$ around the origin is

$$1 + \frac{1}{2}x - \frac{1}{4}\frac{x^2}{2} + \frac{3}{2^3}\frac{x^3}{6} - \frac{3\times5}{2^4}\frac{x^4}{4!} + \cdots$$

or, what is the same,

$$1 + \frac{1}{2}x + \Sigma_{n=2}^{\infty}(-1)^{n-1}\frac{3\times5\times\ldots\times(2n-3)}{2^n n!}x^n.$$

Since

$$\frac{\frac{3\times5\times7\times\ldots\times(2n-1)}{2^{n+1}(n+1)!}}{\frac{3\times5\times\ldots\times(2n-3)}{2^n n!}} = \frac{2n-1}{2(n+1)} \to 1,$$

we can conclude that its radius of convergence is 1. Therefore, for any x in the interval $(-1, 1)$, we expect that

$$\sqrt{1+x} = 1 + \frac{1}{2}x - \frac{1}{4}\frac{x^2}{2} + \frac{3}{2^3}\frac{x^3}{6} - \frac{3\times5}{2^4}\frac{x^4}{4!} + \cdots.$$

Recall that in section 1.13 we found the linearization of $\sqrt{1+x}$, namely $1 + \frac{1}{2}x$. The approximation of second order happens to be $1 + \frac{1}{2}x - \frac{1}{8}x^2$.

Remark

In the five preceding examples we accepted the fact that the functions involved are analytic. The question about the exact nature of the interval of analyticity remained open. In section 3.11 we provided a precise answer for the function e^x. Could we do so for other analytic functions such as $\sin x$ or $\cos x$? It would be necessary, at this stage, to assume that every IVP of the type $y'' + y = 0$, $y(0) = 1$, $y'(0) = 0$, or $y(0) = 0$, $y'(0) = 1$, has a unique solution. This is indeed the case since any IVP $y''(x) + p(x)y'(x) + q(x)y(x) = 0$, $y(0) = \alpha$, $y'(0) = \beta$, has a unique solution ($p(x), q(x)$ are any polynomials) defined over the whole real line; see the fifth application from section 3.13. For instance, if

$$s(x) = 1 - \frac{x^2}{2!} + \frac{x^4}{4!} - \frac{x^6}{6!} + \ldots \quad \text{for all } x,$$

then it is easy to verify that $s''(x) + s(x) = 0$ and $s(0) = 1$, $s'(0) = 0$. But evidently $\cos''(x) + \cos(x) = 0$, $\cos(0) = 1$, $\cos'(0) = 1$. Consequently $\cos x = s(x)$ for all x.

The topic of analytic functions, and intervals of analyticity, will be revisited in section 3.14 from a different perspective.

The binomial series

The Taylor series around the origin of $\sqrt{1+x}$ is a special case of a binomial series, as we will see next. If m is a positive integer, from elementary algebra we know that

$$(1+x)^m = \Sigma_{n=0}^m C_n^m x^n$$

where C_n^m ($n = 0, 1, 2, \ldots, m$) denotes the number $\frac{m!}{(m-n)!n!}$, which is equal to $\frac{m(m-1)\ldots(m-n+1)}{n!}$. It is the number of combinations of m objects taken n at a time and can be easily calculated using the well-known **Pascal triangle**:

$$1$$

$$1\ 2\ 1$$

$$1\ 3\ 3\ 1$$

$$1\ 4\ 6\ 4\ 1$$

$$\cdot\ \cdot\ \cdot\ \cdot\ \cdot\ \cdot$$

For instance, $(1+x)^3 = 1 + 3x + 3x^2 + x^3$. Next we will see that if α is any non-natural number different from zero then

$$(1+x)^\alpha = \Sigma_{n=0}^\infty C_n^\alpha x^n, \quad |x| < 1,$$

where $C_0^\alpha = 1$, $C_n^\alpha = \frac{\alpha(\alpha-1)(\alpha-2)...(\alpha-n+1)}{n!}$, $n \geq 1$. For example, $C_1^\alpha = \alpha$, $C_2^\alpha = \frac{\alpha(\alpha-1)}{2}$, and $C_3^\alpha = \frac{\alpha(\alpha-1)(\alpha-2)}{3!}$. Let us start by defining

$$f(x) = (1+x)^\alpha, \quad |x| < 1.$$

Then $f^{(1)}(x) = \alpha(1+x)^{\alpha-1}$, $f^{(2)}(x) = \alpha(\alpha-1)(1+x)^{\alpha-2}$, $f^{(3)}(x) = \alpha(\alpha-1)(\alpha-2)(1+x)^{\alpha-3}$, ...,

$$f^{(n)}(x) = \alpha(\alpha-1)(\alpha-2)...(\alpha-n+1)(1+x)^{\alpha-n}.$$

In particular, $f(0) = 1$, $f^{(1)}(0) = \alpha$, $f^{(2)}(0) = \alpha(\alpha-1)$, $f^{(3)}(0) = \alpha(\alpha-1)(\alpha-2)$, ...,

$$f^{(n)}(0) = \alpha(\alpha-1)(\alpha-2)...(\alpha-n+1).$$

The Taylor series of $f(x)$, around the origin, is $\Sigma_{n,0} \frac{f^{(n)}(0)}{n!}$; that is,

$$\Sigma_{n,0} \frac{\alpha(\alpha-1)...(\alpha-n+1)}{n!} x^n.$$

In order to find the interval of convergence of this series we will follow the usual path:

$$\left| \frac{\frac{\alpha(\alpha-1)...(\alpha-n+1)(\alpha-n)}{(n+1)!}}{\frac{\alpha(\alpha-1)...(\alpha-n+1)}{n!}} \right| = \left| \frac{\alpha-n}{n+1} \right| \to 1 \text{ as } n \to \infty.$$

Thus, the series converges for $|x| < 1$. That is to say, $\Sigma_{n=0}^{\infty} C_n^\alpha x^n$ is a well-defined number for each x in $(-1, 1)$. In advanced calculus it is proven that $f(x) = (1+x)^\alpha$ is analytic at the origin. Consequently, it sounds reasonable to conclude that

$$(1+x)^\alpha = \Sigma_{n=0}^{\infty} C_n^\alpha x^n, \quad -1 < x < 1. \tag{3.11}$$

Examples

The following three expansions as a series have considerable importance in applications of calculus to problems in the natural sciences.

1. Let $f(x) = \sqrt{1+x}$. Then $C_0^{1/2} = 1$, $C_1^{1/2} = \frac{1}{2}$, $C_2^{1/2} = -\frac{1}{8}$, ... Hence

$$\sqrt{1+x} = 1 + \tfrac{1}{2}x - \tfrac{1}{8}x^2 + ..., \quad -1 < x < 1,$$

a result that had already been found before starting a systematic study of binomial series[10].

---

[10]We do not worry too much about what happens at 1 or -1 because in most applications the interval of interest happens to be $(-a, a)$, where $a < 1$.

2. Let us try to find the expansion of $(1+x)^{-\frac{1}{2}}$ as a power series around the origin. Using the binomial series we get $C_0^{-1/2} = 1$, $C_1^{-1/2} = -\frac{1}{2}$, $C_2^{-1/2} = \frac{(-\frac{1}{2})(-\frac{1}{2}-1)}{2} = \frac{3}{8}$, etc. Therefore

$$\frac{1}{\sqrt{1+x}} = 1 - \frac{1}{2}x + \frac{3}{8}x^2 - ..., \quad -1 < x < 1.$$

3. Our next goal is to find a power series expansion of $\frac{1}{(1+x)^{\frac{3}{2}}}$ around the origin. Right away we get $C_0^{-3/2} = 1$, $C_1^{-3/2} = -\frac{3}{2}$, $C_2^{-3/2} = \frac{(-3/2)(-3/2-1)}{2} = \frac{15}{8}$. Therefore

$$\frac{1}{(1+x)^{3/2}} = 1 - \frac{3}{2}x + \frac{15}{8}x^2 - ..., -1 < x < 1.$$

A short historical note

The theory of series was developed over a period spanning more than a century, starting with Gottfried Leibniz (1646-1716) who dealt with the alternating series test and Isaac Newton (1642-1727) who immersed himself in a deep study of the binomial series. Before them, Pietro Mengoli (1626-1686) had found that the harmonic series diverges.

Jean l'Rond d'Alembert (1717-1783) invented (discovered?) the ratio test and Augustin-Louis Cauchy (1789-1857) developed the root test. Both Colin Maclaurin (1698-1746) and Cauchy are credited with the development of the integral test. By the time that Cauchy's *Cours d'Analyse* was published (1821), the core aspects of the theory of series had reached maturity.

Exercises for Section 3.12

1. Use the fact that $\frac{1}{1+x} = \Sigma_{n=0}^{\infty}(-1)^n x^n$, $-1 < x < 1$, in order to obtain the Taylor series of $\ln(1 + x)$.

2. Find the Taylor series of $f(x) = e^{-x^2}$.

3. Find the Taylor series of $f(x) = \sqrt{1 - x}$.

4. Use the binomial series formula to find the Taylor series of $f(x) = \frac{1}{(1-x)^2}$.

5. Find the Taylor series of $f(x) = \frac{1}{(1-x)^2}$ without using the formula for binomial series[11] (Hint: Observe that $\frac{1}{(1-x)^2} = \frac{d}{dx}(\frac{1}{1-x})$).

6. Use the preceding problem to find a series expansion of $\frac{1}{(1+x)^2}$.

---

[11]The answer to this problem was found, from a different perspective, in section 3.11.

7. Assuming that $\Sigma_{n=0}^{\infty} c_n x^n = 0$ for all x in $\Re$, show that $c_n = 0$ for every n (Hint: Define $f(x) = \Sigma_{n=0}^{\infty} c_n x^n$. Then $c_n = f^{(n)}(0)/n!$).

8. Suppose that $\Sigma_{n=0}^{\infty} c_n x^n = 0$ for all x in the interval $(-r, r)$, where $r > 0$. Could we conclude that $c_n = 0$ for all n ?

9. Calculate the Taylor series of $f(x) = \frac{1}{\sqrt{1+x^2}}$ (Hint: Use the expansion, as a series, of $\frac{1}{\sqrt{1+x}}$).

10. Calculate the first three terms of the Taylor series of $f(x) = \sqrt{1 - \frac{1}{x}}$. What restriction would you impose on x in order to conclude that the resulting quadratic polynomial is a 'good approximation' of $f(x)$?

11. Find the Taylor series of $\frac{1}{\sqrt{1-x^2}}$. Then find the Taylor series of $\arcsin x$ (Hint: Recall the derivative of $\arcsin x$).

12. The Poisson distribution is a probability model that can be used to describe the behavior of a random variable X that represents the number of times that a certain event happens in a unit of time or space. Its probability function is $f(k) = P(X = k) = e^{-\lambda} \frac{\lambda^k}{k!}$ and its corresponding expected value is $E(X) = \Sigma_{k=0}^{\infty} k f(k) = \Sigma_{k=0}^{\infty} k e^{-\lambda} \frac{\lambda^k}{k!}$. Show that $E(X) = \lambda$.

13. Find the Taylor series of $f(x) = \frac{1}{(1-x)^3}$ (Hint: Notice that $\frac{d}{dx} \frac{1}{(1-x)^2} = \frac{2}{(1-x)^3}$ and use exercise 5 above.)

Enrichment Note: Extending the Interval of Convergence

In the third example, at the beginning of the section, we found that

$$\arctan x = \Sigma_{n=0}^{\infty} \frac{(-1)^n}{2n + 1} x^{2n+1}, \quad -1 < x < 1. \tag{3.12}$$

Since both $\Sigma_{n=0}^{\infty} \frac{(-1)^n}{2n+1}$ and $\Sigma_{n=0}^{\infty} \frac{(-1)^{3n+1}}{2n+1}$ are convergent thanks to the alternating series test, and $\arctan x$ is continuous everywhere, we might suspect that (3.12) is valid also at 1 and -1. Let us start with the algebraic identity

$$\tfrac{1}{1-q} = 1 + q + q^2 + ... + q^{n-1} + \tfrac{q^n}{1-q}, \quad q \neq 1,$$

the same identity that we used at the end of section 3.7 when trying to find the exact value of $\Sigma_{n=1}^{\infty} (-1)^{n+1} \frac{1}{n}$. Making $q = -s^2$ we get, for any s,

$$\tfrac{1}{1+s^2} = 1 - s^2 + s^4 + ... + (-1)^{n-1}(s^2)^{n-1} + \tfrac{(-1)^n s^{2n}}{1+s^2}.$$

Hence

$\int_0^x \frac{ds}{1+s^2} = \int_0^x ds - \int_0^x s^2 ds + \int_0^x s^4 ds + ... + \int_0^x (-1)^{n-1} s^{2n-2} ds + (-1)^n \int_0^x \frac{s^{2r} ds}{1+s^2}$.

Therefore

$$\arctan x = x - \frac{x^3}{3} + \frac{x^5}{5} - ... + (-1)^{n-1} \frac{x^{2n-1}}{2n-1} + (-1)^n \int_0^x \frac{s^{2n}}{1+s^2} ds.$$

Consequently,

$$\left| \arctan x - \left(x - \frac{x^3}{3} + \frac{x^5}{5} - ... + (-1)^{n-1} \frac{x^{2n-1}}{2n-1} \right) \right| = \left| \int_0^x \frac{s^{2n}}{1+s^2} ds \right|. \qquad (3.13)$$

Whenever $x \geq 0$ we have

$$\left| \int_0^x \frac{s^{2n}}{1+s^2} ds \right| = \int_0^x \frac{s^{2n}}{1+s^2} ds \leq \int_0^x s^{2n} ds = \frac{x^{2n+1}}{2n+1}, \qquad (3.14)$$

while for $x < 0$

$$\int_x^0 \frac{s^{2n}}{1+s^2} ds \leq \int_x^0 s^{2n} ds = -\frac{x^{2n+1}}{2n+1}.$$

Hence, if $x < 0$

$$\left| \int_0^x \frac{s^{2n}}{1+s^2} ds \right| = \left| \int_x^0 \frac{s^{2n}}{1+s^2} ds \right| = \int_x^0 \frac{s^{2n}}{1+s^2} ds \leq -\frac{x^{2n+1}}{2n+1}. \qquad (3.15)$$

From (3.14) and (3.15) we can conclude that, for any x in $\Re$ and $n = 0, 1, 2, 3, ...$,

$$\left| \int_0^x \frac{s^{2n}}{1+s^2} ds \right| \leq \frac{|x|^{2n+1}}{2n+1}.$$

Therefore, if $|x| \leq 1$ then, as $n \to \infty$,

$$\left| \int_0^x \frac{s^{2n}}{1+s^2} ds \right| \leq \frac{|x|^{2n+1}}{2n+1} \leq \frac{1}{2n+1} \to 0. \qquad (3.16)$$

Due to (3.13) and (3.16) it follows that, whenever $-1 \leq x \leq 1$,

$$\arctan x = \Sigma_{n=0}^{\infty} \frac{(-1)^n}{2n+1} x^{2n+1}.$$

An alternative proof, of the fact that the interval of analiticity of $\arctan x$ can be extended to $[-1, 1]$, stems from **Abel's Limit Theorem**[12], a deep result from Real Analysis (Trench, 2002): Suppose that $f(x) = \Sigma_{n=0}^{\infty} a_n x^n$, $-R < x < R$. If f is continuous from the left at R and $\Sigma_{n,0} a_n R^n$ converges, then $f(R) = \Sigma_{n=0}^{\infty} a_n R^n$.

---

[12]Niels Abel, 1802-1829, was an outstanding Norwegian mathematician.

Similarly, if f is continuous from the right at $-R$ and $\Sigma_{n,0} a_n(-R)^n$ converges, then $f(-R) = \Sigma_{n=0}^{\infty} a_n(-R)^n$.

Since $\arctan x$ is continuous at 1 and -1, and $\Sigma_{n,0} \frac{(-1)^n}{2n+1}$, $-\Sigma_{n,0} \frac{(-1)^n}{2n+1}$ converge, Abel's Limit Theorem implies that $\arctan 1 = \Sigma_{n=0}^{\infty} \frac{(-1)^n}{2n+1}$ and $\arctan(-1) = -\Sigma_{n=0}^{\infty} \frac{(-1)^n}{2n+1}$. On the other hand, since $\ln(1+x)$ is continuous at $x = 1$ and $\Sigma_{n,0}(-1)^n \frac{1}{n+1}$ converges, we can conclude that

$$\ln 2 = \ln(1 + 1) = \Sigma_{n=0}^{\infty}(-1)^n \frac{1}{n+1} = 1 - \frac{1}{2} + \frac{1}{3} - \dots$$

3.13   Applications

As we will see next, the theory of series plays a very important role in the natural sciences as well as in mathematics itself. It is ideally suited to provide approximations.

1. Two electric charges, q and $-q$, are at a distance d from each other; such an arrangement of charges is called a **dipole**. The electric field at a distance x from the left of q is given by

$$E = \frac{q}{x^2} - \frac{q}{(x+d)^2}.$$

Let us assume that $x \gg d$, that is, x is much bigger than d. From exercise 5 (section 3.12) we know that, for $|y| < 1$,

$$\frac{1}{(1+y)^2} = 1 - 2y + 3y^2 - 4y^3 + \dots$$

Thus,

$$\frac{1}{(1+y)^2} \approx 1 - 2y$$

provided that $|y|$ is very small. Therefore

$$\frac{1}{(d+x)^2} = \frac{1}{(x(1+\frac{d}{x}))^2} = \frac{1}{x^2} \frac{1}{(1+\frac{d}{x})^2} \approx \frac{1}{x^2}(1 - 2\frac{d}{x}).$$

Hence

$$E = \frac{q}{x^2} - \frac{q}{(x+d)^2} = q(\frac{1}{x^2} - \frac{1}{(x+d)^2}) \approx q(\frac{1}{x^2} - \frac{1}{x^2} + 2\frac{d}{x^3}) = \frac{2qd}{x^3}.$$

Consequently, the electric field at P, determined by the dipole, is approximately $\frac{2qd}{x^3}$ provided that $x \gg d$.

2. The force exerted by the Earth on a point mass m located at a distance x above it is given by

$$F(x) = \frac{GmM}{(R+x)^2},$$

where R is the radius of the Earth. At $x = 0$ the gravitational force is mg, where g is the acceleration of gravity at a fixed point on the surface of the Earth (g is approximately $32 ft/s^2$). Thus $\frac{GmM}{R^2} = mg$, that is, $GmM = mgR^2$. Hence

$$F(x) = \frac{mgR^2}{(R+x)^2} = \frac{mgR^2}{(R(1+\frac{x}{R}))^2} = \frac{mg}{(1+\frac{x}{R})^2}.$$

But, as in the case of the dipole, $\frac{1}{(1+y)^2} \approx 1 - 2y$ provided that $|y|$ is very small. Therefore, assuming that $x \ll R$, we will have

$$\frac{mg}{(1+\frac{x}{R})^2} \approx mg(1 - 2\frac{x}{R}).$$

Thus, $mg(1 - 2\frac{x}{R})$ is the approximate weight of a point mass m located at a distance x above the Earth (assuming that $x \ll R$).

3. In chapter 4 we will use an important approximation in the context of enzyme kinetics, namely

$$e^{-at} \approx 1 - at + \frac{a^2}{2}t^2 .$$

(t small and $a > 0$), in the quest for mathematical expressions that can be compared with data. Indeed,

$$P(t) = -\frac{a_2}{a_1^2} + \frac{a_2}{a_1^2}e^{-a_1 t} + \frac{a_2}{a_1}t$$

will provide a formula for the amount of product P before the onset of the stationary state, where a_1 and a_2 are parameters. Taking into consideration that during the pre-steady state the values of t are extremely small (sometimes in the order of milliseconds), we can approximate $e^{-a_1 t}$ by $1 - a_1 t + \frac{a_1^2}{2}t^2$. Therefore,

$$P(t) = -\frac{a_2}{a_1^2} + \frac{a_2}{a_1^2}(1 - a_1 t + \frac{a_1^2}{2}t^2) + \frac{a_2}{a_1}t = \frac{a_2}{2}t^2.$$

Thus,

$$\frac{2P(t)}{t} = a_2 t.$$

This expression allows the estimation of a_2 once we have experimental values of $P(t)$.

4. The three previous examples had to do with applications to the natural sciences. This time let us discuss how the theory of series helps in the task of approximating definite integrals. Suppose that we wish to approximate the value of $\int_0^{0.3} e^{-x^2} dx$ in such a way that the error is less than 0.001. Since

$$e^{-x^2} = \Sigma_{n=0}^{\infty} (-1)^n \frac{x^{2n}}{n!}$$

we will have

$$\int_0^{0.3} e^{-x^2} dx = \Sigma_{n=0}^{\infty} (-1)^n \int_0^{0.3} \frac{x^{2n}}{n!} dx = \Sigma_{n=0}^{\infty} (-1)^n \frac{(0.3)^{2n+1}}{n!(2n+1)} =$$
$$0.3 - \frac{0.3^3}{3} + \frac{0.3^5}{5(2!)} - \frac{0.3^7}{7(3!)} + \cdots$$

This is an alternating series. Due to the fact that $\frac{0.3^5}{5(2!)} = 0.000243 < 0.001$ while $\frac{0.3^3}{3} = 0.009 > 0.001$, we can choose $0.3 - \frac{0.3^3}{3} = 0.291$ as the answer.

Next let us see how to find an approximate value of $\int_0^{0.2} \frac{dx}{1+x^3}$. Since

$$\frac{1}{1-x} = \Sigma_{n=0}^{\infty} x^n, \quad |x| < 1,$$

it follows that

$$\frac{1}{1+x^3} = \Sigma_{n=0}^{\infty} (-1)^n x^{3n}, \quad |x| < 1.$$

Consequently,

$$\int_0^{0.2} \frac{dx}{1+x^3} = \Sigma_{n=0}^{\infty} (-1)^n \int_0^{0.2} x^{3n} dx = \Sigma_{n=0}^{\infty} \frac{(-1)^n}{3n+1} \left(\frac{1}{5}\right)^{3n+1} \approx$$
$$\frac{1}{5} - \frac{1}{4(5^4)} + \frac{1}{7(5^7)} - \frac{1}{10(5^{10})} = 0.19960182.$$

The inequality linked to the alternating series test (section 3.6) implies that

$$\left| \int_0^{0.2} \frac{dx}{1+x^3} - 0.19960182 \right| < \frac{1}{13(5^{13})} = 6.3(10^{-11}).$$

We note that the error is exceedingly small. No wonder that a calculator provides the same answer, that is, $\int(\frac{1}{1+x^3}, x, 0, 0.2) = 0.19960182$.

5. Power series play a central role in the solution of differential equations of second or higher order with polynomial coefficients. Let us accept the following theorem, whose proof belongs to the realm of advanced mathematics: Given polynomials $p(x)$, $q(x)$, and arbitrary numbers α, β, the initial value theorem $y''(x) + p(x)y'(x) + q(x)y(x) = 0$, $y(0) = \alpha$, $y'(0) = \beta$, has a unique solution $\phi(x) = \Sigma_{n=0}^{\infty} a_n x^n$ defined over the whole real line, where $a_0 = \alpha$, $a_1 = \beta$ and a_n ($n \geq 2$) are found, through a recursion formula, replacing $\phi(x)$ in the differential equation.

Let us illustrate the methodology to be followed analyzing the initial value problem

$$y'' - xy' + y = 0, \ y(0) = 1, \ y'(0) = 0.$$

According to the above-mentioned theorem, the solution is given by $\phi(x) = \Sigma_{n=0}^{\infty} a_n x^n$ where $a_0 = 1$ and $a_1 = 0$ and the other coefficients can be calculated taking into consideration the fact that $\phi(x)$ is the solution. Indeed, $\phi'(x) = \Sigma_{n=1}^{\infty} n a_n x^{n-1}$, $\phi''(x) = \Sigma_{n=2}^{\infty} n(n-1) a_n x^{n-2}$. Then

$$\Sigma_{n=2}^{\infty} n(n-1) a_n x^{n-2} - \Sigma_{n=1}^{\infty} n a_n x^n + \Sigma_{n=0}^{\infty} a_n x^n = 0,$$

that is,

$$\Sigma_{n=0}^{\infty} (n+2)(n+1) a_{n+2} x^n - \Sigma_{n=1}^{\infty} n a_n x^n + a_0 + \Sigma_{n=1}^{\infty} a_n x^n = 0.$$

Thus

$$(2a_2 + a_0) + \Sigma_{n=1}^{\infty} [(n+2)(n+1) a_{n+2} - n a_n + a_n] x^n = 0 \quad \text{for all } x.$$

Therefore[13]

$$2a_2 + a_0 = 0, \ (n+2)(n+1) a_{n+2} + (1-n) a_n = 0, \ n \geq 1.$$

When $n = 1$ we get $6a_3 = 0$, so $a_3 = 0$. Similarly, when $n = 2$ we get $12a_4 - a_2 = 0$; hence $a_4 = -\frac{1}{24}$, and so on. Therefore, the solution is the power series

$$\phi(x) = 1 - \tfrac{1}{2}x^2 - \tfrac{1}{24}x^4 - ..., \quad \text{for all } x \text{ in } \Re.$$

---

[13]Recall exercise 7, section 3.12

Suppose that we wish to approximate the solution by a polynomial, accepting an error less that 10^{-5} while working on the interval $[-0.1, 0.1]$. Thus, it is necessary to find the smallest value of k so that $|\Sigma_{n=k}^{\infty} a_n x^n| < 10^{-5}$. It is not hard to see that $|a_n| \leq 1$ for all n because $a_0 = 1$, $a_1 = 0$, $a_2 = -1/2$, and $(n+2)(n+1)a_{n+2} + (1-n)a_n = 0$ for every $n \geq 1$. Then

$$|\Sigma_{n=k}^{\infty} a_n x^n| \leq \Sigma_{n=k}^{\infty} |a_n||x^n| \leq \Sigma_{n=k}^{\infty} |x|^n = \tfrac{|x|^k}{(1-|x|)}.$$

Since $|x| < 10^{-1}$ it follows that $\tfrac{|x|^k}{1-|x|} < \tfrac{10^{1-k}}{9}$. Thus, we have to find the smallest value of k such that $\tfrac{10^{1-k}}{9} < 10^{-5}$. A 'guessing and checking' procedure will lead to $k = 6$. Consequently, the answer to the problem at hand is $p(x) = a_0 + a_1 x + ... + a_5 x^5$, that is,

$$p(x) = 1 - \tfrac{1}{2}x^2 - \tfrac{1}{24}x^4.$$

Exercises for Section 3.13

1. Use the fact that $\sin(x) \approx x - \tfrac{x^3}{3!} + \tfrac{x^5}{5!}$ (close to zero), to approximate the value of $\int_0^{0.1} \sin(x^2)dx$. Then use your graphing calculator to compare your answer to that provided by technology.

2. Approximate the value of $\int_0^1 f(x)dx$, where $f(x) = \tfrac{\sin x}{x}$, $0 < x \leq 1$, and $f(0) = 1$, using for this purpose the Taylor expansion of $\sin x$ with $n = 5$ (Hint: Observe that f is continuous on $[0,1]$).

3. Given the initial value problem $y'' - xy = 0$, $y(0) = 0$, $y'(0) = 1$, find a polynomial solution on the interval $[-0.1, 0.1]$ accepting an error less than 10^{-2}.

4. Use the series expansion of $\cos x$ to calculate $\lim_{x \to 0} \tfrac{\cos(x^2)-1}{x^3}$.

5. Suppose that a point mass m is located at a distance y above the center of a thin homogenous circular wire of radius r and mass M. Then (Hahn, 1998) the force of attraction between the ring and the point mass is given by

$$\tfrac{GmMy}{(y^2+r^2)^{3/2}}.$$

If $y \gg r$, using the approximation $(1 + x^2)^{-3/2} = 1$, which is pretty good if $|x| \ll 1$, show that the above-mentioned force of attraction becomes $\tfrac{GmM}{y^2}$. That is, the force acts as if all the mass of the ring were to be concentrated at the center of the circle.

3.14 Taylor's Theorem

As we promised in section 3.12, let us provide a theoretical framework to justify the analyticity of several common functions.

Theorem

Let f be a function defined on an open interval I around the origin, or the whole real line, such that f is infinitely differentiable on its domain. Then, given any x in the domain and any positive integer n, there exists a number θ between 0 and x such that

$$f(x) = T_n(x) + \frac{f^{(n+1)}(\theta)}{(n+1)!}x^{n+1},$$

where $T_n(x) = f(0) + f^{(1)}(0)x + ... + \frac{f^{(n)}(0)}{n!}x^n$.

The polynomial $T_n(x)$ is known as the **Taylor polynomial** of degree n of f around the origin[14]. The theorem itself is known as **Taylor's theorem**. Let us discuss a proof when $n = 1$.

Define the 'error function' $E_1(x)$ by

$$E_1(x) = f(x) - T_1(x) = f(x) - (f(0) + f'(0)x).$$

Using the Generalized Mean Value Theorem we can assert that there exists α between 0 and x such that

$$\frac{E_1(x) - E_1(0)}{x^2 - 0} = \frac{E_1'(\alpha)}{2\alpha}.$$

But $E_1'(x) = f'(x) - f'(0)$, so $E_1'(0) = 0$. Thus

$$\frac{E_1'(\alpha)}{2\alpha} = \frac{1}{2}\frac{E_1'(\alpha) - E_1'(0)}{\alpha - 0}.$$

Applying the Mean Value theorem to the function E_1', it follows that there exists θ between α and 0 such that

$$\frac{E_1'(\alpha) - E_1'(0)}{\alpha - 0} = E_1''(\theta).$$

Hence $\frac{E_1(x)}{x^2} = \frac{1}{2}E_1''(\theta)$. But $E_1'' = f''$, thus $E_1(x) = \frac{x^2}{2}f''(\theta)$. That is,

$$f(x) = f(0) + f'(0)x + f''(\theta)\frac{x^2}{2}.$$

A general proof can be patterned on the case $n = 1$, keeping in mind the definition $E_n(x) = f(x) - T_n(x)$. QED

---

[14]In general, θ depends on f, n, and x.

The analiticity of the cosine and the exponential function

Having made a detour to discuss a proof of Taylor's theorem, let us use the theorem, in its full generality, to confirm that $f(x) = \cos x$ is analytic at the origin - with the whole real line as its interval of analyticity. Thanks to Taylor's theorem, for any fixed

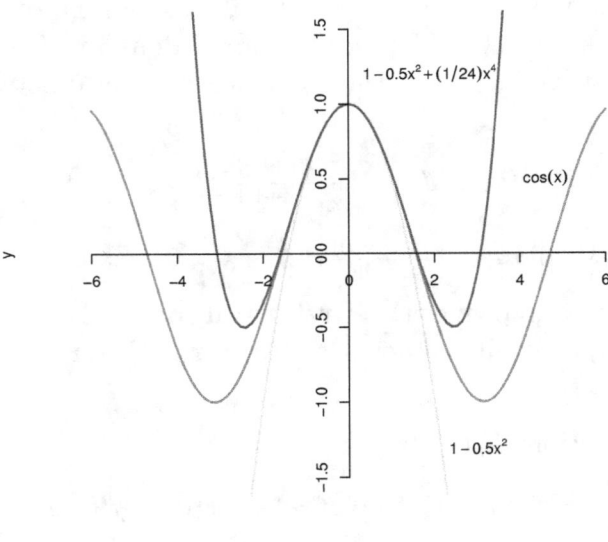

Figure 3.4: The Taylor polynomials $T_1(x)$ and $T_2(x)$, corresponding to $\cos x$.

x and any positive integer n there exists θ such that

$$f(x) = T_n(x) + \frac{f^{(n+1)}(\theta)}{(n+1)!} x^{n+1},$$

where $T_n(x) = 1 - \frac{x^2}{2!} + ... + (-1)^n \frac{x^{2n}}{(2n)!}$. We observe that

$$|f(x) - T_n(x)| = |f^{(n+1)}(\theta)| \frac{|x|^{n+1}}{(n+1)!}.$$

But $|f^{(n+1)}(\theta)| \le 1$ since $f^{(n+1)}$ is either cos or sin and both these functions oscillate between -1 and 1. Therefore

$$0 \le |f(x) - T_n(x)| \le \frac{|x|^{n+1}}{(n+1)!}.$$

Since the ratio test implies that $\Sigma_{n,0} \frac{|x|^{n+1}}{(n+1)!}$ converges, we can assert that

$$\lim_{n \to \infty} \frac{|x|^{n+1}}{(n+1)!} = 0.$$

Hence $\lim_{n \to \infty} |f(x) - T_n(x)| = 0$, so $\lim_{n \to \infty} T_n(x) = f(x)$. Thus

$$\lim_{n \to \infty} 1 - \frac{x^2}{2!} + \dots + (-1)^n \frac{x^{2n}}{(2n)!} = f(x).$$

That is,

$$\Sigma_{n=0}^{\infty} (-1)^n \frac{x^{2n}}{(2n)!} = \cos x.$$

The real number x was fixed, but arbitrary, consequently this last equality is valid for any x. We have succeeded in proving that $\cos x$ is analytic at the origin with the whole real line as its interval of analyticity. Figure 3.4 depicts how the Taylor polynomials $T_1(x)$ and $T_2(x)$, corresponding to the cosine function, are an approximation to $\cos x$ not far from the origin. The higher the degree of the Taylor polynomial, the better the approximation in the sense that $T_n(x)$ is closer to $\cos x$ on a wider interval around the origin.

Next let us provide a new proof of the fact that $f(x) = e^x$ is analytic at the origin with $\Re$ as its interval of analyticity[15]. For any real number x (fixed) and any n there exists θ such that

$$f(x) = T_n(x) + \frac{f^{n+1}(\theta)}{(n+1)!} x^{n+1}$$

where $T_n(x) = 1 + x + \dots + \frac{x^n}{n!}$. Then

$$|f(x) - T_n(x)| = |f^{n+1}(\theta)| \frac{|x|^{n+1}}{(n+1)!} = e^\theta \frac{|x|^{n+1}}{(n+1)!}.$$

Assuming that $x > 0$, since $0 < \theta < x$ we infer that $e^\theta < e^x$. Hence

$$|f(x) - T_n(x)| < e^x \frac{x^{n+1}}{(n+1)!}.$$

But $\lim_{n \to \infty} e^x \frac{x^{n+1}}{(n+1)!} = 0$, consequently $\lim_{n \to \infty} T_n(x) = f(x)$ whenever $x > 0$. What happens if $x < 0$? Under these circumstances $e^\theta < e^0 = 1$. Hence

$$|f(x) - T_n(x)| < \frac{|x|^{n+1}}{(n+1)!}.$$

Since $\lim_{n \to \infty} \frac{|x|^{n+1}}{(n+1)!} = 0$, it follows that

$$\lim_{n \to \infty} T_n(x) = f(x), \ x < 0.$$

Obviously $\lim_{n \to \infty} T_n(0) = 1 = e^0$. We have been able to prove that, for every x,

$$\Sigma_{n=0}^{\infty} \frac{1}{n!} x^n = e^x.$$

Similar techniques allow us to prove that $(1+x)^\alpha$, $\ln(1+x)$, $\arctan x$, $1/(1+x^2)$ and many other functions of common usage are analytic at the origin. Actually, only 'pathological' functions such as $f(x) = e^{-1/x^2}$ $(x \neq 0)$, $f(0) = 0$, are not analytic at the origin.

---

[15]Recall that in section 3.11 we did prove the analyticity at the origin of e^x.

Approximating a function across an interval

Taylor's theorem is also used to approximate a function on a given interval. Suppose that we wish to work on the interval $[0, 0.1]$ and hope to replace e^x by a Taylor polynomial within an accuracy of 0.0001. For any $0 \le x \le 0.1$ we have $e^x \le e^{0.1} < 1.11$. Hence, for any x in $[0, 0.1]$ there exists θ $(0 < \theta < 0.1)$ such that

$$E_n(x) = \frac{e^\theta}{(n+1)!}x^{n+1} < \frac{1.11}{(n+1)!}0.1^{n+1}.$$

Thereafter we have to choose the smallest value of n such that

$$\frac{(1.11)0.1^{n+1}}{(n+1)!} < 0.0001.$$

For $n = 2$ we get

$$\frac{(1.11)0.1^3}{3!} = 0.000185.$$

We have to choose a bigger n because $0.000185 > 0.0001$. Trying $n = 3$ we get

$$\frac{(1.11)0.1^4}{4!} = 0.00000463 < 0.000185.$$

Thus

$$T_3(x) = 1 + x + \tfrac{1}{2}x^2 + \tfrac{1}{6}x^3$$

can replace e^x in the interval $[0, 0.1]$, and we are confident that the error, for any x in $[0, 0.1]$, is less than 10^{-4}. It is to be expected that we will need a Taylor polynomial of higher order if the interval is larger, say $[0, 0.3]$. The idea of replacing e^x by a polynomial, in a small interval around the origin, is widely used in science (see section 4.6).

A different vision of Taylor's theorem

Under the same hypotheses of Taylor's Theorem it can be shown that

$$\lim_{x \to 0} \frac{E_n(x)}{x^n} = 0 \tag{3.17}$$

where $E_n(x) = f(x) - T_n(x)$. Thus, E_1 approaches zero 'faster' than x, ..., E_n approaches zero 'faster' than x^n. No wonder then that $T_n(x)$ becomes a better approximation of $f(x)$ insofar as n increases. A proof of (3.17) is not out of reach of a calculus student. Indeed, since

$$E_1(x) = f(x) - (f(0) + f^{(1)}(0)x)$$

we will have

$$\lim_{x \to 0} \frac{E_1(x)}{x} = \lim_{x \to 0} \frac{f(x) - f(0)}{x - 0} - f^{(1)}(0) = f^{(1)}(0) - f^{(1)}(0) = 0.$$

On the other hand,

$$E_2(x) = f(x) - \left(f(0) + f^{(1)}(0)x + \frac{f^{(2)}(0)}{2}x^2 \right).$$

Hence

$$E_2'(x) = f^{(1)}(x) - f^{(1)}(0) - f^{(2)}(0)x$$

and $E_2(0) = 0$. Then L'Hôpital's rule implies that

$$\lim_{x \to 0} \frac{E_2(x)}{x^2} = \frac{1}{2} \lim_{x \to 0} \frac{E_2'(x)}{x}.$$

But

$$\lim_{x \to 0} \frac{E_2'(x)}{x} = \lim_{x \to 0} \frac{f^{(1)}(x) - f^{(1)}(0)}{x - 0} - f^{(2)}(0) = f^{(2)}(0) - f^{(2)}(0) = 0.$$

Therefore $\lim_{x \to 0} \frac{E_2(x)}{x^2} = 0$, as we wished to prove. A pattern emerges and the same techniques can be applied for any n.

Exercises for Section 3.14

1. Show that $\sin x$ is analytic at the origin with $\Re$ as its interval of analyticity.

2. Replace $\cos x$ by a Taylor polynomial on $[-0.1, 0.1]$, within an accuracy of 0.001.

3. Show that $\lim_{x \to 0} \frac{E_3(x)}{x^3} = 0$.

4. Replace $\sin x$ by a Taylor polynomial on $[-0.5, 0.5]$, within an accuracy of 0.01.

5. Assume that f is defined on a neighborhood around the origin and is twice differentiable there, with its second derivative continuous. Prove that

$$f(x) = f(0) + f^{(1)}(0)x + \int_0^x f^{(2)}(t)(x - t)dt.$$

(Hint: Integrate by parts $\int_0^x f^{(2)}(t)(x - t)dt$).

6. Assume that f is defined on a neighborhood around the origin and is three times differentiable there, with its third derivative continuous. Prove that

$$f(x) = f(0) + f^{(1)}(0)x + \frac{f^{(2)}(0)}{2}x^2 + \frac{1}{2}\int_0^x f^{(3)}(t)(x - t)^2 dt.$$

(Hint: Integrate by parts $\int_0^x f^{(3)}(t)(x - t)^2 dt$).

7. Assume that f is defined and differentiable $n + 1$ times on a neighborhood of the origin, and $f^{(n+1)}$ is continuous. On the basis of the two previous exercises we might suspect that

$$f(x) = T_n(x) + \frac{1}{n!} \int_0^x f^{(n+1)}(t)(x - t)^n dt.$$

Integrating by parts the integral, show that the equality holds true for every x in the neighborhood . Thus

$$E_n(x) = \frac{1}{n!} \int_0^x f^{(n+1)}(t)(x - t)^n dt.$$

Quite appropriately, this new way of expressing E_n is called the 'error in integral form.'

8. Defining $f(x) = \cos x$ show, independently of Taylor's theorem, that

$$\lim_{n \to \infty} \frac{1}{n!} \int_0^x f^{(n+1)}(t)(x - t)^n dt = 0.$$

This fact provides an alternative way of proving the analiticity of the cosine function.

Enrichment Note: How the Exact Value of a Series was Found

In section 3.9 we mentioned that in 1734 Leonhard Euler presented the surprising equality $\sum_{n=1}^{\infty} \frac{1}{n^2} = \frac{\pi^2}{6}$. How did he reach it? He tried different approaches. One of them is essentially as follows (Kalman, 1993): The Taylor series of $\sin x$ is

$$\sin x = x - \frac{x^3}{3!} + \frac{x^5}{5!} - \frac{x^7}{7!} + \dots.$$

Therefore, for any $x > 0$,

$$\sin \sqrt{x} = \sqrt{x} - \frac{(\sqrt{x})^3}{3!} + \frac{(\sqrt{x})^5}{5!} - \frac{(\sqrt{x})^7}{7!} + \dots.$$

Consequently, for any $x > 0$,

$$\frac{\sin \sqrt{x}}{\sqrt{x}} = 1 - \frac{1}{3!}x + \frac{x^2}{5!} - \frac{x^3}{7!} + \dots.$$

The roots of $\frac{\sin \sqrt{x}}{\sqrt{x}}$ are precisely the positive roots of $\sin \sqrt{x}$, that is, π^2, $4\pi^2$, $9\pi^2$, $16\pi^2$, ... Euler surely knew that given any polynomial $x^n + a_{n-1}x^{n-1} + \dots + a_1 x + a_0$, $a_0 \neq 0$, its n roots $r_1, r_2, \dots, r_n$ have two basic properties, namely $r_1 r_2 \dots r_n = a_0$ and $\sum r_{i_1} \dots r_{i_{n-1}} = -a_1$ where $\{i_1, \dots, i_{n-1}\}$ runs over all subsets of size $n-1$ of $\{1, \dots, n\}$. For instance, given $x^2 + a_1 x + a_0$ we have $r_1 r_2 = a_0$ and $r_1 + r_2 = -a_1$ while given the third order polynomial $x^3 + a_2 x^2 + a_1 x + a_0$ we have $r_1 r_2 r_3 = a_0$ and $r_1 r_2 + r_1 r_3 + r_2 r_3 = -a_1$. Consequently

$$\frac{1}{r_1} + \frac{1}{r_2} + \dots + \frac{1}{r_n} = -\frac{a_1}{a_0}.$$

At this stage Euler assumed that this property is also valid for power series, in particular for the power series that stems from $\frac{\sin\sqrt{x}}{\sqrt{x}}$. Therefore

$$\frac{1}{\pi^2} + \frac{1}{4\pi^2} + \frac{1}{9\pi^2} + \frac{1}{16\pi^2} + \dots = -(-\frac{1}{3!}) = \frac{1}{3!}.$$

Then $\frac{1}{\pi^2}\sum_{n=0}^{\infty}\frac{1}{n^2} = \frac{1}{6}$, thus $\sum_{n=0}^{\infty}\frac{1}{n^2} = \frac{\pi^2}{6}$. Of course, Euler realized that his finding needed a rigorous justification because there is no certainty that what is a valid property for polynomials has to be also valid for power series. Many different proofs, acceptable under modern standards of rigor, have been developed since Euler's time. The interested reader can consult the above-mentioned work by Kalman.

Chapter 4

Enzyme Kinetics

4.1 Introduction

Enzymes are mainly proteins that catalyze many organic chemical reactions which otherwise would proceed very slowly. They are essential to life. Their kinetics began to be understood at the beginning of the 20th century. It was observed that a typical enzyme (E) converts a substrate (S) into a product (P) according to the chemical formula $S + E \longrightarrow E + P$. Assuming that we are dealing with a single-step reaction, the rate at which the concentration of a product increases is proportional to the concentration of reactants[1], thus

$$\frac{d[P]}{dt} = k[S][E].$$

This is due to the law of mass action, which asserts that the rate at which an elementary chemical reaction proceeds is proportional to the product of the concentration of reactants. By increasing $[S]_o$, the initial concentration of substrate, and keeping the amount of enzyme concentration constant, we could increase, without limits, the initial rate v_o at which the product is formed. It is very interesting to point out that this conclusion is not in agreement with observations: v_o reaches a value beyond which the addition of more substrate does not increase the rate of initial formation of the product. In other words, the enzyme is saturated with substrate. To circumvent this and other difficulties, scientists postulated the existence of an intermediate compound, which achieved rapidly an equilibrium with the reactants and decomposed gradually producing a molecule of the product and regenerating a molecule of enzyme. That is to say,

$$S + E \leftrightarrow C \rightarrow E + P.$$

---

[1]The symbol [ ] is often used by chemists to denote concentration.

Let us assume that the reversible process has rate constants k_1 and k_{-1} for the forward and backward reaction, respectively, while the irreversible process is governed by the rate constant k_2. Due to the above-mentioned equilibrium, we have

$$k_1[S][E] = k_{-1}[C],$$

so $[E] = \frac{k_{-1}[C]}{k_1[S]}$. But the enzyme exists either as free enzyme or part of the intermediate compound, thus $[E]_T = [E] + [C]$ where $[E]_T$ is the total concentration of enzyme. Therefore $[C] = [E]_T - [E] = [E]_T - \frac{k_{-1}[C]}{k_1[S]}$, which in turn leads to $[C](1 + \frac{k_{-1}}{k_1[S]}) = [E]_T$. Finally we get

$$[C] = \frac{[E]_T[S]}{K+[S]}$$

where $K = \frac{k_{-1}}{k_1}$. Since the rate of the reaction is given by $v = \frac{d[P]}{dt}$ and $\frac{d[P]}{dt} = k_2[C]$, we reach the expression $v = \frac{k_2[E]_T[S]}{K+[S]}$. In particular

$$v_o = \frac{k_2[E]_T[S]_o}{K + [S]_o}. \tag{4.1}$$

A close look at this expression allows us to conclude that, if we increase $[S]_o$ while keeping $[E]_T$ constant, eventually it will be much greater than K. So v_o will tend to the limiting rate $k_2[E]_T$, which we denote V_{max} following common usage among biochemists. Thus

$$v = \frac{V_{max}[S]}{K + [S]}. \tag{4.2}$$

This relationship is known as the **Michaelis-Menten equation**, honoring Leonor Michaelis and Maud Menten, who in 1913 published a groundbreaking paper on enzyme kinetics. They were two early pioneers in a relatively new field.

If we consider v_o as a function of $[S]_o$ (keeping $[E]_T$ constant), a graph, shared by all functions of the form $f(x) = \frac{ax}{b+x}$, can be drawn (Figure 4.1).

The Michaelis-Menten equation (MM) predicts the appearance of the phenomenon of saturation because, no matter how much substrate we add, the initial rate cannot surpass the limiting value V_{max}. By the early 1920s, solid experimental evidence supporting the Michaelis-Menten equation had accumulated. But the existence of an equilibrium between reactants and the intermediate compound was challenged by George Briggs and John Haldane in a remarkable two-page paper (Briggs and Haldane, 1925). Rather than accepting the equilibrium between substrate, enzyme, and the intermediate compound, they claimed that the rate at which the concentration of the intermediate compound varies is practically zero, except at the very beginning of the reaction. This alternative hypothesis led them also to the Michaelis-Menten equation, as we will see next.

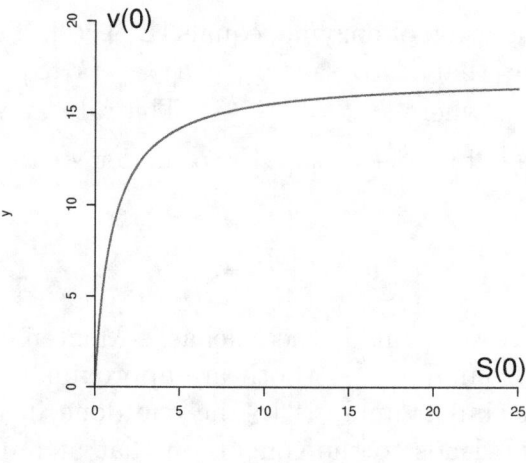

Figure 4.1: The initial rate as a function of $[S]_o$

4.2 The Steady-State Hypothesis

Let us recall that the basic model of enzyme kinetics is given by

$$S + E \leftrightarrow C \rightarrow E + P$$

with rate constants k_1, k_{-1} for the reversible part of the reaction and k_2 for the irreversible part. The substrate S combines with the enzyme E giving birth to an intermediate compound C through a reversible reaction. C decomposes into the product P and regenerates the enzyme E. It should be noted that one works with a much higher concentration of substrate than of enzyme.

The rate at which C varies is given by

$$\frac{d[C]}{dt} = k_1[S][E] - (k_{-1} + k_2)[C].$$

At the beginning of the experiment, substrate and enzyme combine quite rapidly yielding the intermediate compound C. Thereafter, a steady-state ensues, during which the concentration of C remains practically constant. This is because each time a molecule of P is formed by a rearrangement of C, a molecule of the enzyme is regenerated and combines rapidly with a molecule of substrate (there is a high affinity between both of them and during most part of the process there are many more molecules of substrate than enzyme). This mechanism lasts as long as there is substrate. Thus, we should expect that $\frac{d[C]}{dt} = 0$ under steady-state conditions, which in turn implies

$$k_1[S][E] - (k_{-1} + k_2)[C] = 0.$$

But $[E]_T$, the total concentration of enzyme, equals $[E] + [C]$. Therefore $k_1[S]([E]_T - [C]) - (k_{-1} + k_2)[C] = 0$; that is to say, $k_1[S][E]_T - (k_1[S] + k_{-1} + k_2)[C] = 0$. Consequently $[C] = \frac{[E]_T[S]}{[S]+k_m}$, where $K_m = \frac{k_{-1}+k_2}{k_1}$. The rate at which the product is formed is given by $v = \frac{d[P]}{dt}$; but $\frac{d[P]}{dt} = k_2[C]$. So, under steady-state conditions we will have

$$v = \frac{k_2[E]_T[S]}{[S]+K_m}.$$

The reader may observe that this is the Michaelis-Menten equation, except that $K_m = \frac{k_{-1}+k_2}{k_1}$ is different from $K = \frac{k_{-1}}{k_1}$ (both are approximately equal when k_{-1} is much bigger than k_2). A similar analysis to the one done in the previous section, right after displaying (4.1), leads to the conclusion that at any instant, during the steady-state, the maximum rate that can be achieved is $k_2[E]_T$. Thus, we can write $V_{max} = k_2[E]_T$ as we did before.

In particular, this formula is valid at the beginning of the steady-state when we can measure the rate v_o for a certain concentration of substrate $[S]_o$. For most reactions catalyzed by enzymes, the steady-state is reached very quickly, on the order of milliseconds, so we may assume that $[S]_o$ is the concentration of substrate at the beginning of the experiment. Let us pay close attention to the formula

$$v_o = \frac{V_{max}[S]_o}{[S]_o+K_m}.$$

For each value of $[S]_o$ we should expect a different value of v_o. How could we calculate v_o? So far we do not have information about K_m or V_{max}, hence v_o has to be found from experiments. Indeed, under steady-state conditions

$$k_1[E][S] = (k_{-1} + k_2)[C].$$

Therefore

$$\frac{d[S]}{dt} = -k_1[E][S] + k_{-1}[C] = -(k_{-1} + k_2)[C] + k_{-1}[C] =$$

$$= -k_2[C] = -\frac{d[P]}{dt} = -v.$$

Thus, $v = -\frac{d[S]}{dt}$, which in turn implies that v can be approximated by the slope of the tangent line to the $[S]$ curve. In actual practice, to estimate v_o we would have to calculate

$$\frac{S(t_2)-S(t_1)}{t_2-t_1}$$

where t_1 and t_2 are very close to each other and measurements are made at the beginning of the experiment[2]. Later, once we learn more about $[S]$ as a function of time, a practical method will be analyzed.

Biochemists prefer to perform measurements at the beginning of the steady-state, in other words measure $[S]_o$ and v_o rather than $[S]$ and v at a later time, because some enzymes may be denatured as the process is under way or an appreciable amount of product may inhibit the catalytic role of the enzyme.

4.3 The Lineweaver-Burk Plot

How can we estimate V_{max} and K_m? One approach is to perform experiments and choose different concentrations of substrate $[S]_o$, then measure the corresponding initial rate v_o. Having a $[S]_o, v_o$ table (e.g., Table 4.1) of experimentally determined values, we could then fit a curve as best as possible. The horizontal asymptote would be V_{max} while K_m is the value of $[S]_o$ at which the initial rate becomes $V_{max}/2$. The latter assertion follows from the fact that, using the Michaelis-Menten equation,

$$\frac{V_{max}}{2} = \frac{V_{max}[S]_o}{[S]_o + K_m} \text{ if and only if } [S]_o = K_m.$$

However, it is not an easy task to fit by hand the above-mentioned curve to experimental values of initial substrate concentrations and initial velocities. Fortunately, there is a simple alternative, which we will study next. Taking the converse of the Michaelis-Menten equation, we get $\frac{1}{v_o} = \frac{[S]_o + K_m}{V_{max}[S]_o}$ which, in turn, is equivalent to

$$\frac{1}{v_o} = \frac{1}{V_{max}} + \frac{K_m}{V_{max}}\frac{1}{[S]_o}.$$

If we have $\frac{1}{[S]_o}$ on the x-axis and $\frac{1}{v_o}$ on the y-axis, the experimental values should cluster around a straight line. We use linear regression to find the least squares linear regression line. In turn, this will allow us to calculate $1/V_{max}$ as the intersection with the y-axis and K_m/V_{max} as the slope. From these values, we can easily obtain V_{max} and K_m.

This plot is known as a Lineweaver-Burk plot (Figure 4.2). Hans Lineweaver and Dean Burk introduced this way of calculating V_{max} and K_m in 1934. An example will help to understand the procedure. Let us consider the kinetic data related to the hydration of CO_2 utilizing the enzyme carbonic anhydrase (McQuarrie and Simon, 1977), shown in Table 4.1.

We may calculate the least squares linear regression line using a graphing calculator (TI-89 or similar calculators), placing $1/[S]_o$ values in the first column and $1/v_o$ values in the second column (Table 4.2). We can work with the Data/Matrix Editor of the graphing calculator, found under Applications. Once data are loaded

---

[2]Sometimes it is convenient to write $S(t)$ instead of $[S]$; in particular, $S(0) = [S]_o$

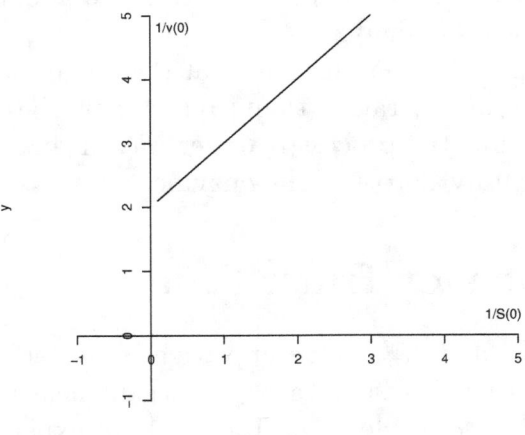

Figure 4.2: The Lineweaver-Burk plot

Table 4.1: Hydration of CO_2

$[S]_o(mol.dm^{-3})$	$v_0(mol.dm^{-3}.s^{-1})$
1.25×10^{-3}	2.78×10^{-5}
2.5×10^{-3}	5.00×10^{-5}
5×10^{-3}	8.33×10^{-5}
20×10^{-3}	1.66×10^{-4}

in columns one and two, LinReg (F5 option, with c1 for x and c2 for y) does the job. Minitab, **R**, and many other software packages have linear regression capabilities too. The following line should appear on the screen: $\hat{y} = 39.934042x + 4023.940015$. Hence $1/V_{max} = 4023.940015$, which in turn leads to $V_{max} = 0.00024851265$, while $K_m/V_{max} = 39.934042$. Replacing the value of V_{max}, we get $K_m = 0.009924$. Therefore, $V_{max} = 2.4851265 \times 10^{-4}$ while $K_m = 9.924 \times 10^{-3}$. It should be noted that the correlation coefficient happened to be 1. Of course, once V_{max} has been estimated, it is not difficult to estimate k_2 since $V_{max} = k_2 E_T$.

Why is it so important to calculate K_m and V_{max}? The answer is that each enzyme has a specific, unique value for these constants; thus, calculating the above-mentioned parameters may help to identify an enzyme.

It should be noted that there are several alternatives to the Lineweaver-Burk plot (Helfgott and Seier, 2007).

Table 4.2: Table of transformed values

$1/[S]_o$	$1/v_0$
$10^3/1.25$	$10^5/2.78$
$10^3/2.5$	$10^5/5$
$10^3/5$	$10^5/8.33$
$10^3/20$	$10^4/1.66$

4.4 Integration of the Michaelis-Menten Equation

We have learned how to calculate K_m and V_{max} on the basis of experimental values for S_o and v_0. But, under certain circumstances, it might be difficult to measure initial velocities because often the reaction proceeds quite rapidly. What can be done? There is an alternative way if we can measure $S(t)$ at different values of t during the steady state.

Let us recall that Michaelis-Menten equation asserts that, at any instant t during the steady-state, $v = \frac{V_{max}S(t)}{S(t)+K_m}$ where $v = P'(t) = -S'(t)$. Thus $-S'(t) = \frac{V_{max}S(t)}{S(t)+K_m}$. Multiplying by $\frac{S(t)+K_m}{S(t)}$, we arrive at $-S'(t) - K_m\frac{1}{S(t)}S'(t) = V_{max}$. Integrating with respect to time between 0 and t (considering 0 as the instant when the steady-state begins), we get

$$-\int_0^t S'(u)du - K_m\int_0^t \frac{S'(u)}{S(u)}du = \int_0^t V_{max}du.$$

Consequently,

$$-(S(t) - S(0)) - K_m\ln\frac{S(t)}{S(0)} = V_{max}t. \tag{4.3}$$

Therefore,

$$\frac{1}{t}\ln\frac{S(0)}{S(t)} = -\frac{1}{K_m}\frac{S(0)-S(t)}{t} + \frac{V_{max}}{K_m}.$$

This is a remarkable identity, known as the integrated form of the Michaelis-Menten equation, because it does not involve rates but only experimental values of $S(t)$ during the course of an enzymatic reaction. Also, it predicts the appearance of a line if we have $\frac{S(0)-S(t)}{t}$ on the horizontal axis and $\frac{1}{t}\ln\frac{S(0)}{S(t)}$ on the vertical axis (Figure 4.3). It is a line with slope $-1/K_m$ and vertical intersection V_{max}/K_m.

So, in order to calculate V_{max} and K_m we need a table of $S(t)$ values obtained at different times t. We can build a table of two columns with $\frac{S(0)-S(t)}{t}$ in the first column and $\frac{1}{t}\ln\frac{S(0)}{S(t)}$ in the second column. A linear regression analysis will provide us with an approximation to the slope $\frac{-1}{K_m}$ and the vertical intersection $\frac{V_{max}}{K_m}$. Finally, a simple arithmetical procedure will lead to the corresponding values of K_m and V_{max}.

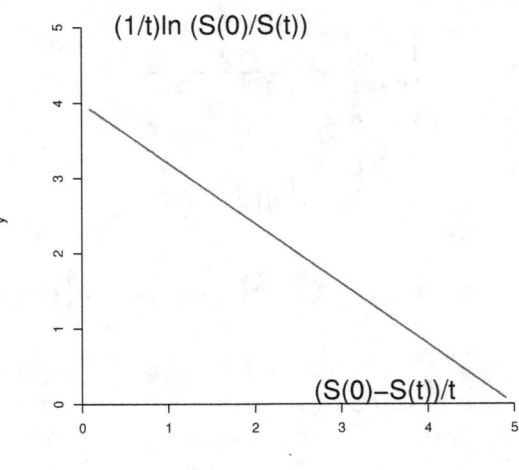

Figure 4.3: A prediction that stems from the integrated form of MM

The advantage of this approach to the estimation of K_m and V_{max} is that it does not require measuring rates (often a challenging experimental procedure). A difficulty that may appear when working with the integrated equation is that it requires measuring values of S across time, a situation that could lead to its own set of problems because diverse factors may eventually distort the reaction after it has gotten under way. For instance, as we mentioned before, the product might inhibit enzyme activity or the enzyme might become unstable.

4.5 The Variation of Substrate at the Beginning

The integrated form of the Michaelis-Menten equation provides a useful way of calculating V_{max} and K_m, but it does not provide an explicit formula for $S(t)$. Interestingly enough, it is possible to find an approximate formula for $S(t)$ at the beginning of the steady state. For this purpose we have to keep in mind a mathematical approximation, namely that $\ln(1 + x)$ is approximately equal to x whenever x is a very small positive number. Hence, at the beginning of the steady state, when $S(t)$ does not differ much from S_o,

$$\ln \frac{S(t)}{S_o} = \ln(1 + \frac{S(t)}{S_o} - 1) = \ln(1 + \frac{S(t) - S_o}{S_o}) \approx \frac{S(t) - S_o}{S_o}.$$

Multiplying both sides of (4.3) by -1, we get

$$S(t) - S_o + K_m \ln \frac{S(t)}{S_o} = -V_{max}t,$$

which can be replaced by $S(t) - S_o + K_m \frac{S(t)-S_o}{S_o} = -V_{max}t$ due to the above-mentioned approximation. Therefore,

$$S(t) = S_o - \frac{V_{max}}{1+\frac{K_m}{S_o}}t = S_o - \frac{V_{max}S_o}{S_o+K_m}t.$$

Hence $S(t) = S_o - v_o t$. Having t on the horizontal axis and S on the vertical axis, we are then dealing with a straight line with slope $-v_o$ and vertical intercept S_o. In other words, at the beginning of the steady state, the variation of S with respect to time is linear. The fact that $-v_o$ is the slope of this line has practical implications because it suggests a way to estimate the initial rate: make several measurements of S at the beginning of the experiment and apply linear regression. The slope of the least squares linear regression line will be an approximation of v_o.

4.6 Estimating Parameters

Let us go back to the enzymatic reaction $S + E \leftrightarrow C \to E + P$, with kinetic constants k_1, k_{-1} for the reversible stage and k_2 for the irreversible stage. In section 4.3 we were able to estimate k_2 under steady-state conditions because V_{max} can be estimated through Lineweaver-Burk or similar plots (recall that $V_{max} = k_2 E_T$, where E_T is the total amount of enzyme). Our next task is to estimate k_1 and k_{-1}.

We have $C'(t) = k_1 S(t) E(t) - (k_{-1} + k_2)C(t)$, $P'(t) = k_2 C(t)$, and $E_T = E(t) + C(t)$. Before the onset of the steady-state it is a good approximation to replace $S(t)$ by S_o. Therefore,

$$P''(t) = k_2 C'(t) = k_2(k_1(E_T - C(t))S_o - (k_{-1} + k_2)C(t) =$$

$$= k_2 k_1 E_T S_o - (k_1 S_o + k_{-1} + k_2)k_2 C(t)$$

$$= k_2 k_1 E_T S_o - (k_1 S_o + k_{-1} + k_2)P'(t).$$

Hence $P''(t) + a_1 P'(t) = a_2$, where $a_1 = k_1 S_o + k_{-1} + k_2$ and $a_2 = k_1 S_o V_{max}$. Then the general solution[3] is given by

$$P(t) = d_1 + d_2 e^{-a_1 t} + \frac{a_2}{a_1}t.$$

Since $P(0) = 0$ and $P'(0) = k_2 * 0 = 0$ we can conclude that $d_1 + d_2 = 0$ and $-a_1 d_2 + \frac{a_2}{a_1} = 0$, which lead to $d_2 = \frac{a_2}{a_1^2}$, $d_1 = -\frac{a_2}{a_1^2}$. Then

$$P(t) = -\frac{a_2}{a_1^2} + \frac{a_2}{a_1^2}e^{-a_1 t} + \frac{a_2}{a_1}t.$$

---

[3]Recall the last subsection of section 2.9.

Taking into consideration that during the pre-steady state the values of t are extremely small (sometimes in the order of milliseconds), we can approximate $e^{-a_1 t}$ by $1 - a_1 t + \frac{a_1^2}{2} t^2$. Therefore,

$$P(t) = -\frac{a_2}{a_1^2} + \frac{a_2}{a_1^2}(1 - a_1 t + \frac{a_1^2}{2} t^2) + \frac{a_2}{a_1} t = \frac{a_2}{2} t^2 = \frac{k_1 V_{max} S_o}{2} t^2.$$

That is to say,

$$\frac{2P(t)}{t V_{max} S_o} = k_1 t.$$

Thus, we can predict that if it is possible to measure $P(t)$ before the steady-state, and plot points on a graph with time on the horizontal axis and $\frac{2P(t)}{t V_{max} S_o}$ on the vertical axis, the points should be spread around a line that passes not far from the origin. Thereafter we estimate k_1 through linear regression.

It is to be noted that rapid reaction techniques, developed in the 1950's, allowed the measurement of $P(t)$ before the onset of the steady state. A great success of the basic model of enzyme kinetics was to make the above-mentioned prediction, which was later found to be in agreement with observations (Roughton, 1954). Indeed, any scientific theory should have the capacity to make testable predictions.

Having developed a method to estimate k_1, it is an easy task to estimate k_{-1} because $k_{-1} = k_1 K_m - k_2$ where, let us recall, K_m is found through Lineweaver-Burk or similar plots.

4.7   The Pseudo-Steady-State Hypothesis

Biochemists have found that the steady-state hypothesis is very fruitful. Many consequences of it are in agreement with experimental results. However, it has been challenged by several scientists who prefer to call it the 'pseudo-steady-state hypothesis' because, strictly speaking, the derivative of $[C]$ equals zero at just one instant. Thereafter, the curve that describes $[C]$ is only approximately constant while there is enough substrate left. Starting in the 1960's, a new approach, based on the perturbation theory of differential equations, was developed. This is an advanced branch of mathematics that allows the construction of a framework that supersedes previous theories. But, at an introductory level, the steady-state hypothesis is still widely used in enzyme kinetics, and with reasonably good outcomes.

4.8   Non-Linear Regression

Before computers became powerful and widely available, linearization of a non-linear model through the application of non-linear transformations (Lineweaver-Burk plot and related ones) was the practical way to estimate V_{max} and K_m. However, nowadays there are computer programs available to estimate the parameters of a non-linear model without transforming the variables. In particular, **R** is a freeware (available from http://www.r-project.org) that contains a command to do non-linear regression; step by step instructions to download the program can be found at

http://faculty.etsu.edu/seier/CalcBioBook.htm

We will illustrate its use with the same example to which we applied the traditional model of linearization (Table 4.1, section 4.3). Its scatter plot can be seen in Figure 4.4.

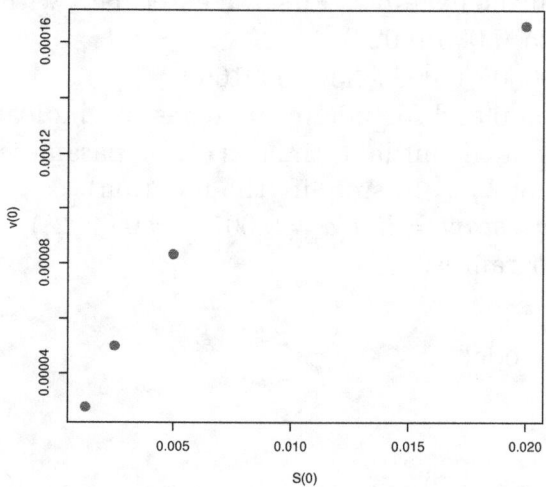

Figure 4.4: Scatterplot of initial rate versus initial concentration

The procedure of doing non-linear regression can be summarized in two steps:

1. First, we need to come up with initial estimates of the parameters. For this purpose we need to recall the role of the parameters of the curve. If we write the model as $y = \frac{V_{max}x}{x+K}$, we have to keep in mind that V_{max} is the maximum rate. So it would just make sense to have, as the initial estimate, the maximum rate attained in the experiment or a value close to it. In the example, the maximum value of v_o was 0.000166, so we could use its rounded version, 0.0002, as an initial estimate for V_{max}.

To get an initial estimate for K_m we must remember that K_m is the value of the variable x (substrate concentration) that corresponds to half of V_{max}. In the example, $\frac{V_{max}}{2}$ is approximately 0.0001. Locating the value 0.0001 on the vertical axis and going to the right to guess a value of x, we see that $x = 0.007$ when $V = 0.0001$. Thus, we will take 0.007 as our initial estimate for K_m.

2. We input the initial estimates, and the program will calculate the sum of squares of residuals using those initial estimates. Then the software will change the values of the parameters a little bit and will recalculate the sum of squares of residuals. The process continues until the change in the sum of squares of residuals is negligible. The sum of squares of residuals can be thought of as a function of the values of the parameters; we can think of it as a surface and we want to reach the minimum of that surface. Imagine a valley that has hills and slopes and we want to reach the location in the valley that has the minimum altitude. In order to arrive there soon, it is important to choose the 'correct direction.' The algorithm that searches for the minimum uses mathematical tools (usually the Gauss-Newton method) to walk on the route of the steepest slope.

Next we include the commands we need to type in **R** to perform the non-linear estimation for the data in the example. First we enter data with:

$x <- c(0.00125, 0.0025, 0.005, 0.020)$

$y <- c(0.0000278, 0.00005, 0.0000833, 0.000166)$

Then we write the command for nonlinear regression indicating the equation of the model and the values of the initial estimates of the parameters (we are using the symbols a for K_m and b for V_{max} to simplify the notation):

$nls(y \sim (b*x)/(x+a), start = list(a = 0.007, b = 0.0002))$

The output of the program is:

Nonlinear regression model
model: $y \sim (b*x)/(x+a)$
data: parent.frame()
$a = 9.9041861615, b = 0.0002482121$
residual sum-of-squares: $1.895619e - 15$

The final estimated value for K_m and V_{max} are 0.0099041861 and 0.0002482121, respectively. That is to say, $K_m = 9.9041861 \times 10^{-3}$ and $V_{max} = 2.482121 \times 10^{-4}$, quite close to the values found before using linear regression.

When data are such that the curve fits quite perfectly, it is likely that the estimated values using transformed variables and linear regression are going to be very similar to the estimated values through non-linear regression. However, when there is a lot of experimental error and the points look scattered around a hypothetical curve, or the experiment has been replicated and there is a lot of variability from replicate to replicate also producing points that look scattered around the curve, the two methods could give different estimated values. We also need to remember that linear regression makes several assumptions, such as linearity, constant error variance, and normality of errors. If those assumptions are violated by the transformed data, we might be better off working with non-linear estimation using appropriate software. A detailed

Table 4.3: Segel's data

x	y
0.00000833	$1.3E - 08$
0.00001000	$1.60E - 08$
0.00001250	$1.90E - 08$
0.00001670	$2.36E - 0.8$
0.00002500	$3.08E - 08$
0.00003000	$3.43E - 08$
0.00003330	$3.63E - 08$
0.00004000	$4.00E - 08$
0.00005000	$4.44E - 08$
0.00006000	$4.80E - 08$
0.00008000	$5.33E - 08$
0.00010000	$5.71E - 08$
0.00020000	$6.67E - 08$

discussion of these and related ideas can be found in the literature (Helfgott and Seier, 2007).

Exercises for chapter 4

1. Multiplying by $[S]_o$ the Lineweaver-Burk expression $\frac{1}{v_o} = \frac{1}{V_{max}} + \frac{K_m}{V_{max}}\frac{1}{[S]_c}$ we get $\frac{[S]_o}{v_o} = \frac{1}{V_{max}}[S]_o + \frac{K_m}{V_{max}}$. A plot of $\frac{[S]_o}{v_o}$ versus $[S]_o$ will be linear. Using Table 3.1 perform a linear regression with $[S]_o$ on the x-axis and $\frac{[S]_o}{v_o}$ on the y-axis. Estimate the parameters V_{max} and K_m and compare your results with those obtained before.

2. The data in Table 4.4 come from experiments with an enzyme (Segel. 1976). The x-values are concentrations of substrate while the y-values are corresponding initial velocities. Apply both linear regression and non-linear regression to estimate the parameters K_m and V_{max} (in the case of linear regression we have in mind the corresponding Lineweaver-Burk plot). Compare both approaches, keeping in mind that, as usual, $E - 08$ is another way of writing 10^{-8}; for instance, $1.38E - 08$ means $1.38 * 10^{-8}$.

3. Use the integrated form of Michaelis-Menten equation to do a linear regression analysis, and estimate the values of K_m and V_{max} from Stern's data portrayed in Table 4.5 (Stern, 1936).

Table 4.4: Stern's data

$t(min)$	$S(t)(mol \times cm^{-3})$
0	10.27
3	7.98
6	5.20
9	2.86
12	1.19
15	0.32

4. It can happen that besides the substrate and the enzyme there is a substance, called the inhibitor, that binds with the enzyme in competition with the substrate. This phenomenon causes a decrease in the rate of formation of the product P. For instance, insecticides inhibit very important enzymes in insect metabolism. In addition to the scheme $S+E \leftrightarrow C \to E+P$, we need to consider the reversible chemical reaction $E+I \leftrightarrow IE$ between the enzyme, the inhibitor, and the compound IE formed by the union of I and E. We can assume that the latter chemical reaction soon reaches equilibrium, thus $k_i[E][I] = k_{-i}[IE]$, where k_i and k_{-i} are the forward and backward rate constants, respectively. So $[IE] = \frac{[E][I]}{K}$, where $K = \frac{k_{-i}}{k_i}$. On the other hand $[E]_T = [E] + [C] + [IE]$ since the enzyme exists either as free enzyme, or has combined either with the substrate or the inhibitor. Therefore $[E] = \frac{[E]_T - [C]}{1 + \frac{I}{K}}$. Show that under steady-state conditions

$$v = \frac{V_{max}[S]}{[S] + K_m(1 + \frac{[I]}{K})}.$$

This expression looks very much like Michaelis-Menten equation, except that K_m is multiplied by the factor $1 + \frac{[I]}{K}$ (Hint: Follow the steps that led to Michaelis-Menten equation).

5. At the beginning of the steady-state, we will have $v_o = \frac{V_{max}[S]_o}{[S]_o + K_m(1 + \frac{[I]_o}{K})}$. Thus

$$\frac{1}{v_o} = \frac{1}{V_{max}} + \frac{K_m(1 + \frac{[I]_o}{K})}{V_{max}} \frac{1}{[S]_o}.$$

Keeping $[I]_o$ constant, a linear equation will appear if we put $1/[S]_o$ on the x-axis and $1/v_o$ on the y-axis (this is the Lineweaver-Burk plot when inhibition is present). What will be the intersection of this line with both axes? Finally, make a rough sketch of the lines that correspond to $[I]_o$, $2[I]_o$, and $3[I]_o$. What do you notice with regard to the intersection of these three lines with each axis?

Chapter 5

Transport Across Cell Membranes

5.1   Introduction

The molecules of solute in a solution exhibit a particular behavior when a concentration gradient is present. They move around due to their kinetic energy in such a way that, after a certain time, they are uniformly distributed in the solution. We then say that the substance diffuses and does so from a higher to a lower concentration.

Diffusive processes across membranes were studied in the 18th century, but no quantitative law could be found to explain the phenomenon of diffusion. In 1855 Adolf Fick, a physician, conducted experiments and, in close analogy to the theory of heat elaborated by Joseph Fourier around 1820, put forward a mathematical relationship that nowadays is called **Fick's law**.

5.2   Diffusion Across Cell Membranes

For the case of diffusion across a cell membrane, Fick's law leads to

$$\frac{dS_{in}}{dt} = kA(C_{out} - C_{in}(t)) \tag{5.1}$$

where $S_{in}(t)$ denotes the mass of solute inside the cell, A is the area of the membrane, $C_{in}(t)$ is the concentration of solute inside the cell and C_{out} is the constant concentration outside the cell (the volume outside the cell is usually much larger than the volume of the cell, thus we can assume that C_{out} is constant); k is a positive parameter known as the 'permeability of the membrane,' which depends on the structure and width of the cell membrane (a typical width of a cell membrane is $5 * 10^{-7} cm$).

Since $C_{in}(t) = \frac{S_{in}(t)}{V}$, where V is the volume of the cell, we can conclude that

$$\frac{dC_{in}}{dt} = \frac{kA}{V}(C_{out} - C_{in}(t)). \tag{5.2}$$

We may observe that if $C_{out} > C_{in}(t)$ then $\frac{dC_{in}}{dt} > 0$, that is, the concentration of solute inside the cell is increasing due to the fact that molecules of solute are migrating from a higher concentration to a lower concentration. We wish to point out that some molecules of solute will cross the membrane in the other direction, but the net flow of molecules of solute will be from the outside to the inside of the cell.

Similarly, if $C_{out} < C_{in}(t)$ then $\frac{dC_{in}}{dt} < 0$; thus, the concentration of solute inside the cell is diminishing: the net flow of molecules is from the inside to the outside of the cell.

5.3   The Quotient A/V

From the equation (5.2) we observe that the rate at which the concentration of solute changes inside the cell depends not only on the difference $C_{out} - C_{in}$ but on the quotient A/V. The larger the quotient A/V, the higher the rate at which the concentration inside the cell is changing.

Suppose that the cell has a spherical shape with radius r. Then

$$\frac{A}{V} = \frac{4r^2\pi}{(4/3)r^3\pi} = \frac{3}{r}.$$

Hence, the quotient A/V is inversely proportional to the radius. A typical spherical bacteria may have a radius of $1\mu m$ ($1\mu m = 10^{-6}m$), thus $A/V = 3$, while an spherical eukaryotic cell may have a radius of $10\mu m$. For the latter we have $A/V = 0.3$, hence the above-mentioned bacteria does much better in terms of diffusion!

What happens if the cell has the shape of a cylinder? Suppose that the radius is r and the height is r too. Then $A = 2\pi r^2 + 2\pi r * r = 4\pi r^2$ and $V = \pi r^3$, thus $A/V = 4/r$. If the volume is fixed we would have $r = (V/\pi)^{1/3}$, which in turn leads to $A = 4\pi^{1/3}V^{2/3}$; therefore $A/V = 4\pi^{1/3}/V^{1/3} \approx 5.86/V^{1/3}$.

A natural question is whether, for a fixed volume V, we would get a better ratio A/V if the height is $r/2$ rather than r. Indeed,

$$V = \pi r^2 \frac{r}{2} = \frac{\pi r^3}{2}$$

that in turn leads to $r = (\frac{2V}{\pi})^{1/3}$. But $A = 2\pi r^2 + 2\pi r(\frac{r}{2}) = 3\pi r^2$, consequently

$$A = 3\pi(\tfrac{2V}{\pi})^{2/3} = 3 * 2^{2/3}\pi^{1/3}V^{2/3} \approx 6.97V^{2/3}.$$

Thus $\frac{A}{V} \approx \frac{6.97}{V^{1/3}}$. We note that the flattened cylinder with radius $r/2$ has a better A/V ratio than the cylinder of the same volume but radius r. It is not hard to show that for a cylinder-shaped cell, keeping the volume fixed, the ratio A/V increases insofar as the height decreases. To be more precise, if the height is r/n, where n is a natural number, then

$$\frac{A}{V} = \frac{2\pi^{1/3}}{V^{1/3}}\left(n^{2/3} + \frac{1}{n^{1/3}}\right).$$

Thus A/V becomes bigger and bigger when n increases. For instance, $A/V \approx 14.96/V^{1/3}(n = 10)$, $A/V \approx 63.74/V^{1/3}(n = 100)$, $A/V \approx 293.21/V^{1/3}$ $(n = 1000)$.

On the other hand, one would expect that, for a fixed volume V, if the height is given by nr (instead of r/n) then A/V will also increase when n becomes bigger and bigger. This is so when $n > 2$ (when the height is $2r$ the quotient $A/V \approx 5.54/V^{1/3}$, the smallest quotient among all cylinders of fixed volume V). In other words, an elongated cell in the shape of a cylinder will have a greater quotient A/V than a more regular cylinder.

For a sphere of fixed volume V we can prove that $A/V \approx 4.84/V^{1/3}$, worse than the cylinder of the same volume and height r. Actually, the sphere has the smallest A/V ratio among all solids of the same volume because it has the smallest area. This latter fact stems from the famous isoperimetric problem of higher mathematics.

What happens when a spherical cell divides in two? The new volume of each cell will be $V/2$, thus $V/2 = 4s^3\pi/3$ where s is the new radius. Therefore $s = (3V/8\pi)^{1/3}$, which allows us to conclude that the new area of each cell is $A = 4\pi(3V/8\pi)^{2/3}$. Hence

$$\tfrac{A}{V/2} = (8\pi)^{1/3}3^{2/3}/V^{1/3}.$$

Consequently, the quotient *area/volume* for each new cell increases from the original $4.84/V^{1/3}$ to $6.09/V^{1/3}$, that is, by a factor of almost 1.26. Thus, the offspring do better than the parent with regard to the ratio A/V.

From a biological perspective it is understandable why a high value for the quotient A/V is good for the cell. A cell gets what it needs and excretes waste through the surface area, thus a higher value of the above-mentioned quotient will make the cell more efficient.

5.4 The Integrated Form of the Diffusion Equation

The diffusion equation cannot be used readily because it involves rates, which usually are much harder to measure than concentrations. Thus, let us try to integrate it in order to find an expression that is easier to compare with data.

Suppose that $C_{out} > C_{in}(0)$. Integrating (5.2) we obtain

$$\int \tfrac{-dC_{in}}{C_{out}-C_{in}} = \int \tfrac{-kA}{V}dt.$$

Thus $\ln(C_{out} - C_{in}(t)) = -\tfrac{kA}{V}t + L$ for a certain constant L. We should emphasize that $C_{out} - C_{in}(t) > 0$ because when $C_{out} - C_{in}(t) = 0$ the process of diffusion comes to an end. Moreover, $C_{out} - C_{in}(t)$ cannot be negative at any time t because it would imply that, at some time t_1, $C_{out} - C_{in}(t_1) = 0$; the Intermediate Value Theorem, for continuous functions, implies that such a t_1 exists.

We note that $\ln(C_{out}-C_{in}(0)) = L$, thus $\ln(C_{out}-C_{in}(t))-\ln(C_{out}-C_{in}(0)) = -\frac{kA}{V}t$. That is to say,

$$\ln\frac{C_{out}-C_{in}(0)}{C_{out}-C_{in}(t)} = \frac{kA}{V}t. \tag{5.3}$$

If we put $\ln\frac{C_{out}-C_{in}(0)}{C_{out}-C_{in}(t)}$ on the vertical axis and time on the horizontal axis, the experimental points should gather around a straight line that goes through the origin. In other words, we should get a correlation coefficient close to 1. A linear regression analysis will lead to the line of best fit and an estimation of the slope kA/V. From there we can approximate the permeability constant k.

Once the model has been validated and an approximation to the permeability constant has been calculated, it is time to see whether an explicit expression for $C_{in}(t)$ can be found. Applying the exponential function to both sides of (5.3) we obtain

$$\frac{C_{out}-C_{in}(0)}{C_{out}-C_{in}(t)} = e^{kAt/V}.$$

An elementary algebraic manipulation finally leads to

$$C_{in}(t) = C_{out} - (C_{out} - C_{in}(0))e^{-kAt/V}.$$

As expected, $\lim_{t\to\infty} C_{in}(t) = C_{out}$. The function $C_{in}(t)$ is depicted in Figure 5.1.

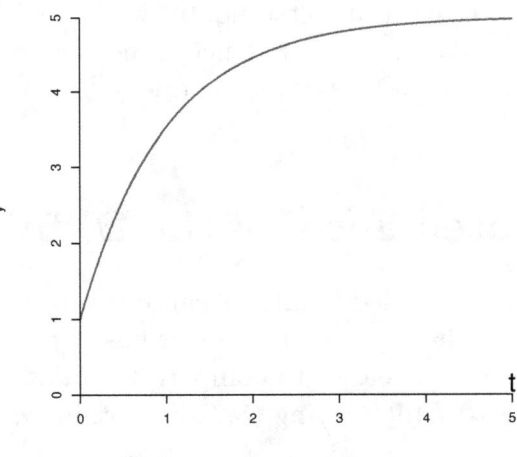

Figure 5.1: The function $C_{in}(t)$ when $C_{out} > C_{in}(0)$

What happens if $C_{out} < C_{in}(0)$? Equation (5.2) can be written

$$\frac{dC_{in}}{dt} = -\frac{kA}{V}(C_{in}(t) - C_{out}). \tag{5.4}$$

Integrating this differential equation we get $\int \frac{dC_{in}}{C_{in}(t)-C_{out}} = \int -\frac{kA}{V} dt$. Thus $\ln(C_{in}(t) - C_{out}) = -\frac{kA}{V}t + M$ for a certain constant M. In particular, for $t = 0$ we will have $\ln(C_{in}(0) - C_{out}) = M$. Consequently

$$\ln \frac{C_{in}(0) - C_{out}}{C_{in}(t) - C_{out}} = \frac{kA}{V}t. \tag{5.5}$$

This time the prediction is that the experimental values will gather around a line that goes through the origin if we put $\ln \frac{C_{in}(0)-C_{out}}{C_{in}(t)-C_{out}}$ on the vertical axis and time on the horizontal axis. Thereafter, we use linear regression to estimate the value of the slope of the line of best fit, which then happens to be an approximation of kA/V. A simple arithmetic calculation finally leads to an approximation of the permeability constant k.

We can apply the exponential function to both sides of (5.5) and obtain

$$C_{in}(t) = C_{out} - (C_{out} - C_{in}(0))e^{-kAt/V}.$$

This time, the function $C_{in}(t)$ is depicted in Figure 5.2.

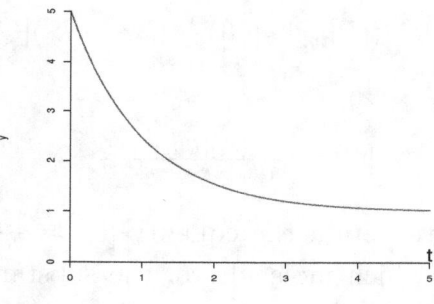

Figure 5.2: The function $C_{in}(t)$ when $C_{out} < C_{in}(0)$

5.5 Facilitated Diffusion

Scientists noticed that several substances cross the cell membrane at a much higher rate than that predicted by simple diffusion. Furthermore, the phenomenon of saturation appears, that is to say, the rate at which the substance crosses the cell membrane cannot be increased beyond a certain threshold. Addition of more substance does not lead to a higher rate. Experimental evidence pointed toward an expression of the form $v = a[S_e]/(b + [S_e])$, where $[S_e]$ is the concentration of the substance in the exterior of the cell (a, b are certain parameters). Thus, an alternative explanation to simple diffusion had to be found. The existence of proteins inside the membrane led to a coherent theory, in close analogy to enzyme kinetics.

We may hypothesize that membrane proteins play the role of carriers, shuttling molecules from the outside to the inside of the cell and vice versa. Let us first assume that the mechanism proceeds in only one direction, from the outside to the inside of the cell. A molecule of S (the substance) combines with a molecule of P (the protein) to produce an intermediate compound C through a reversible reaction which, in turn, deposits a molecule of S in the interior of the cell through an irreversible reaction and a molecule of protein is free to act once more as a carrier. Schematically, we have $S_e + P \leftrightarrow C \rightarrow P + S_i$, where S_e denotes a molecule of S in the exterior of the cell and S_i denotes a molecule of S in the interior of the cell.

Let k_1, k_{-1} be the kinetic parameters of the reversible process and k_2 the kinetic parameter of the irreversible process. We have

$$\frac{d[C]}{dt} = k_1[S_e][P] - (k_{-1} + k_2)[C]$$

where, as usual, the brackets denote concentration.

The protein is either free or forming part of the intermediate compound, thus $[P]_{tot} = [P] + [C]$. Under steady-state conditions $\frac{d[C]}{dt} = 0$; therefore,

$$0 = k_1[S_e]([P]_{tot} - [C]) - (k_{-1} + k_2)[C].$$

Consequently,

$$[C] = \frac{k_1[S_e][P]_{tot}}{k_1[S_e]+k_{-1}+k_2}.$$

Defining $K_m = \frac{k_{-1}+k_2}{k_1}$, we obtain the equality $[C] = \frac{[S_e][P]_{tot}}{K_m+[S_e]}$. But $v = \frac{d[S_i]}{dt} = k_2[C]$, the transport rate at which molecules of the substance are deposited in the interior of the cell. Hence

$$v = \frac{k_2[P]_{tot}[S_e]}{K_m+[S_e]}.$$

As $[S_e]$ increases, the rate will tend to the limiting value $k_2[P]_{tot}$; we denote the latter by the symbol V_{max}. Thus

$$v = \frac{V_{max}[S_e]}{K_m + [S_e]}, \tag{5.6}$$

an expression for the rate of transport that is in agreement with experimental data. We should not forget that simple diffusion and facilitated diffusion are two types of passive transport, wherein no metabolic energy is expended.

The analogy with enzyme kinetics suggests transforming (5.6) into a linear expression in order to estimate V_{max} and K_m. Lineweaver-Burk type plots, or similar ones, can be used in practice to validate the model and estimate the above-mentioned parameters. It should be noted that nowadays scientists think that membrane proteins not only act as carriers but some of them play the role of channels across the membrane, allowing only certain specific substances to pass through the membrane and enter the cell. Either as carriers or channels, it is understandable why saturation takes place: after a short period of time all the proteins are busy doing their job and a dynamic equilibrium is established.

5.6 A Closer Look at Facilitated Diffusion

We have to take into account the fact that some molecules of the substance are transported by the protein from the inside to the outside of the cell. Thus, to the equation $\frac{dS_i}{dt} = \frac{V_{max}S_e}{S_e + K_m}$ we have to subtract something (from now on the brackets will be dropped whenever concentration is meant). It seems reasonable to subtract the expression $\frac{V_{max}S_i}{S_i + K_m}$, assuming that the parameters V_{max} and K_m are the same for the forward and backward transport process. Hence

$$\frac{dS_i}{dt} = \frac{V_{max}S_e}{S_e + K_m} - \frac{V_{max}S_i}{S_i + K_m}. \tag{5.7}$$

We have to obtain mathematical consequences of (5.7), and compare them with experimental data if we wish to validate the model of facilitated diffusion that we are discussing. With this idea in mind, let us start by assuming that the concentration of substance on the outside is much bigger than in the inside of the cell; thus, S_e in (5.7) can be considered to be a constant. We are dealing with a separable variables differential equation, consequently

$$\int \frac{dS_i}{a - \frac{S_i}{S_i + b}} = \int V_{max} dt$$

where $a = \frac{S_e}{S_e + K_m}$ and $b = K_m$. Using a graphing calculator or a computer we get

$$\int \frac{1}{a - \frac{x}{x+b}} dx = \frac{x}{a-1} - \frac{b}{(a-1)^2} \ln |(a-1)x + ab|.$$

Thus

$$\frac{S_i}{\frac{S_e}{K_m + S_e} - 1} - \frac{K_m}{(\frac{S_e}{K_m + S_e} - 1)^2} \ln[(\frac{S_e}{K_m + S_e} - 1)S_i + \frac{K_m S_e}{K_m + S_e}] = V_{max}t + L$$

for a certain constant L. A little bit of arithmetic leads to

$$S_i(K_m + S_e) + (K_m + S_e)^2 \ln[\frac{K_m}{K_m + S_e}(S_e - S_i)] = -K_m V_{max}t + H$$

where $H = -LK_m$. If we assume that $S_i = 0$ at $t = 0$, right away we get

$$H = (K_m + S_e)^2 \ln \frac{K_m S_e}{K_m + S_e}.$$

Replacing this value in the preceding equation, we finally reach the equality

$$S_i(K_m + S_e) + (K_m + S_e)^2 \ln \frac{S_e - S_i}{S_e} = -K_m V_{max} t. \qquad (5.8)$$

The challenge is to estimate the values of K_m and V_{max} from (5.8). For this purpose we will use the half-equilibrium time method (Stein, 1986). Let $t_{1/2}$ be the time it takes the internal concentration S_i to adopt the value $\frac{S_e}{2}$. Thus,

$$\frac{S_e}{2}(K_m + S_e) - (K_m + S_e)^2 \ln 2 = -K_m V_{max} t_{1/2}.$$

Therefore,

$$(K_m + S_e)((\ln 2 - 0.5)S_e + K_m \ln 2) = K_m V_{max} t_{1/2}. \qquad (5.9)$$

Next we make two experiments, starting from two different values of S_e (S_I and S_{II}) and measure the corresponding half-equilibrium times (t_I and t_{II}). Then, from (5.9), we get the following two equations:

$$(K_m + S_I)(\ln 2 - 0.5)S_I + K_m \ln 2) = K_m V_{max} t_I$$

and

$$(K_m + S_{II})((\ln 2 - 0.5)S_{II} + K_m \ln 2) = K_m V_{max} t_{II}.$$

Dividing both equations, we obtain

$$\frac{(K_m + S_I)(\ln 2 - 0.5)S_I + K_m \ln 2)}{(K_m + S_{II})((\ln 2 - 0.5)S_{II} + K_m \ln 2)} = \frac{t_I}{t_{II}}$$

The only unknown in this equation is precisely K_m. Once K_m has been found through a simple algebraic procedure, we use (5.9) to calculate V_{max}.

5.7   Fick's Law in a More General Context

Let us consider two compartments (with volumes V_1 and V_2), separated by a membrane, each of which has a certain amount $S_1(t)$, $S_2(t)$ of solute. Fick's law asserts that

$$S_1'(t) = K(C_2(t) - C_1(t))$$

$$S_2'(t) = K(C_1(t) - C_2(t))$$

where $C_1(t)$, $C_2(t)$ are the concentrations of solute in the first and second compartment, respectively, and K is a parameter (actually, $K = kA$, where k is the permeability constant of the membrane and A is its area). Therefore

$$C_1'(t) = \frac{K}{V_1}(C_2(t) - C_1(t))$$

$$C_2'(t) = \frac{K}{V_2}(C_1(t) - C_2(t)).$$

Defining $a = \frac{K}{V_1}$, $b = \frac{K}{V_2}$ we arrive to the following system of differential equations:

$$C_1'(t) = -aC_1(t) + aC_2(t) \tag{5.10}$$

$$C_2'(t) = bC_1(t) - bC_2(t). \tag{5.11}$$

How can we solve it? Taking the derivative of equation (5.10) and using equations (5.11) and (5.10), we get

$$C_1''(t) = -aC_1'(t) + aC_2'(t) = -aC_1'(t) + a(bC_1(t) - bC_2(t)) =$$

$$= -aC_1'(t) + abC_1(t) - b(C_1'(t) + aC_1(t)) = -(a + b)C_1'(t).$$

That is to say,

$$C_1''(t) + (a + b)C_1'(t) = 0. \tag{5.12}$$

This is a second order differential equation, which can be transformed into a first order differential equation by defining $v = C_1'$. Indeed,

$$v'(t) + (a + b)v(t) = 0,$$

whose general solution, as we know, is $v(t) = Le^{-(a+b)t}$; consequently $C_1'(t) = Le^{-(a+b)t}$. But $L = C_1'(0) = -aC_1(0) + aC_2(0)$, hence

$$C_1'(t) = a(C_2(0) - C_1(0))e^{-(a+b)t}.$$

The solution of this differential equation can be found by just one integration, thus

$$C_1(t) = \frac{a(C_1(0) - C_2(0))}{a+b}e^{-(a+b)t} + H$$

for some constant H. But $C_1(0) = \frac{aC_1(0) - aC_2(0)}{a+b} + H$, therefore

$$C_1(t) = \frac{aC_1(0) - aC_2(0)}{a + b}e^{-(a+b)t} + \frac{bC_1(0) + aC_2(0)}{a + b}. \tag{5.13}$$

Equation (5.11) is identical to (5.10), with C_1 interchanged by C_2 and a by b, therefore

$$C_2(t) = \frac{bC_2(0) - bC_1(0)}{a + b}e^{-(a+b)t} + \frac{aC_2(0) + bC_1(0)}{a + b}. \tag{5.14}$$

We observe that

$$\lim_{t\to\infty} C_1(t) = \lim_{t\to\infty} C_2(t) = \frac{bC_1(0) + aC_2(0)}{a+b} =$$

$$= \frac{V_1 C_1(0) + V_2 C_2(0)}{V_1 + V_2}.$$

As expected, in the long run, the concentration of solute in both compartments will be the same (see Figure 5.3).

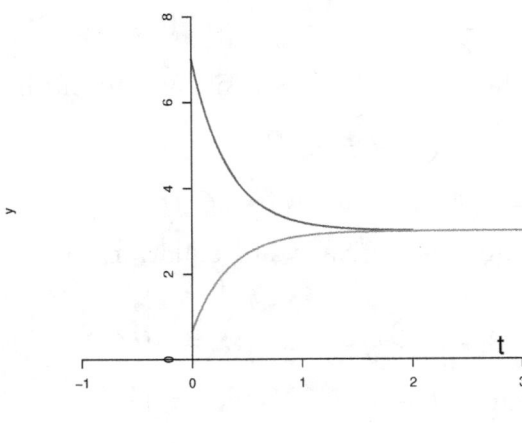

Figure 5.3: Variation of concentration in each compartment

Exercises for chapter 5

1. It could happen that some molecules of the substance cross the cell membrane by simple diffusion and others by facilitated diffusion. Thus

$$\frac{dC_{in}}{dt} = \frac{kA}{V}\left(C_{out} - C_{in}\right) + \frac{V_{max}C_{out}}{K_m + C_{out}}.$$

Assuming that C_{out} is constant and $C_{in}(0) = 0$, solve the differential equation and calculate the limit when $t \to \infty$.

2. Transform the equation $v = \frac{V_{max}[S_e]}{K_m + [S_e]}$ for facilitated diffusion in such a way that $v = V_{max} - K_m \frac{v}{[S_e]}$. If we put $\frac{v}{[S_e]}$ on the x-axis and v on the y-axis, we should obtain a straight line with negative slope $-K_m$ and y-intersection V_{max}. Use Table 5.1 (Christensen and Liang, 1966) and linear regression to estimate the values of K_m and V_{max}.

3. A cell in the shape of a cylinder has radius r and height nr, where n is a natural number. Assuming that the volume V is fixed, show that

$$\frac{A}{V} = \frac{2\pi^{1/3}}{V^{1/3}}\left(\frac{1}{n^{2/3}} + n^{1/3}\right).$$

Interpret what happens when n becomes bigger and bigger.

4. Solve $y''(t) + 3y'(t) = 0$, with initial conditions $y(0) = 1$, $y'(0) = 0$.

5. Solve $y'' + ay' = b$, $a \neq 0$, with $y(0) = \alpha$, $y'(0) = \beta$.

Table 5.1: Uptake of benzylamine by *Ehrlich ascites carcinoma cells*

$[S_e]$ mM/lt	v $mM/lt/min$
0.05	0.1
0.24	0.27
0.58	0.42
1.2	0.53
1.69	0.54
2.38	0.57

Bibliography

[1] Adler, F.R. (2005), *Modeling the Dynamics of Life* (2nd edition), Thomson Brooks/Cole, California.

[2] Altman, P.L., D.A. and Dittmer, D.S., editors (1964), *Biology Data Book*, Federation of American Societies for Experimental Biology, Washington, D.C.

[3] Barrow, G.M. (1979), *Physical Chemistry*, McGraw-Hill.

[4] Bartle, R.G. and Ionescu-Tulcea, C.(1970), *Honors Calculus*, Scott, Foresman and Company, Glenview, Illinois.

[5] Batschelet, E. (1975), *Introduction to Mathematics for Life Scientists* (2nd edition), Springer Verlag, New York.

[6] Burghes, D.N. and Borrie, M.S. (1981), *Modelling with Differential Equations*, John Wiley and Sons, New York.

[7] Briggs, G.E. and Haldane, J.B.S. (1925), A note on the kinetics of enzyme action, *Biochemical Journal* **39**: 338-339

[8] Causton, D.R. (1977), *A Biologist's Mathematics*, Edward Arnold Publishers, London.

[9] Christensen, H.N. and Liang, M. (1966) Modes of Uptake of Benzylamine by the Ehrlich Cell, *The Journal of Biological Chemistry* **241**, No. 23, 5552-5556.

[10] Cullen, M.R. (1983) *Mathematics for the Biosciences*, PWS-Kent Publishing Company.

[11] Frank, P. (1957), *Philosophy of Science*, Prentice Hall, Englewood Cliffs, N.J.

[12] Fulling, S.A. (2005), How to Avoid the Inverse Secant (and Even the Secant Itself), *College Mathematics Journal*, November, 381-387.

[13] Gordon, S.P. and Gordon, F.S. (2004), Deriving the Regression Equations Without Using Calculus, *Mathematics and Computer Education* **38**, No. 1, 64-68.

[14] Hahn, A.J. (1998), *Basic Calculus: From Archimedes to Newton to its Role in Science*, Springer, New York.

[15] Helfgott, M. (2004), Placing the Natural Logarithm and The Exponential Function on an Equal Footing, *College Mathematics Journal* **35**, No. 5, 390-393.

[16] Helfgott, M. and Seier, E. (2007), Some mathematical and statistical aspects of enzyme kinetics, *Journal of Online Mathematics and its Applications* **7**, October, 1611.

[17] Helfgott, H. and Helfgott, M. (2009), A Modern Vision of the Work of Cardano and Ferrari on Quartics, *Convergence*, MAA MathDL, 3312 July.

[18] Helfgott, M. and Moore, D. (2011), *Introductory Calculus for the Natural Sciences*, ISBN 978 1453880838.

[19] Jenson, V.G. and Jeffreys, G.V. (1963), *Mathematical Methods in Chemical Engineering*, Academic Press, New York.

[20] Kalman, D. (1993), Six Ways to Sum a Series, *The College Mathematics Journal*, Vol.24, No. 5, 402-421.

[21] Katz, M.A. (1976), *Calculus for the Life Sciences*, Marcel Dekker, New York.

[22] Lislevand, T. et al. (2007) Avian body sizes in relation to fecundity, mating system, display behavior, and resource sharing, *Ecology* **88**: 1605.

[23] Lomen, D. and Lovelock, D. (1999), *Differential Equations*, Wiley.

[24] Miller, I. and Miller, M. (2004), *John E. Freund's Mathematical Statistics with Applications*, Pearson Prentice Hall, N.J.

[25] McCarthy, M.A. (1997), Competition and Dispersal from Multiple Nests, *Ecology* **78**(3): 873-883.

[26] McQuarrie, D.A. and Simon, J.D. (1997), *Physical Chemistry*, University Science Books, Sausalito, California.

[27] Reed, H.S. and Holland, R.H. (1919), The Growth Rate Of An Annual Plant Helianthus, *Proc. Nat. Acad. of Sciences*, 135-144.

[28] Rosenlicht, M. (1968), *Introduction to Analysis*, Scott, Foresman and Company, Glenview, Illinois.

[29] Roughton, F.J.W. (1954), Rapid Reactions in Biology, *Disc. Faraday Soc.* **17**: 116-120.

[30] Sacher, G.A. and Staffeldt, E.F. (1974), Relation of Gestation Time to Brain Weight for Placental Mammals; Implications for the Theory of Vertebrate Growth, *American Naturalist* 108: 593-613.

[31] Segel, I.H.(1976), *Biochemical Calculations* (2nd edition), John Wiley and Sons.

[32] Seier, E. and Joplin, K. (2011), *Introduction to Statistics in a Biological Context*, ISBN 978 1463613372.

[33] Stein, W.D. (1986), *Transport and Diffusion Across Cell Membranes*, Academic Press, Orlando.

[34] Stern, K.G. (1936), A study of the decomposition of monoethyl hydrogen peroxide by catalase and of an intermediate enzyme-substrate compound, *Journal of Biological Chemistry* **114**: 473-494.

[35] Sved, J.A. and Mayo, O. (1970), The Evolution of Dominance, in *Mathematical Topics in Population Genetics*, edited by Ken-ichi Kojima, Springer Verlag, Berlin, p. 294.

[36] Tazawa, H. et al. (2001), Allometric relationships between embryonic heart rate and fresh egg mass in birds, *The Journal of Experimental Biology* **204**: 165-174.

[37] Thomas, G.B., Weir, M. and Hass, J. (2010), *Thomas' Calculus (Early Transcendentals)*, 12th edition, Addison-Wesley, Boston, pp. 488-490.

[38] Thornley, J.H.M. and Johnson, I.R. (1990) *Plant and Crop Modelling*, Clarendon Press, Oxford, UK.

[39] Thornton, H.G. (1922) On the development of a standardised agar medium for counting soil bacteria, with special regard to the repression of spreading colonies, *Annals of Biology* **9**: 241-274.

[40] Trench, W.F. (2002), *Introduction to Real Analysis*, Pearson Education.

Appendix A

A.1   List of Enrichment Notes

1. The Range of a Continuous Function (Section 1.6).

2. Harmonic Functions (Section 1.8).

3. A Proof of Fermat's Theorem (Section 1.18).

4. A Proof of the Second Derivative Test (Section 1.21).

5. The Weighted Average and Simpson's Method (Section 2.3).

6. A Proof of the Fundamental Theorem of Calculus (Section 2.7).

7. More about the Logistic Equation (Section 2.16).

8. Proving an Inequality (Section 3.6).

9. Extending the Interval of Convergence (Section 3.12).

10. How the Exact Value of a Series was Found (Section 3.14).

Appendix B

B.1   Some proofs about convergent sequences

1. Every convergent sequence is **bounded**. That is to say, if $a_n \to L$ for some L, there exists $M > 0$ such that $|a_n| \le M$ for every n. Why is this so? Since $a_n \to L$ there exists N such that $|a_n - L| < 1$ whenever $n \ge N$. Thus, if $n \ge N$ we will have $|a_n| < L + 1$. Define M as the maximum of the N numbers $|a_1|, ..., |a_{N-1}|, L + 1$. Then $|a_n| \le M$ for every n. We should keep in mind that the converse of this proposition is not true. That is to say, boundedness does not imply convergence. For instance, the sequence (a_n), where $a_n = 1$ if n is odd and $a_n = -1$ if n is even, is bounded but it certainly is not convergent since its values are oscillating between 1 and -1, never clustering around any number.

2. It is to be expected that if $a_n \to L$ and $L > 0$ then $a_n > 0$ '**eventually**,' that is, there exists a natural number N such that $a_n > 0$ for every $n \ge N$. Indeed, $a_n \to L$ implies that there exists N such that $-\frac{L}{2} < a_n - L < \frac{L}{2}$ whenever $n \ge N$. Thus, if $n \ge N$ we will have $a_n > L - \frac{L}{2} = \frac{L}{2} > 0$. In a similar fashion, it can be proven that if $a_n \to L$ and $L < 0$ then $a_n < 0$ eventually. Thus, if $a_n \to L$ and $L \ne 0$ we can assert that $a_n \ne 0$ eventually.

3. The fifth property of convergent sequences, mentioned in section 2.1, establishes that if $a_n \le b_n$ for all n and $a_n \to L_1$, $b_n \to L_2$, then $L_1 \le L_2$. It looks intuitively obvious, so the proof might be short. Indeed, given any $\epsilon > 0$, the convergence of both sequences implies the existence of a natural number N such that

$$L_1 - \tfrac{\epsilon}{2} < a_N \le b_N < L_2 + \tfrac{\epsilon}{2}$$

Therefore $L_1 < L_2 + \epsilon$ for every $\epsilon > 0$, which in turn implies $L_1 \le L_2$ (otherwise $L_1 < L_2 + (L_1 - L_2)$, that is, $L_1 < L_1$).

In section 2.1 we mentioned a particular case of this property: if $a_n \to L$ and $a_n \geq 0$ for all n, then $L \geq 0$. Similarly, if $a_n \to L$ and $a_n \leq 0$ for all n, then $L \leq 0$.

4. Let us prove that if $a_n \to L_1$ and $b_n \to L_2$ then $a_n b_n \to L_1 L_2$. We have

$$|a_n b_n - L_1 L_2| = |a_n b_n - a_n L_2 + a_n L_2 - L_1 L_2| \leq$$

$$|a_n||b_n - L_2| + |L_2||a_n - L_1| \leq M|b_n - L_2| + |L_2||a_n - L_1|,$$

where $M > 0$ is a bound of (a_n) (recall that every convergent sequence is bounded). Since $b_n \to L_2$ there exists N_1 such that $|b_n - L_2| < \frac{\epsilon}{2M}$ provided that $n \geq N_1$, and since $a_n \to L_1$ there exists N_2 such that $|a_n - L_1| < \frac{\epsilon}{2(|L_2|+1)}$. Among N_1 and N_2, let N denote the bigger. Thus, if $n \geq N$ then

$$M|b_n - L_2| + |L_2||a_n - L_1| < \frac{\epsilon}{2} + \frac{\epsilon}{2} = \epsilon$$

We have been able to prove that given any $\epsilon > 0$ there exists a natural number N such that if $n \geq N$ then $|a_n b_n - L_1 L_2| < \epsilon$. Consequently, $a_n b_n \to L_1 L_2$.

5. Next let us show that if $b_n \to L_2$ and $L_2 \neq 0$, then $\frac{1}{b_n} \to \frac{1}{L_2}$. Indeed,

$$\left|\frac{1}{b_n} - \frac{1}{L_2}\right| = \frac{|b_n - L_2|}{|b_n||L_2|}$$

Since $b_n \to L_2$, there exists a natural number N_1 such that if $n \geq N_1$ then $||b_n| - |L_2|| < \frac{|L_2|}{2}$ and, in particular, $|b_n| > \frac{|L_2|}{2}$. Hence $\frac{1}{|b_n|} < \frac{2}{L_2}$ provided that $n \geq N_1$, which in turn implies

$$\frac{1}{|b_n||L_2|} < \frac{2}{|L_2|^2}$$

whenever $n \geq N_1$.

On the other hand, given any ϵ, since $b_n \to L_2$ there exists N_2 such that

$$|b_n - L_2| < \frac{\epsilon|L_2|^2}{2}$$

whenever $n \geq N_2$. Let N denote the bigger of the numbers N_1 and N_2. Then, if $n \geq N$, multiplying the two previous inequalities we get

$$\frac{|b_n - L_2|}{|b_n||L_2|} < \epsilon$$

In summary, we have succeeded in showing that given any ϵ there exists N such that if $n \geq N$ then $\left|\frac{1}{b_n} - \frac{1}{L_2}\right| < \epsilon$; that is, $\frac{1}{b_n} \to \frac{1}{L_2}$.

Having proven that if $b_n \to L_2$ then $\frac{1}{b_n} \to \frac{1}{L_2}$, it immediately follows that $\frac{a_n}{b_n} \to \frac{L_1}{L_2}$ provided that $a_n \to L_1$ and $b_n \to L_2$ ($L_2 \neq 0$). We just have to keep in mind that the limit of the product of two convergent sequences is the product of the corresponding limits.

B.2 The equivalence between two statements

Let a function f be defined in a neighborhood of c, but not necessarily at c. In section 1.4 we saw the definition of $\lim_{x \to c} f(x) = L$, namely:

1. For every $\epsilon > 0$ there exists $\delta > 0$ such that for any x, if $0 < |x - c| < \delta$, then $|f(x) - L| < \epsilon$.

   Interestingly, this statement is equivalent to the assertion that, for any sequence (x_n) in the domain of f,

2. If $x_n \to c$ then $f(x_n) \to L$

Let us first show that the definition of limit implies the assertion about sequences. Assume that $x_n \to c$. Given any $\epsilon > 0$, since $x_n \to c$, there exists a natural number N such that if $n \geq N$, then $|x_n - c| < \delta$ (δ is the positive number whose existence is guaranteed by the definition). Therefore $|f(x_n) - L| < \epsilon$. We have succeeded in showing that $f(x_n) \to L$.

Next we have to prove that the assertion by sequences implies the definition of limit. Proceeding by contradiction[1], suppose that there exists an $\epsilon > 0$ such that, for every $\delta > 0$, there exists x such that $0 < |x - c| < \delta$ and $|f(x) - L| \geq \epsilon$. In particular, for every natural number n, there exists x_n such that $0 < |x_n - c| < \frac{1}{n}$ and $|f(x_n) - L| \geq \epsilon$. A sequence (x_n) has been built and, since $\frac{1}{n} \to 0$, the squeeze property implies that $x_n \to c$. Then, thanks to the hypothesis, $f(x_n) \to L$. We have reached an impossibility because $|f(x_n) - L| \geq \epsilon$ for every n. QED

What is the importance of the equivalence that we have just proven? Properties of convergent sequences lead without much effort to proofs of properties of limits of functions. For instance, property (iii) (section 1.4) can be proven as follows: let $x_n \to c$. Since $\lim_{x \to c} f(x) = L_1$ we will have $f(x_n) \to L_1$. Similarly $g(x_n) \to L_2$. Then $f(x_n)g(x_n) \to L_1 L_2$ due to the fact that the limit of the product

---

[1] We could as well develop a direct proof by using the logical contrapositive of the implication.

of two convergent sequences equals the product of the corresponding limits. In a similar fashion we can prove property (iv) (section 1.4).

B.3　A proof of the chain rule

Let us recall the statement of the chain rule: Suppose f is differentiable at x_o and g is differentiable at $f(x_o)$, then $g \circ f$ is differentiable at x_o and $(g \circ f)'(x_o) = g'(f(x_o)f'(x_o)$. As expected, we are assuming that f is defined on an interval I, g is defined on an interval J, and the image of I under f is contained in J.

Right away, we could write

$$\frac{g(f(x)) - g(f(x_o))}{x - x_o} = \frac{g(f(x)) - g(f(x_o))}{f(x) - f(x_o)} \frac{f(x) - f(x_o)}{x - x_o}$$

provided that $f(x) \neq f(x_o)$ in a neighborhood of x_o, excluding the point x_o itself. This restriction does not seem to be outlandish because in an introductory calculus course many functions are increasing or decreasing in a neighborhood of each point of their domain.

Taking the limit, when $x \to x_o$, we would obtain that

$$\lim_{x \to x_o} \frac{g(f(x)) - g(f(x_o))}{x - x_o} = g'(f(x_o))f'(x_o),$$

that is, $(g \circ f)'(x_o) = g'(f(x_o))f'(x_o)$, and the proof is done. But we have implicitly accepted that

$$\lim_{x \to x_o} \frac{g(f(x)) - g(f(x_o))}{f(x) - f(x_o)} = \lim_{y \to f(x_o)} \frac{g(y) - g(f(x_o))}{y - f(x_o)}.$$

The differentiability of g at $f(x_o)$ and the continuity of f at x_o will allow us to justify the last equality. Indeed, given any $\epsilon > 0$, since g is differentiable at $f(x_o)$, there will exist a $\delta > 0$ such that

$$0 < |y - f(x_o)| < \delta \implies |\tfrac{g(y) - g(f(x_o))}{y - f(x_o)} - g'(f(x_o))| < \epsilon.$$

But f is continuous at x_o. Consequently, there exists $\mu > 0$ such that

$$|x - x_o| < \mu \implies |f(x) - f(x_o)| < \delta.$$

Therefore,

$$0 < |x - x_o| < \mu \implies |\tfrac{g(f(x)) - g(f(x_o))}{f(x) - f(x_o)} - g'(f(x_o))| < \epsilon.$$

The proof has come to an end.

What could be done if every neighborhood of x_o has a point x such that $f(x) = f(x_o)$? We would have to modify the preceding proof to avoid any division by the number zero. Instead of following this path, we will present a different, more straightforward, approach that avoids any difficulties. It starts with the observation that an arbitrary function h is differentiable at c if and only if there exists a function ϕ, continuous at c, such that

$$h(x) = h(c) + (x - c)\phi(x)$$

in a neighborhood of c. Moreover, $\phi(c) = h'(c)$.

Let us prove the above-mentioned observation. Suppose that such a function ϕ exists. Then

$$\frac{h(x) - h(c)}{x - c} = \phi(x)$$

for every x in a neighborhood of c, $x \neq c$. Since ϕ is continuous at c we will have

$$\lim_{x \to c} \frac{h(x) - h(c)}{x - c} = \lim_{x \to c} \phi(x) = \phi(c).$$

Hence h is differentiable at c and $h'(c) = \phi(c)$. On the other hand, assume that h is differentiable at c and define

$$\phi(x) = \frac{h(x) - h(c)}{x - c}, \qquad x \neq c$$

$$\phi(c) = h'(c).$$

Then $h(x) = h(c) + (x - c)\phi(x)$ and

$$\lim_{x \to c} \phi(x) = h'(c) = \phi(c).$$

Thus, ϕ is continuous at c.

We are now ready to develop a proof of the chain rule without restrictions. Since g is differentiable at $f(c)$, there exists a neighborhood V of $f(c)$ and there exists a function ψ, continuous at $f(c)$, such that

$$g(y) = g(f(c)) + (y - f(c))\psi(y)$$

for every y in V. Besides, $\psi(f(c)) = g'(f(c))$. However, f is continuous at c. Consequently, there exists a neighborhood U of c such that $f(x)$ is in V for every x in U. Therefore,

$$g(f(x)) = g(f(c)) + (f(x) - f(c))\psi(f(x))$$

for every x in U. But, since f is differentiable at c, there exists a neighborhood W of c and there exists a function ϕ, continuous at c, such that

$$f(x) = f(c) + (x - c)\phi(x)$$

for every x in W (we choose W in such a way that W is a subset of U). Then

$$g(f(x)) = g(f(c)) + (x - c)\phi(x)\psi(f(x))$$

for every x in W. Evidently, the function $\phi(x)\psi(f(x))$ is continuous at c, consequently $g \circ f$ is differentiable at c and

$$(g \circ f)'(c) = \phi(c)\psi(f(c)) = f'(c)g'(f(c)).$$

The proof has come to an end.

B.4   The continuity of the inverse

Let $f : I \to \Re$ be continuous and increasing (decreasing). Then $g = f^{-1} : J \to \Re$ is continuous and increasing (decreasing), where $J = f(I)$. We will develop a proof when f is increasing. It can be easily adapted if f is decreasing.

The function f is continuous and non-constant, so $f(I)$ is an interval (a fact proven in section 1.6). Since g is the inverse of f, and f is increasing, right away it follows that g is increasing (section 1.2, exercise 6). Let a be an arbitrary interior point of J and define $r = g(a)$. Is r an interior point of I? Since a is an interior point of J there exist y_1, y_2 in J such that a belongs to the interval (y_1, y_2), which is a subset of J. But g is increasing, so $g(y_1) < g(a) < g(y_2)$, that is, $g(y_1) < r < g(y_2)$. Moreover, since I is an interval it follows that $(g(y_1), g(y_2))$ is a subset of I. Thus r happens to be an interior point of I.

Given any $\epsilon > 0$ choose $\alpha > 0$ so that $[r - \alpha, r + \alpha]$ is a subset of I and $\alpha < \epsilon$. Defining $a_1 = f(r - \alpha), a_2 = f(r + \alpha)$ we can assert that a_1, a_2 belong to J and $a_1 < a < a_2$. Next choose $\delta > 0$ so that $\delta < min\{a_2 - a, a - a_1\}$. Hence $a_1 < a - \delta < a < a + \delta < a_2$. Let us assume that $|x - a| < \delta$. Then $a - \delta < x < a + \delta$, which in turn implies $a_1 < x < a_2$. Consequently $g(a_1) < g(x) < g(a_2)$, so $r - \alpha < g(x) < r + \alpha$. Finally we get $|g(x) - g(a)| < \alpha < \epsilon$.

What happens if a is a left endpoint of J? Firstly we note[2] that $r = g(a)$ is a left endpoint of I. Given $\epsilon > 0$ choose α such that $0 < \alpha < \epsilon$ and $[r, r + \alpha]$ is a subset of I and define $a_2 = f(r + \alpha)$. Hence a_2 belongs to J and $a_2 > a$ ($a \geq a_2$ would imply $g(a) \geq g(a_2)$, that is, $r \geq r + \alpha.$). Let $\delta = a_2 - a$ and assume that $a \leq x < a + \delta$. Then $a \leq x < a_2$, so $g(a) \leq g(x) < g(a_2)$. Thus $r \leq g(x) < r + \alpha$, which in turn leads to $0 \leq g(x) < r + \alpha$, so $0 \leq g(x) - r < \epsilon$. Hence $|g(x) - g(a)| < \epsilon$. The proof is almost the same if a is a right endpoint.

B.5   The derivative of the inverse

In section 1.11 we stated the following result: Let f be defined and differentiable on an open interval I, where it is increasing or decreasing, and $f'(y) \neq 0$ for every y in I. Then g, the inverse of f, is differentiable on the interval J, where J is the image of I under f. A proof goes as follows:

In the first place, let us keep in mind that $g(x)$ is continuous on J because $f(y)$ is continuous on I (recall the third property of continuous functions, section 1.6). Besides, g inherits from f the property of being increasing or decreasing on J. It should be noted that $J = f(I)$ is an open interval because I is an open interval and

---

[2]We have that $a \leq x$ for all x in J. Let z be an arbitrary point in $g(J) = I$, thus $z = g(u)$ for a certain point u in J. But $a \leq u$, consequently $g(a) \leq g(u) = z$.

f is an increasing (or decreasing) continuous function (recall the enrichment note at the end of section 1.6).

Let c be an arbitrary point of J. Given any $\epsilon > 0$, since $g(c)$ belongs to I and f is differentiable at $g(c)$, there exists $\mu > 0$ such that

$$0 < |y - g(c)| < \mu \implies \left| \tfrac{f(y) - f(g(c))}{y - g(c)} - f'(g(c)) \right| < \epsilon.$$

However, g is continuous at c. Hence, there exists $\delta > 0$ such that

$$|x - c| < \delta \implies |g(x) - g(c)| < \mu.$$

Assume $0 < |x - c| < \delta$. Then[3]

$$\left| \tfrac{f(g(x)) - f(g(c))}{g(x) - g(c)} - f'(g(c)) \right| < \epsilon.$$

We have been able to show that

$$\lim_{x \to c} \tfrac{f(g(x)) - f(g(c))}{g(x) - g(c)} = f'(g(c)).$$

But $f(g(x)) = x$ and $f(g(c)) = c$. Consequently,

$$\lim_{x \to c} \tfrac{x - c}{g(x) - g(c)} = f'(g(c)).$$

Therefore,

$$\lim_{x \to c} \tfrac{g(x) - g(c)}{x - c} = \tfrac{1}{f'(g(c))}$$

We have just shown that $g(x)$ is differentiable at c. Moreover, $g'(c) = \tfrac{1}{f'(g(c))}$.

Remark

The preceding proof has a drawback. It relies on the continuity of g (the inverse of the function f), a proposition that we have not proven although we discussed it in section 1.6.

B.6 A proof of the Mean Value Theorem

The theorem states that if f is continuous on $[a, b]$ and differentiable on (a, b), there exists θ in (a, b) such that

$$f'(\theta) = \tfrac{f(b) - f(a)}{b - a}.$$

---

[3]Since g is increasing or decreasing, the possibility that $g(x) = g(c)$, for some x different from c, has to be ruled out. Thus, there is no danger of dividing by zero in the next expression.

We will divide the proof in two parts, first when $f(a) = f(b)$ and thereafter when $f(a) \neq f(b)$.

1. Let us first consider the case[4] when $f(a) = f(b)$. If $f(x) = f(a)$ for every x in (a, b) then $f'(x) = 0$ for every x in (a, b) and the proof comes to an end (we can choose $\theta = (a + b)/2$). Suppose that there exists c in (a, b) such that $f(c) \neq f(a)$, say $f(c) > f(a)$. Let θ be the point in $[a, b]$ where f attains its maximum. Therefore,

$$f(\theta) \geq f(c) > f(a) = f(b).$$

Hence $a < \theta < b$, otherwise we would have $f(a) > f(a)$ or $f(b) > f(b)$. Then Fermat's theorem implies that $f'(\theta) = 0$, consequently

$$f'(\theta) = \tfrac{f(b)-f(a)}{b-a}.$$

What happens if $f(c) < f(a)$? Let μ be the point in $[a, b]$ where f adopts its minimum. Then

$$f(\mu) \leq f(c) < f(a) = f(b).$$

Hence $a < \mu < b$ because otherwise we would have $f(a) < f(a)$ or $f(b) < f(b)$. Fermat's theorem implies then that $f'(\mu) = 0$, consequently

$$f'(\mu) = \tfrac{f(b)-f(a)}{b-a}.$$

2. Next we will consider the case when $f(b) \neq f(a)$. The graph of the function

$$g(x) = f(a) + \tfrac{f(b)-f(a)}{b-a}(x - a), \qquad a \leq x \leq b,$$

is the segment that joins the points $(a, f(a))$ and $(b, f(b))$. We build a new function on the domain $[a, b]$, namely $h(x) = f(x) - g(x)$. Obviously, $h(x)$ is continuous on $[a, b]$ and differentiable on (a, b). Moreover, $h(a) = 0 = h(b)$. The first part of the theorem, applied to the function $h(x)$, implies that there exists θ in (a, b) such that $h'(\theta) = 0$. But

$$h'(x) = f'(x) - \tfrac{f(b)-f(a)}{b-a}.$$

Consequently $f'(\theta) = \tfrac{f(b)-f(a)}{b-a}$, which is precisely what we wanted to prove.

---

[4]This case is known in the mathematical literature as Rolle's Theorem, in honor of Michel Rolle (1652-1719).

B.7   A proof of the fourth rule of L'Hôpital

Let us assume that f and g are continuous functions defined on the interval $[c, c+\delta_o)$, and differentiable on $(c, c+\delta_o)$. Furthermore, $\lim_{x\to c^+} f(x) = \lim_{x\to c^+} g(x) = 0$, and $\lim_{x\to c^+} \frac{f'(x)}{g'(x)} = L$. Then $\lim_{x\to c^+} \frac{f(x)}{g(x)} = L$. Let us provide a proof of this particular case of the fourth rule. We need to show that given any $\epsilon > 0$ there exists $\delta > 0$ such that if $c < x < c + \delta$ then $|\frac{f(x)}{g(x)} - L| < \epsilon$. Indeed, since $\lim_{x\to c^+} \frac{f'(x)}{g'(x)} = L$ there exists $\delta > 0$ such that

$$\text{if } c < z < c+\delta \text{ then } |\tfrac{f'(z)}{g'(z)} - L| < \epsilon.$$

Let x be an arbitrary (fixed) point in $(c, c+\delta)$, that is $c < x < c+\delta$. The Generalized Mean Value Theorem implies that there exists θ, $c < \theta < x$, such that

$$\tfrac{f(x)-f(c)}{g(x)-g(c)} = \tfrac{f'(\theta)}{g'(\theta)}.$$

However $f(c) = g(c) = 0$ since f and g are right-continuous at c and $\lim_{x\to c^+} f(x) = \lim_{x\to c^+} g(x) = 0$. Thus

$$\tfrac{f(x)}{g(x)} = \tfrac{f'(\theta)}{g'(\theta)}.$$

But $c < \theta < c + \delta$, consequently $|\frac{f'(\theta)}{g'(\theta)} - L| < \epsilon$. Then $|\frac{f(x)}{g(x)} - L| < \epsilon$. The proof has come to an end.

Instead of working with the interval $[c, c + \delta_o)$ we could have worked with the interval $(c - \delta_o, c]$. The proof is almost identical. Thus the fourth rule is fully proven when the interval under consideration is $(c - \delta_o, c + \delta_o)$ and the limit L is a real number. What happens if $L = \infty$ or $L = -\infty$?

Assume that f and g are defined and continuous on $(c - \delta_o, c]$, and differentiable on $(c - \delta_o, c)$. Furthermore $\lim_{x\to c^-} f(x) = \lim_{x\to c^-} g(x) = 0$ and $\lim_{x\to c^-} \frac{f'(x)}{g'(x)} = \infty$. Then $\lim_{x\to c^-} \frac{f(x)}{g(x)} = \infty$.

We need to show, under the given hypotheses, that given any $M > 0$ there exists $\delta > 0$ such that if $c - \delta < x < c$ then $\frac{f(x)}{g(x)} > M$. Right away we can assert that there exists $\delta > 0$ such that

$$\text{If } c - \delta < z < c \text{ then } \tfrac{f'(z)}{g'(z)} > M.$$

Assume $c - \delta < x < c$ and fix x. The Generalized Mean Value Theorem implies that there exists θ, $x < \theta < c$, such that $\frac{f(c)-f(x)}{g(c)-g(x)} = \frac{f'(\theta)}{g'(\theta)}$. Since $f(c) = g(c) = 0$ we can conclude that $\frac{f(x)}{g(x)} = \frac{f'(\theta)}{g'(\theta)}$. Evidently $c - \delta < \theta < c$, consequently $\frac{f'(\theta)}{g'(\theta)} > M$. Therefore $\frac{f(x)}{g(x)} > M$.

The proof of the case $L = -\infty$ is quite similar, as well as when the interval under consideration is $[c, c+\delta_0)$. Thus, the conclusion is valid when working with an interval of the type $(c - \delta_o, c + \delta_o)$.

B.8   A proof of the Comparison Test

Let us recall the statement of the Comparison test, which was used in section 2.29. Suppose that f and g are continuous functions defined on $[a, \infty)$, such that $0 \leq f(x) \leq g(x)$ for every $x \geq a$. The following implications are true:

(i) If $\int_a^\infty g(x)dx$ converges then $\int_a^\infty f(x)dx$ converges;

(ii) If $\int_a^\infty f(x)dx$ diverges then $\int_a^\infty g(x)dx$ diverges.

Moreover, in case (i) the inequality $\int_a^\infty f(x)dx \leq \int_a^\infty g(x)dx$ holds.

It is time to provide a proof. Define, for $x \geq a$, the following functions:

$$F(x) = \int_a^x f(t)dt \quad \text{and} \quad G(x) = \int_a^x g(t)dt.$$

Obviously, F and G are increasing, continuous[5]and $F(x) \leq G(x)$ on $[a, \infty)$ since $0 \leq f(x) \leq g(x)$, $x \geq a$. Let us assume that $\int_a^\infty g(x)dx$ converges. Thus, $\lim_{x\to\infty} G(x)$ exists, say $\lim_{x\to\infty} G(x) = L$. Therefore, there exists $p \geq a$ such that $|G(x) - L| < 1$ whenever $x \geq p$. Consequently,

$$G(x) \leq |G(x)| < 1 + |L| \text{ whenever } x \geq p.$$

Since G is continuous on $[a, p]$, the Extreme Value Theorem (section 1.6) implies that there exists x_o in $[a, p]$ such that $G(x) \leq G(x_o)$ for any x in $[a, p]$. Hence,

$$G(x) \leq M \text{ for any } x \geq a,$$

where M is the maximum of the numbers $G(x_o)$ and $1 + |L|$. Therefore, $F(x) \leq M$ for any $x \geq a$.

Since F is increasing and bounded above, it follows that $\lim_{x\to\infty} F(x)$ exists (proposition 1a, section 3.1). That is, $\int_a^\infty f(x)dx$ converges. It only remains to prove that $\int_a^\infty f(x)dx \leq \int_a^\infty g(x)dx$ is true. Let $L_1 = \lim_{x\to\infty} F(x)$ and $L_2 = \lim_{x\to\infty} G(x)$. Given any $\epsilon > 0$ there exists r such that $L_1 - \epsilon/2 < F(x)$ whenever $x \geq r$ and there exists s such that $G(x) < L_2 + \epsilon/2$ whenever $x \geq s$. Then, if q is the maximum of the two numbers r and s, it follows that:

$$L_1 - \tfrac{\epsilon}{2} < F(q) \leq G(q) < L_2 + \tfrac{\epsilon}{2}$$

Therefore, $L_1 < L_2 + \epsilon$ for every $\epsilon > 0$, which in turn implies that $L_1 \leq L_2$. The proof of (i) has come to an end. How about a proof of (ii)? It is pretty simple: the implication (ii) is nothing but the logical contrapositive of (i)!

---

[5]Actually, thanks to FTC, we can assert that F and G are differentiable on the common domain of definition of f and g.

Appendix C

C.1 Answers to odd-numbered exercises

Section 1.1

1) $y = \frac{1}{2}x - \frac{1}{2}$. 3) $-1 < x < 4$. 5) $x > 4$ or $x < -5$. 7) $y = 3x - 1$. 9) $y = -\frac{1}{3}x + \frac{7}{3}$.
11) 2 seconds.

Section 1.2

1) $(-\infty, -2]$. 3) It defines a relation, namely a horizontal parabola open to the left.
5) $(f \circ g)(x) = \sin(x^2 - 5)$, $(g \circ f)(x) = \sin^2(x) - 5$. 7) Let $f : I \to \Re$ be decreasing
and let $g : ran(f) \to \Re$ be its inverse, where we are accepting that $ran(f)$ is an
interval. Let y_1 and y_2 be arbitrary elements of $ran(f)$, $y_1 < y_2$. Then $y_1 = f(x_1)$,
$y_2 = f(x_2)$ for some points x_1, x_2 in I. We claim that $x_1 > x_2$ because otherwise
$x_1 \leq x_2$ implies that $f(x_1) \geq f(x_2)$ (recall that f is decreasing). But $g(y_1) = x_1$ and
$g(y_2) = x_2$, consequently $g(y_1) > g(y_2)$. 9) Its inverse is $g(x) = -x^2$, $x \geq 0$. Note
that g has domain $[0, \infty)$ and range $(-\infty, 0]$.

Section 1.3

1) $x = \frac{\log 5}{\log 3} - 2$ 3) $\frac{1}{3.9810717}$ less intense. 5) Yes, it is correct. Why? $10^{\log a - \log b} =$
$10^{\log a} \times 10^{\log b} = ab$. 7) $[H^+] = 10^{-11.8} \approx 1.5848932 \times 10^{-12}$. 9)$x = \log 5 / \log 2$. The
equation $x2^x = 5$ cannot be solved by hand; it is equivalent to $\log x + x \log 2 = \log 5$,
but this equation cannot be solved in terms of elementary functions. A graphing
calculator provides the answer: $Solve(x * 2^x = 5, x) \approx 1.6231403$.

Section 1.4

1) The limit under consideration has the value 2. Hence, an equation of the tangent
line at (1,1) is $y - 1 = 2(x - 1)$. 3) 0. 5) The limit from the left is -1 while the

limit from the right is 1. 7) The limit from the left is 1 while the limit from the right is 0. 9) The vertical asymptote is the line $x = 4$. We note that the curve intersects the horizontal axis at the point $(-6, 0)$, the vertical axis at $(0, -3/2)$, and gets closer and closer to 1 as $|x|$ becomes larger and larger. 11) Given $\epsilon > 0$ choose $\delta = \epsilon^2$. If $0 < x < \epsilon^2$ then $\sqrt{x} < \epsilon$, i.e., $|\sqrt{x} - 0| < \epsilon$. 13) Assume that $|x - c| < 1$. Then $||x| - |c|| < 1$, which in turn implies $|x| < |c| + 1$. Thus, $|x^2 - c^2| < (2|c| + 1)|x - c|$ whenever $|x - c| < 1$. Given $\epsilon > 0$ choose $\delta =$ minimum of the two numbers 1, $\frac{\epsilon}{2|c|+1}$. This δ 'will do the job.'

Section 1.5

1) 0. 3) The vertical asymptote is given by $x = 2$, while its horizontal asymptote is $y = 1$. The curve intersects the vertical axis at $(0, -3/2)$ and intersects the horizontal axis at $(-3, 0)$. 5) The vertical asymptote is $x = 1$ and the oblique asymptote is $y = x + 1$. 7) Let $f(x) = \frac{x^2 + 3x - 10}{x^2 - 4}$. Then $\lim_{x \to -2+} = \infty$ and $\lim_{x \to -2-} = -\infty$; thus, $\lim_{x \to -2}$ does not exist. But $\lim_{x \to \pm\infty} f(x) = 1$. 9) Define $f(x) = 1/x^2$ and $g(x) = 1/x$. We have $f(x) < g(x)$ on $(1, \infty)$, but $\lim_{x \to \infty} f(x) = \lim_{x \to \infty} g(x) = 0$.

Section 1.6

1) It has a removable discontinuity at $x = 0$. Redefining $h(0) = 0$ we will achieve continuity everywhere. 3) Let $f(x) = x^2$ and $g(x) = \sin x$; then $h(x) = (f \circ g)(x)$. 5) $y = 3/2$ will do the job. 7)$[1, 0)$. 9) Let $f : [a, b] \to \Re$ be continuous and decreasing. We will show that $ran(f) = [f(b), f(a)]$. First let us prove that $[f(b), f(a)]$ is a subset of $ran(f)$. Indeed, if $y = f(a)$ or $y = f(b)$ then obviously y belongs to $ran(f)$. Assuming that $f(b) < y < f(a)$, the Intermediate Value Theorem implies that there exists x, $a < x < b$, such that $f(x) = y$; thus, y belongs to $ran(f)$. It remains to show that $ran(f)$ is a subset of $[f(b), f(a)]$. With this purpose in mind, let y be an arbitrary element of $ran(f)$. Then $y = f(x)$ for some x, $a \leq x \leq b$. Since f is decreasing we will have $f(a) \geq f(x) \geq f(b)$, consequently $f(b) \leq y \leq f(a)$. That is to say, y belongs to the interval $[f(b), f(a)]$. 11) One such pair is $f(x) = 1$, $x \geq 0$, $f(x) = -1$, $x < 0$, and $g(x) = -1$, $x \geq 0$, $g(x) = 1$, $x < 0$. Note that $f + g$ is the identically zero function, which is obviously continuous everywhere.

Section 1.7

1) $\frac{-x^2 - 6x - 3}{(x^2 + x)^2}$, $\frac{5}{2}x^{3/2}$. 3)163.95 ft/s 5) $2 \pm \sqrt{3}$ 7) Yes, it is differentiable at $x = 0$. The derivative happens to be zero. 9) $\frac{1 - x^2}{(1 + x^2)^2}$. 11)$v'(r) = -2kr$. 13) $-\frac{nRT}{P^2}$. 15) For any point x where f and g are differentiable and $g(x) \neq 0$, we have: $(\frac{f}{g})'(x) = (f\frac{1}{g})'(x) = \frac{f'(x)}{g(x)} + f(x)(-\frac{g'(x)}{g^2(x)}) = \frac{f'(x)g(x) - f(x)g'(x)}{g^2(x)}$. 17) Let $h(x) = x^2/2$, $x < 0$ and

$h(x) = 0$, $x \geq 0$. Then $h'(x) = x$, $x < 0$, and $h'(x) = 0$, $x \geq 0$. We can observe that h' is not differentiable at $x = 0$. 19)Let $f(x) = |x|$ and $g(x) = x^2$. We can see that $f(x)g(x) = x^3$ when $x \geq 0$ and $f(x)g(x) = -x^3$ when $x < 0$, which happens to be a function that is differentiable everywhere (including the origin). Of course, there are many other pairs of functions that could be defined.

Section 1.8

1) $2x \sin x + x^2 \cos x$. 3) $2 \sin x \cos x$. 5) $\sec x \tan^2 x + \sec^3 x$. 7) $2 \sin x \cos^2 x - \sin^3 x$. 9) $2 \sec^2 x \tan^2 x + \sec^4 x$.

Section 1.9

1) $e^x(\sin x + \cos x)$. 3) $-\frac{e^x+1}{(e^x+x)^2}$. 5) $e^{2x}(2x+2(x^2+1))$. 7) $3e^{3x}$. 9) $e^{3x}(3 \cos x - \sin x)$. 11) $(-0.112, e^{-0.112})$, $(-3.58, e^{-3.58})$.

Section 1.10

1) $-3x^2 \sin(x^3+7)$, $x/\sqrt{x^2+1}$. 3) $2xe^{x^2}$, $2xe^{x^2} \cos(e^{x^2})$. 5) $\frac{2\pi}{15} \cos(\frac{2\pi}{15})$, $\pi/15$, $\frac{2\pi}{15} \cos(\frac{8\pi}{15})$. 7) $-kI(0)e^{-kz}$. 9) $\frac{G'(t)}{G(t)} = 2.06\frac{H'(t)}{H(t)}$. 11) $y - \frac{\sqrt{5}}{3} = -\frac{1}{\sqrt{5}}(x - \frac{4}{3})$, $y + \frac{\sqrt{5}}{3} = \frac{1}{\sqrt{5}}(x - \frac{4}{3})$. 13) $-\pi/6$. 15) $\frac{dP}{dt} = 15$.

Section 1.11

1) $-\tan x$. 3) $\frac{2x}{x^2+1}$. 5) $\frac{1-\ln x}{x^2}$. 7) $c = e$. 9) Taking the derivative to the given expression, we get $\frac{c(0)}{c(t)} \frac{c'(t)}{c(0)} = -k$. Therefore $c'(t) = -kc(t)$.

Section 1.12

1) $\frac{2x}{(x^2+1)^2+1}$. 3) $\frac{1}{x\sqrt{x^2-1}}$ when $x > 1$, $\frac{-1}{x\sqrt{x^2-1}}$ when $x < -1$. 5) $(0,0)$. 7) $\pm\sqrt{\frac{1-\alpha}{\alpha}}$. 9) Just apply the chain rule to the function $\arccos(1/x)$.

Section 1.13

1) $L(x) = 1$, $L(x) = -\frac{\sqrt{2}}{2}(x - \frac{\pi}{4}) + \frac{\sqrt{2}}{2}$. 3) $L(x) = x - 1$. 5)$L(x) = -ax + 1$.

Section 1.14

1) Let $f(x) = x^3 - x + 1$. Since $f(-2) < 0$ and $f(0) > 0$ we can choose $x_o = -1$. Then $x_1 = -1.5$, $x_2 = -1.3478261$, $x_3 = -1.3252004$. On the other hand, using

the program *Newt*, with $n = 5$ and $x_0 = -1$, we get the following five numbers: -1.5, -1.3478261, -1.3252004, -1.3247182,-1.324718. So, we are confident that a root is -1.32471, where the five decimals are 'true.' 3) Since $f(0) > 0$ and $f(-1) < 0$ we can choose $x_o = -1/2$ as our starting point. With $n = 4$ the program *Newt* provides four numbers, namely -0.44642857, -0.44504291, -0.44504187, -0.44504187. Thus, an approximation to a root is -0.445041 with six 'true' decimals. Next we note that $f(-1) < 0$ and $f(-2) > 0$. Thus, there is a negative root between -2 and -1. Choosing $x_o = -3/2$ and $n = 6$ we get -2.125, -1.8928302, -1.8119037, -1.8020762, -1.8019378, -1.8019377. We can then confidently assert that -1.801937 is an approximation to a root, with all its decimals 'true.' 5) Since $f(0) > 0$ and $f(\sqrt{2\pi}) < 0$, it makes sense to select $x_o = \sqrt{\pi}$. Using *Newt*, with $n = 3$, we get three numbers on the screen: 1.7524826, 1.7523603, 1.7523603. Thus, 1.75236 is an approximation to a root. We notice that $f(-x) = f(x)$. Therefore, another root is -1.75236.

Section 1.15

1) The given function is increasing on $(-\infty, -1/\sqrt{3})$ and $(1/\sqrt{3}, \infty)$, while it is decreasing on $(-1/\sqrt{3}, 1/\sqrt{3})$. 3) The given function is increasing over the whole real line. 5) Just note that $h'(x) = 1/x > 0$ for every $x > 0$, hence $h(x)$ is increasing on its domain of definition. On the other hand, f is increasing on $(0, e)$ and decreasing on (e, ∞). 7) $\lim_{t \to \infty} z(t) = 1$ whenever $\alpha > \beta$. 9) Solving exercise 8 we reach the expression $x(t) = x(0)e^{-\alpha t}$. Then $x(\tau) = x(\frac{\ln 2}{\alpha}) = x(0)e^{\frac{-\alpha \ln 2}{\alpha}} = x(0)e^{-\ln 2} = \frac{x(0)}{e^{\ln 2}} = \frac{x(0)}{2}$. 11) $f(t) = \sin t - \cos t + 2$. 13) Multiplying the differential equation by e^{2t}, we obtain $e^{2t}(x'(t) + 2x(t)) = 0$; thus, $(e^{2t}x(t))' = 0$. Therefore, there exists a constant C such that $e^{2t}x(t) = C$ for every t, that is, $x(t) = Ce^{-2t}$ for every t. But $5 = x(0) = C$, hence $x(t) = 5e^{-2t}$ for every t. 15) We observe that $f'(x) = 3x^2 + 6x + 1$; thus, the critical points are $-3 \pm \sqrt{6}$. The graph of $y = 3x^2 + 6x + 1$ is a parabola that opens upwards. Then f is increasing when $x > -3 + \sqrt{6}$ or $x < -3 - \sqrt{6}$, and decreasing on $(-3 - \sqrt{6}, -3 + \sqrt{6})$.

Section 1.16

1) $t = 5/8$ seconds and $v \approx -85.66$ ft/s. 3) 0.8125 miles. 5) 40.65 meters. 7) $F(x) = -\frac{5}{3}\cos(3x) + C$, $G(x) = \frac{1}{\pi}\sin(\pi x) + \frac{x^3}{3} + 2x + C$. 9) $y(x) = \arctan x + 3x + \frac{\pi}{4}$. 11) $h\alpha^2$, where h is the starting height. 13) $F(x) = -e^{-\nu x} + 1$ 15) $F(x) = x^2/2$ when $0 \le x < 1$, $F(x) = -x^2/2$ when $-1 < x < 0$. 17) $F(x) = -\frac{\cos x}{\sin x} + C$.

Section 1.17

1) $-1/x^2$, $x > 0$; $9e^{3x} - 4\cos(2x)$. 3) $y(t) = e^{-t} - \frac{1}{\pi^2}\cos(\pi t) + C_1 t + C_2$. 5) $y(x) = 1 + \frac{1}{2}x^2$. 7) $y(t) = \frac{1}{2}t^3 - \cos t + C_1 t + 1$. 9) $\frac{1}{2}y'^2(t) = \frac{K}{y(t)+R} + \frac{1}{2}v_o^2 - \frac{K}{R}$. 11) Let $f(x) = x^3/6$, $x < 0$ and $f(x) = 0$, $x \geq 0$. Then $f'(x) = x^2/2$, $x < 0$ and $f'(x) = 0$, $x \geq 0$, while $f''(x) = x$, $x < 0$ and $f''(x) = 0$, $x \geq 0$. The function f'' is not differentiable at $x = 0$ because its left derivative at the origin is 1 and its right derivative at the origin is 0.

Section 1.18

1) The maximum is adopted at -1 and the minimum is attained at 1 and -2. 3) Minimum at π, maximum at $\pi/4$. 5) Minimum at $5\pi/4$, maximum at $\pi/4$. 7) She should swim to a point situated 106.066 meters down the coast, and run from there to her destination. 9) The best option is to install the pump at a distance of 400 ft down the river from the closest cottage. 11) It should land at a point $\frac{1}{\sqrt{3}}$ Km down the coast from its original position.

Section 1.19

1) $r = (\frac{V}{2\pi})^{1/3}$, $h = 2(\frac{V}{2\pi})^{1/3}$. 3) $r = \sqrt{\frac{S}{3\pi}} = h$. 5) $x = e$ 7) $t = \sqrt{2}$. 9) The right triangle has to be isosceles, with their legs having the common length $3/\sqrt{2}$.

Section 1.20

1) The given function is concave up on $(-\infty, -1/\sqrt{6})$ and $(1/\sqrt{6}, \infty)$, while it is concave down on $(-1/\sqrt{6}, 1/\sqrt{6})$. 3) No, f has no points of inflection (it is concave up everywhere). 5) $h = \frac{\sqrt{2}}{2}r$. 7) $L/6$ by $L/8$. 9) Approximately 11.15 ft from the strongest source, answer that is found using the first version of the second derivative test. 11) According to the first derivative test, all we need to do is show that $f'(x) < 0$ for $x < c$ and $f'(x) > 0$ for $x > c$. Let us prove the latter. Suppose that it were not to be true, then there exists d, $d > c$, such that $f'(d) \leq 0$. We cannot have $f'(d) = 0$ since c is the only critical point, thus $f'(d) < 0$. But the hypothesis $f''(x) > 0$ for all x in I implies that $f'(x)$ is increasing, consequently $f'(d) > f'(c) = 0$. This is an impossibility because previously we reached the inequality $f'(d) < 0$.

Section 1.21

1) It has a local maximum at 0, and a local minimum at 4/3. For the purpose of sketching a graph, we may note that the given function is concave up on $(\frac{2}{3}, \infty)$ and concave down on $(-\infty, \frac{2}{3})$. 3) It has a local minimum at $x = -1$, which happens to

be a global minimum (the function is concave up everywhere). 5) Using the practical form of the first derivative test for local extrema, we find that the function has a local maximum at 0.8212, a local minimum at 2.6692, and a local maximum at 4.51715.

Section 1.22

1) The correlation coefficient happens to be 0.908473 and the corresponding least squares regression line leads to the estimation A = 0.02983569, $\alpha = 0.734382$. Thus, $y = 0.02983569x^{0.734382}$ happens to be an allometric model that agrees with data reasonably well.

Section 2.1

1) $520(1.05)^{10} \approx 847$ bears. 3) We observe that $|1 + \frac{3}{n} - 1| = \frac{3}{n}$. Given $\epsilon > 0$ choose a natural number N such that $N > \frac{3}{\epsilon}$. Assuming $n \geq N$ we will have $n > \frac{3}{\epsilon}$, hence $\frac{1}{n} < \frac{\epsilon}{3}$, which is equivalent to $\frac{3}{n} < \epsilon$. In summary, given any $\epsilon > 0$ we have been able to find N such that if $n \geq N$ then $|1 + \frac{3}{n} - 1| < \epsilon$. 5) The quotient under consideration has the values 3/3, 5/3, 7/3, 9/3, 11/3,... (corresponding to $n = 1, 2, 3, 4, 5, ...$), hence a pattern emerges: $\frac{2n+1}{3}$, $n = 1, 2, 3, 4, 5, ...$. Since $\Sigma_{i=1}^{n} i = \frac{n(n+1)}{2}$ we can conclude that $\Sigma_{i=1}^{n} i^2 = \frac{n(n+1)}{2} \frac{2n+1}{3} = \frac{n}{6}(n+1)(2n+1)$. 7) We note that $R(1+i)^{-1} + ... + R(1+i)^{-n} = R(1+i)^{-1}[1 + (1+i)^{-1} + ... + (1+i)^{-n+1}] = \frac{R}{1+i} \frac{1-(1+i)^{-n}}{1-(1+i)^{-1}} = \frac{R}{i}[1 - (1+i)^{-n}]$. 9) 132.86 dollars. 11) Given any $\epsilon > 0$, since $a_n \to c$ there exists N such that if $n \geq N$ then $|a_n - c| < \epsilon\sqrt{c}$. Assuming $n \geq N$ we will have $\frac{1}{\sqrt{c}}|a_n - c| < \epsilon$. Hence $|\sqrt{a_n} - \sqrt{c}| < \epsilon$ provided that $n \geq N$. 13) Proceeding by contradiction, assume that $L < 0$. Since $a_n \to L$, there exists N such that $|a_n - L| < -L$ whenever $n \geq N$. In particular, $|a_N - L| < -L$, thus $L < a_N - L < -L$. Therefore $a_N < 0$, which is an impossibility. 15) Let $b_n = a_{n-1}$. We need to show that given any ϵ there exists a natural number N_1 such that if $n \geq N_1$ then $|b_n - L| < \epsilon$. Well, given ϵ, since $a_n \to L$ there exists N_2 such that if $n \geq N_2$ then $|a_n - L| < \epsilon$. Choose $N_1 = N_2 + 1$ and assume that $n \geq N_1$. Then $n - 1 \geq N_2$ and consequently $|a_{n-1} - L| < \epsilon$, that is, $|b_n - L| < \epsilon$. 17) It follows from the inequality $||a_n| - |L|| \leq |a_n - L|$ for all n.

Section 2.3

1) The corresponding weighted averages are: 1.0000034 ($n = 5$), 1.0000002 ($n = 10$), and 1 ($n = 15$). Thus, we might conjecture that the area has dimension 1. 3) 128. 5) 1/4.

Section 2.4

1) 50. 3) -2.5 5)130 7) 3 9) Recall that $\int_a^b h = \lim_{n\to\infty} \Sigma_{i=1}^n h(c_i)\frac{b-a}{n}$. But $a_n = \Sigma_{i=1}^n h(c_i)\frac{b-a}{n} \geq 0$ since $h(c_i) \geq 0$ and $\frac{b-a}{n} > 0$. Therefore $\int_a^b h \geq 0$. 11) 3/2.

Section 2.5

1) $\int_0^\pi \cos x dx = 0$, $\int_{\pi/2}^\pi \cos x dx = -1$. 3)$\frac{1}{12}$. 5) 1 7) $\frac{1}{V(t)}\frac{dV}{dt} = 4\frac{1}{R(t)}\frac{dR}{dt}$. 9) $2\ln 2$. 11)$[\arctan x]_{-a}^a - \frac{1}{3}[x^3]_{-a}^a$, where $a = 0.7861514$.

Section 2.6

1) $\frac{4}{7}x^{7/4} + C$. 3) $\frac{1}{3}\tan^{-1}\frac{x}{3} + C$. 5) $3\tan\frac{x}{3} + C$. 7) $\frac{1}{2}[e^{x^2+2x+1}]_1^2$.

9) $\frac{1}{\sqrt{2}}[\arcsin(x\sqrt{2})]_0^{1/3}$. 11) $\frac{x^2}{2} + \arctan x + \frac{1}{2}\ln(x^2 + 1)$.

Section 2.7

1) $2x\cos(x^2)$ 3) We notice that $F_1'(x) = \sin(x^2) = F_2'(x)$. Hence, there exists a constant C such that $F_1(x) = F_2(x) + C$ for every x. In particular, $F_1(3) = F_2(3) + C$. Therefore $C = -F_2(3) = -\int_5^3 \sin(u^2)du = \int_3^5 \sin(u^2)du$ 5) $xe^x - e^x + 2$. 7) We wish to show that $\int_a^b f(t)dt = [F(x)]_a^b$, where $F(x)$ is any antiderivative of $f(x)$. Let $G(x) = \int_a^x f(t)dt$. The Fundamental Theorem of Calculus implies that $G'(x) = f(x)$. Hence, there exists a constant C such that $G(x) = F(x) + C$ for every x. In particular $G(a) = F(a) + C$. But $G(a) = 0$, thus $C = -F(a)$. Finally, we observe that $G(b) = F(b) - F(a)$. That is to say, $\int_a^b f = F(b) - F(a)$. 9) $F(x) = x^3/3$ for $x < 0$, $F(x) = x^2/2$ for $x \geq 0$. This function is differentiable everywhere, as a graph can confirm. 11) If $0 \leq x < 1$ we have $y(x) = \int_0^x udu \frac{1}{2}x^2$, while if $x \geq 1$ we get $y(x) = \int_0^1 udu + \int_1^x \frac{1}{2}du = \frac{1}{2}x$. Implicit in our argument is the certainty that although f is not continuous on $[0, 1]$, we can define on $[0, 1]$ a function $g(x)$ such that $g(1) = 1$ and $g(x) = f(x)$, $0 \leq x < 1$. Evidently, g is continuous on $[0, 1]$ and it makes a lot of sense to define $\int_0^1 f(x)dx = \int_0^1 g(x)dx$. 13) Dividing $\alpha x = T_2\sin\theta$ by $T_1 = T_2\cos\theta$ we arrive at $\tan\theta = \frac{\alpha}{T_1}x$, that is, $f'(x) = \frac{\alpha}{T_1}x$. Integrating, we get $f(x) = \frac{\alpha}{2T_1}x^2 + C$. In particular $0 = f(0) = 0 + C$, therefore $f(x) = \frac{\alpha}{2T_1}x^2$. 15)$I(t) = e^{-t}\int_0^t e^u \cos(2u)du$. 15) $I(t) = e^{-2t}\int_0^t e^{2u}\cos u du$. Tools from section 2.19 are needed to write the integral in terms of elementary functions.

Section 2.8

1) $\frac{1}{2}\ln\frac{11}{2}$. 3) $\frac{d}{dx}\ln(x^2 + 1) = \frac{2x}{x^2+1}$. Thus, the function is increasing for $x > 0$ and decreasing for $x < 0$. Taking the second derivative, we can assert that the function is concave up on $(-0.5, 0.5)$, concave down on $(0.5, \infty)$ and $(-\infty, -0.5)$. 5) $-\tan x$. 7)

Assume that x is a fixed number less than 1. Then $\frac{1}{t} \geq 1$ for every t in $[x, 1]$. Therefore $\int_x^1 \frac{1}{t} dt \geq \int_x^1 dt$, which in turn leads to $-\int_1^x \frac{1}{t} dt \geq -\int 1x dt$. Hence $-\ln x \geq -(x-1)$, that is, $\ln x \leq x - 1$. 9) $\frac{1}{2} \arctan^2 x + C$. 11) $\int \frac{dx}{(\arctan x)(x^2+1)} = \ln|\arctan x| + C$. Therefore, the answer is $\ln|\arctan(-1)| - \ln|\arctan(-3)| \approx -0.46394435$.

Section 2.9

1) Let $f(x) = e^{x^2}$. This function is increasing on $(0, \infty)$ and decreasing on $(-\infty, 0)$; moreover, it is concave up everywhere. On the other hand, the function $g(x) = e^{-x^2}$ is increasing on $(-\infty, 0)$ and decreasing on $(0, \infty)$. It is concave up on $(-\infty, -\frac{\sqrt{2}}{2})$ and $(\frac{\sqrt{2}}{2}, \infty)$, while it is concave down on $(-\frac{\sqrt{2}}{2}, \frac{\sqrt{2}}{2})$. 3) A linear regression analysis leads to the line of best fit $y = -0.405139x + 4.218385$. Since the correlation coefficient happens to be -0.971722, we are pretty confident about the chosen model and proceed to estimate k as 0.405139. 5) All the time, the process is governed by the differential equation $c'(t) = -kc(t)$. After ingesting the first pill we will have $c(t) = c_o e^{-kt}$, $0 < t \leq T$. Right after the second dosis, the amount is $c_o e^{-kT} + c_o$. Thus, between T and $2T$ we have to deal with the IVP $c'(t) = -kc(t)$, $c(T) = c_o e^{-kT} + c_o$. The general solution of the differential equation under study is $c(t) = He^{-kt}$. In particular, $c_o e^{-kT} + c_o = c(T) = He^{-kT}$; hence $H = c_o + c_o e^{kT}$. Consequently $c(t) = (c_o + c_o e^{kT})e^{-kt}$, $T < t \leq 2T$. Therefore $c(2T) = (c_o + c_o e^{kT})e^{-2kT}$. Right after the third dosis, we will have $(c_o + c_o e^{kT})e^{-2kT} + c_o$, that is, $c_o + c_o e^{-kT} + c_o e^{-2kT}$. A pattern emerges: right after the nth dosis, we will have $c_o + c_o e^{-kT} + ... + c_o e^{-(n-1)kT}$. In the long run, it is a matter of dealing with the geometric series $\Sigma_{n,0}(e^{-kT})^n$. Then $\Sigma_{n=0}^\infty (e^{-kT})^n = \Sigma_{n=0}^\infty (1/e^{kT})^n = 1/(1 - 1/e^{kT}) = \frac{c_o}{1-e^{-kT}}$. 7) Since h and W are inverse of each other, we will have $W(h(x)) = x$ for every $x \geq -1$ and $h(W(x)) = x$ for every $x \geq -1/e$, that is, $W(xe^x) = x$ for every $x \geq -1$ and $W(x)e^{W(x)} = x$ for every $x \geq -1/e$. 9) $x = \frac{1}{c}W(\frac{dc}{b}e^{\frac{ca}{b}}) - \frac{a}{b}$ (if $c = 0$ or $b = 0$, obviously it is not necessary to use the W function). 11) $y(t) = \frac{7}{2}t + \frac{7}{4}e^{-2t} - \frac{3}{4}$. 13) $y(t) = te^{-t}$.

Section 2.10

1) $3^{x^2+2x}(2\ln(3)x + 2\ln(3))$. 3) $\frac{7}{\ln 2}$. 5) $\frac{1}{2\ln 3}(\ln x)^2 + C$.

Section 2.11

1) $A = \frac{y_1}{x_1^\alpha}$, $\alpha = \ln\frac{y_1}{y_2}/\ln\frac{x_1}{x_2}$. 3) $e^{\frac{2}{3}\ln x}$, $x > 0$; 0 when $x = 0$; $e^{\frac{2}{3}\ln(-x)}$, $x < 0$. 5) We obtain the correlation coefficient -0.973761 and the line of best fit $\hat{y} = -0.126367x + 6.087279$. Then we can proceed to estimate α and A: $\alpha = -0.126367$, $\ln A = 6.087279$, which in turn leads to $A = e^{6.087279} = 440.22194$. The estimated model becomes $y = 440.22194x^{-0.126367}$. 7) From $R'(I) = k\frac{R(I)}{I}$ we get $\int \frac{R'(I)}{R(I)} dI = \int \frac{k}{I} dI$,

thus $\ln R(I) = k \ln I + C$. Hence $\ln R(I) = \ln I^k + C$, which in turn leads to $R(I) = AI^k$. We would use the expression $\ln R = k \ln I + C$ to do a linear regression analysis.

Section 2.12

1) $\frac{\ln 2}{k}$. 3) We have $c'(t) = -kc^2(t)$, thus $\int_0^t c^{-2}(s)c'(s)ds = -kt$. Hence $-[c^{-1}(s)]_0^t = -kt$, i.e., $-\frac{1}{c(t)} + \frac{1}{c_o} = -kt$. By the definition of half-life we get $-\frac{1}{c_o/2} + \frac{1}{c_o} = -k\tau$, therefore $\tau = \frac{1}{kc_o}$. In particular, $\tau_1 = \frac{1}{kc_1}$ and $\tau_2 = \frac{1}{kc_2}$. So $\frac{\tau_1}{\tau_2} = \frac{c_2}{c_1}$. There is reasonable agreement between this prediction and the given data: the three possible combinations of quotients of initial concentrations and quotients of half-lives provide the values 1.9270833 and 2, 3.89474 and 4, 2.02105 and 2. 5) A linear regression analysis, accepting first order kinetics, provides the line of best fit $\hat{y} = 0.053883x - 0.024896$. Since the correlation coefficient happens to be 0.997913, we are quite confident that the first-order model of chemical kinetics is acceptable and thereafter we proceed to estimate k as 0.053883.

Section 2.13

1) $\tau = 8.63575$ days. 3) 4.4784 mg 5) $8,633.4056$ years.

Section 2.14

1) $\frac{1}{\ln(40/31)}\ln 4$. 3) $58.67^\circ C$ 5) A linear regression analysis leads to $\hat{y} = -0.018363x + 4.173424$. Since the correlation coefficient is -0.998899 we are pretty confident about the validity of the model and proceed to estimate k as 0.018363.

Section 2.15

1) $64,824,380$ 3) $P(t) = 44,633e^{0.045162t}$ 5) $\frac{(t_1-t_2)\ln 2}{\ln \frac{N_1}{N_2}}$. 7) Integrating the given differential equation, between 0 and t, we obtain $\int_0^t \frac{P'(s)}{P(s)}ds = \int_0^t(\alpha s + \beta)ds$. Hence $\ln \frac{P(t)}{P(0)} = \frac{\alpha t^2}{2} + \beta t$. Then we divide by t and obtain $\frac{1}{t}\ln \frac{P(t)}{P(0)} = \frac{\alpha}{2}t + \beta$. A linear regression analysis, of t versus $\frac{1}{t}\ln \frac{P(t)}{P(0)}$, will allow us to test the model with data. From the line of best fit, $\hat{y} = ax + b$, we estimate $\alpha/2$ as a and β as b. Of course, we expect a correlation coefficient close to 1.

Section 2.16

1) $y(x) = \frac{1}{1+e^{-x}}$ 3) $y(t) = \frac{N}{(N-1)e^{-kNt}+1}$, thus $\lim_{t\to\infty} y(t) = N$. 5) We observe that $N'(t) = -\frac{a}{K}(N^2 - KN + \frac{K}{a}h)$. A 'completion of squares' leads to $N^2 - KN + \frac{K}{a}h =$

$(N - \frac{K}{2})^2 + (-\frac{K^2}{4} + \frac{Kh}{a})$. This is a parabola with vertex at $(\frac{K}{2}, -\frac{K^2}{4} + \frac{Kh}{a})$. Since $-\frac{a}{K} < 0$, in order to have $N'(t) < 0$ it is necessary that the parabola adopts only positive values. For this to happen, the second coordinate of the vertex has to be positive, that is, $-\frac{K^2}{4} + \frac{Kh}{a} > 0$, which is equivalent to $h > \frac{aK}{4}$. 7) $W(t) = a - (a - W_o)e^{-kt}$ 9) The inflection point happens to be $t = \frac{1}{\beta} \ln \frac{\alpha_o}{\beta}$. The solution to Gomperz's equation is $W(t) = W_o exp(\frac{\alpha_o}{\beta} - \frac{\alpha_o}{\beta}e^{-\beta t})$. A typical solution, for each of the four models, is portrayed in Figure C.1. 11) Recall that the solution of Gompertz's equation is $W(t) = W_o \exp(\frac{\alpha_o}{\beta} - \frac{\alpha_o}{\beta}e^{-\beta t})$. When t is very small, we will have $e^{-\beta t} \approx 1 - \beta t$. Hence $W(t) = W_o \exp(\frac{\alpha_o}{\beta} - \frac{\alpha_o}{\beta}(1 - \beta t)) = W_o e^{\beta t}$.

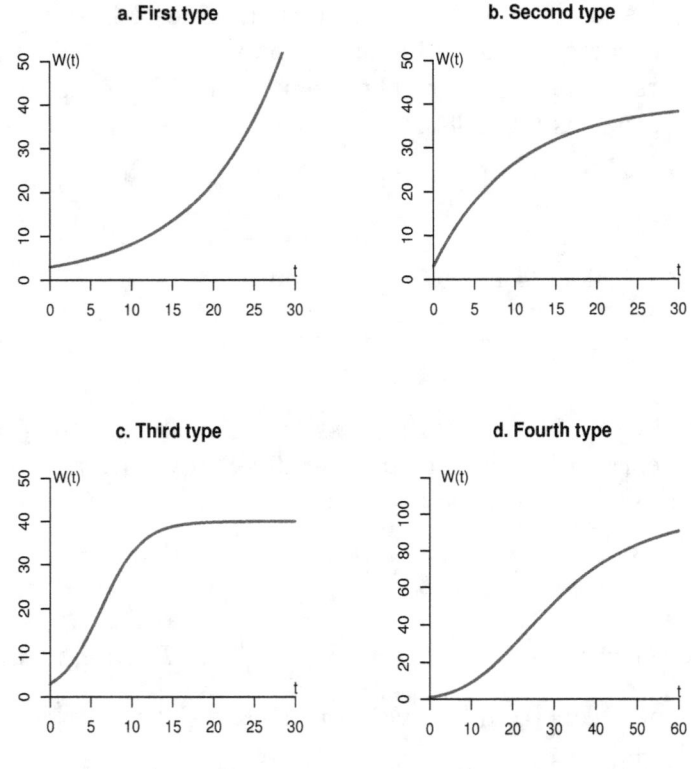

Figure C.1: Four types of plant growth

Section 2.17

1) $x(t) = Vp + (x_o - Vp)e^{-\frac{b}{V}t}$ 3) $x(t) = x_o(\frac{V-(b-a)t}{V})^{\frac{b}{b-a}}$ 5) 6.93 minutes. 7) The fourth corollary of MVT (section 1.15) leads to the conclusion that there exists a

constant C such that $e^{\frac{b}{V}t}x(t) = \frac{rV}{b}e^{\frac{b}{V}t} + C$ for every t. Then $x(t) = \frac{rV}{b}e^{\frac{b}{V}t} + Ce^{-\frac{b}{V}t}$. In particular, $x_o = \frac{rV}{b} + C$; consequently $x(t) = \frac{rV}{b} + (x_o - \frac{rV}{b})e^{-\frac{b}{V}t}$, exactly the same expression portrayed before (2.33). 9) The median, because it is 'more resistant' to outliers.

Section 2.18

1) $0.5(\cos 10 - \cos 2)$. 3) $-0.5(\ln|\cos(\frac{\pi^2}{16})| - \ln|\cos(\frac{\pi^2}{9})|)$. 5) $\ln|\sec x + \tan x| + C$ 7) $\ln(e^4 + 5) - \ln 6$ 9) $\frac{1}{3}\ln(\ln 6) - \frac{1}{3}\ln(\ln 2)$ 11) $\int \cos^2 x dx = \frac{1}{4}\sin 2x + \frac{1}{2}x + C$; $\int \sin^2 x dx = \frac{1}{2}x - \frac{1}{4}\sin(2x) + C$. 13) $-\frac{1}{4}\sin(2\pi - 2x) + \frac{1}{2}x + 1$. 15) $\frac{1}{3-x(t)} = kt + C$. Applying linear regression, we obtain $\hat{y} = 0.008112t + 0.329427$; thus, we estimate k as 0.008112. By the way, the correlation coefficient is 0.991096. 17) $2\sqrt{x} - 6\ln(\sqrt{x} + 3) + C$.

Section 2.19

1) $\frac{x}{3}\sin(3x) + \frac{1}{9}\cos(3x) + C$ 3) $x\arctan(3x) - \frac{1}{6}\ln(9x^2 + 1) + C$ 5) $\frac{\pi}{4} - \frac{1}{2}$ 7) $\frac{1}{13}(3\cos(2x) + 2\sin(2x))e^{3x} + C$ 9) $4\cos(x/2) + 2x\sin(x/2) + C$ 11) $\frac{x}{2}(\cos(\ln x) + \sin(\ln x)) + C$. 13) $2\sqrt{r}(\ln(r) - 2) + 2\sqrt{2}(2 - \ln 2) + C$. 15) $12{,}347{,}844$. 17) $\int \sec^3 \theta d\theta = \frac{1}{2}\sec\theta\tan\theta + \frac{1}{2}\ln|\sec\theta + \tan\theta| + C$. 19) $y(t) = \frac{1}{2}t^2 - t - 2e^{-t} + 4$.

Section 2.20

1) $-\frac{1}{3}\cos^3 x + \frac{1}{5}\cos^5 x + C$. 3) $\frac{1}{6}\cos x \sin^5 x - \frac{1}{24}\cos x \sin^3 x - \frac{1}{8}\sin x \cos x + \frac{1}{8}x + C$. 5) $\frac{1}{2}\sin x - \frac{1}{18}\sin(9x) + C$.

Section 2.21

1) $\frac{1}{3}\ln|x - 2| - \frac{1}{3}\ln|x + 1| + C$. 3) $\frac{2}{\sqrt{3}}\tan^{-1}(\frac{2x-1}{\sqrt{3}}) + C$. 5) $\frac{5}{2}\ln|x^2 + x + 1| - \frac{1}{\sqrt{3}}\tan^{-1}(\frac{2x+1}{\sqrt{3}}) + C$. 7) $\frac{\sqrt{3}}{18}\tan^{-1}(\frac{x}{\sqrt{3}}) + \frac{x}{6(x^2+3)} + C$ 9) $\frac{1}{2}\ln|x - 1| - \frac{1}{2}\ln|x + 1| + C$.

Section 2.22

1) $\frac{1}{2}x\sqrt{x^2 + 9} + \frac{9}{2}\ln(\sqrt{x^2 + 9} + x) + C$. 3) $\ln|x + \sqrt{x^2 - 16}| + C$. 5) $-\frac{1}{25}\frac{x}{\sqrt{x^2-25}} + C$. 7) $2\sin^{-1}(\frac{x}{2}) - \frac{x}{2}\sqrt{4 - x^2} + C$. 9) $\sqrt{x^2 + 4}(\frac{1}{3}x^2 - \frac{8}{3}) + C$. 11) $\ln|x| - \ln(\sqrt{1 - x^2} + 1) + C$.

Section 2.23

1) $2x\cosh x + x^2\sinh x$. 3) $\frac{2x}{\sqrt{x^4-1}}$. 5) $\ln(7 + \sqrt{48}) - \ln(3 + \sqrt{8})$. 7) $\frac{1}{\cosh^2 x}$. 9) $\frac{2x}{\cosh^2(x^2+1)}$. 11) $\cosh^{-1}(-x) = \ln(-x + \sqrt{x^2 - 1})$ for any $x \leq -1$.

Section 2.24

1) Proceeding by hand, we get $\int \cos(ax)dx = \frac{1}{a}\sin(ax) + C$. Using a calculator, $\int (\cos(ax), x) = \cos(ax)x + C$. What happens is that the calculator receives the information that x is the variable; thus, it interprets $\cos(ax)$ as a constant. To avoid this difficulty, one has to write $\int (\cos(a * x), x)$. 3) $\frac{1}{3}\ln(\frac{1+y(t)}{2-y(t)}) = kt + C$. 5) We used the expression $\frac{1}{a-b}\ln\frac{A(t)}{B(t)} = kt + C$ to do a linear regression analysis with data from Table 2.17. The least squares regression line happened to be $\hat{y} = 0.001973t + 0.038781$ and we got the information that the correlation coefficient was 0.999994. Thus, we can confidently write $\hat{k} = 0.001973$. 7) Using technology we get $\int(\frac{\sin x}{x}, x, 1, 2) = 0.65932991$. With $n = 4$, the program TMA provides the approximation 0.65932999 for the weighted average. It is the smallest such n. 9) From exercise 14 (section 1.15) we know that $\sec^{-1} x = \tan^{-1}\sqrt{x^2 - 1}$, thus $\frac{1}{b}\sec^{-1}(\frac{x}{b}) = \frac{1}{b}\tan^{-1}\sqrt{\frac{x^2}{b^2} - 1} = \frac{1}{b}\tan^{-1}(\frac{\sqrt{x^2-b^2}}{b})$.

Section 2.25

1) $\tan(t^2/2)$, $-\frac{\pi}{2} < t < \frac{\pi}{2}$. 3) $y(t) = 5e^{-t^2}$ when $y(0) = 5$, $y(t) = -e^{-t^2}$ when $y(0) = -1$. Both functions are defined everywhere on the real line. 5) $\tau = \frac{2^{n-1}-1}{(n-1)kc_o^{n-1}}$, where $c_o = c(0)$. 7) $v(t) = a\frac{e^{at}-e^{-at}}{e^{at}+e^{-at}}$, where $a = \sqrt{\frac{mg}{k}}$. We have $\lim_{t\to\infty} v(t) = \sqrt{\frac{mg}{k}}$. 9) Since $c > a/b$ and the constant function a/b is a solution, we can conclude that $y(t) > a/b$ for every t, that is, $a - by(t) < 0$ for every t. From $\int \frac{-b}{a-by}dy = \int -bdt$ it follows that there exists a constant D such that $\ln|a - by(t)| = -bt + D$ for every t. Thus $\ln(by(t) - a) = -bt + D$ and in particular $\ln(bc - a) = D$. Therefore $\ln\frac{by(t)-a}{bc-a} = -bt$, and consequently $\frac{by(t)-a}{bc-a} = e^{-bt}$. Elementary algebra then leads to $y(t) = \frac{a}{b} + (c - \frac{a}{b})e^{-bt}$.

Section 2.26

1) $\frac{\sqrt{2}}{2} + \ln(\sqrt{2} + 1)$. 3) $\sinh(3)$. 5) 2.14662.

Section 2.27

1) $\frac{\pi}{2}$. 3) $\pi(\frac{\pi^2}{4} - 2)$. 5) $\frac{\pi}{2}(e^2 - 1)$. 7) $\frac{16\pi}{3}$. Revolving around the y-axis we obtain the same answer.

Section 2.28

1) ∞ 3) 1 5) 0 7) $-1/2$ 9) 1 11) ∞ 13) 1. 15) 0.

Section 2.29

1) $1/2$ 3) $\pi^2/8$ 5) We have $0 < \frac{1}{x^3+2} < \frac{1}{x^3}$. Since $\int_1^\infty \frac{1}{x^3}dx$ is convergent, we can conclude that $\int_1^\infty \frac{1}{x^3+2}dx$ is convergent. Using a graphing calculator we get $\int_1^{30} \frac{dx}{x^3+2} = 0.31037033$, $\int_1^{35} \frac{dx}{x^3+2} = 0.31051771$. Hence $\int_1^\infty \frac{dx}{x^3+2} \approx 0.31$. 7) We have $0 < \frac{1}{e^x+e^{-x}} < \frac{1}{e^x} = e^{-x}$. Since $\int_0^\infty e^{-x}dx$ converges, we can conclude that $\int_0^\infty \frac{1}{e^x+e^{-x}}dx$ converges. The use of technology leads to $\int_0^{10} \frac{1}{e^x+e^{-x}}dx = 0.78535276$ and $\int_0^{15} \frac{1}{e^x+e^{-x}}dx = 0.78539786$. Hence, the approximate value of the improper integral is 0.7853. Interestingly, it is possible to find the *exact* value of the integral under consideration: $\int \frac{dx}{e^x+e^{-x}} = \int \frac{e^x}{e^{2x}+1} = \int \frac{du}{u^2+1} = \arctan(e^x) + C$. Therefore $\int_0^r \frac{dx}{e^x+e^{-x}} = [\arctan(e^x)]_0^r = \arctan(e^r) - \arctan(1) = \arctan(e^r) - \frac{\pi}{4} \to \frac{\pi}{2} - \frac{\pi}{4} = \frac{\pi}{4}$. 9) Obviously, $0 < \frac{1}{x} < \frac{e^x}{x}$ if $x > 0$ if $x > 0$. Since $\int_1^\infty \frac{dx}{x}$ is divergent, we can assert that $\int_1^\infty \frac{e^x}{x}dx$ is divergent. 11) $\pi/2$.

Section 2.30

1) $S'(x) = \sqrt{\frac{2}{\pi}}\sin(x^2)$, $S''(x) = \sqrt{\frac{2}{\pi}}\cos(x^2)2x$. Since $\int_0^x \sin(t^2)dt$ appears again on the screen when we write $\int(\sin(t^2), t, 0, x)$, we conjecture that $S(x)$ is not an elementary function. 3) $y(x) = \int_0^x te^{t^3}dt+1$. In particular, $y(0.1) = \int_0^{0.1} te^{t^3}dt+1 \approx 0.005002+1 = 1.005002$ (the approximation was found using the TMA program with $n = 30$). 5) $\int(\frac{\sin x}{x}, x, 1, 2) = 0.65932991$ while $\int_1^2 \frac{\sin x}{x}dx \approx 0.65932991$. For the latter we used the TMA program with $n = 10$. Thus, the approximation given by the calculator is the same as the approximation achieved through the use of the TMA program (weighted average option).

Section 2.31

1) $23.111362^\circ C$ 3) 11.623675. 5) Since $x'(t) = (\alpha - \beta y(t))x(t)$ it follows that $\frac{x'(t)}{x(t)} = \alpha - \beta y(t)$. Therefore $\frac{1}{p}\int_0^p \frac{x'(u)}{x(u)}du = \frac{1}{p}\int_0^p \alpha du - \frac{\beta}{p}\int_0^p y(u)du$, which in turn implies $\frac{1}{p}(\ln x(p) - \ln x(0)) = \frac{1}{p}\alpha p - \beta av(y)$. But $x(p) = x(0)$, hence $\beta av(y) = \alpha$. Finally, $av(y) = \frac{\alpha}{\beta}$. Analogously, $y'(t) = (\gamma x - \delta)y(t)$ implies $\frac{1}{p}\int_0^p \frac{y'(u)}{y(u)}du = \frac{\gamma}{p}\int_0^p x(u)du - \frac{1}{p}\int_0^p \delta du$. Thus, $\frac{1}{p}(\ln y(p) - \ln y(0)) = \gamma av(x) - \frac{1}{p}\delta p$. Since $y(p) = y(0)$ we get $0 = \gamma av(x) - \delta$, that is, $av(x) = \frac{\delta}{\gamma}$.

Section 3.1

1) Converges to zero 3) Converges to e^{-2} 5) Converges to zero 7) Converges to zero 9) Converges to 1 11) Diverges to infinity 13) Converges to zero 15) Diverges to infinity 17) Converges to zero. 19) The converse does not have to be true. For instance,

define $f(x) = \sin(\pi x)$. We have $f(n) = \sin(n\pi) = 0$ for every natural number, so $\lim_{n\to\infty} f(n) = 0$. However, $\lim_{x\to\infty} f(x)$ does not exist; the function $f(x)$ is oscillating all the time as $x \to \infty$. 21)By hypothesis A is bounded below, that is, there exists a number m such that $m \leq a$ for every a in A. Thus $-a \leq -m$ for every a in A, so $b \leq -m$ for every b in B. Since B is bounded above we can assert that l.u.b B exists. Next we will prove that $g.l.b.A$ exists. As a matter of fact, we will show that $g.l.bA = -l.u.b.B$. Indeed, $b \leq l.u.b.B$ for every b in B. Hence $-l.u.b.B \leq a$ for every a in A, that is, $-l.u.b.B$ is a lower bound of A. Is it the greatest lower bound? Given any $\epsilon > 0$, since $l.u.b.B - \epsilon$ cannot be an upper bound of B we can conclude that there exist b in B such that $b > l.u.b.B - \epsilon$. Consequently $-b < -l.u.b.B + \epsilon$, hence there exists a in A such that $a < -l.u.b.B + \epsilon$ (namely $a = -b$).

Section 3.2

1) Converges to 21/4 3) Converges to $\frac{x}{x-1}$ whenever $|x| > 1$ 5) Converges to 2/3 7) Observe that $1 + x + \ldots + x^{k-1} + \sum_{n=k}^{\infty} x^n = \frac{1}{1-x}$. Thus, $\frac{1-x^k}{1-x} + \sum_{n=k}^{\infty} x^n = \frac{1}{1-x}$; consequently, $\sum_{n=k}^{\infty} x^n = \frac{1}{1-x} - \frac{1-x^k}{1-x} = \frac{x^k}{1-x}$.

Section 3.3

1) 2 3) 0 5) $\frac{1}{e^3(e-1)}$.

Section 3.4

1) Diverges 3) $(1 + \frac{4}{n})^n \to e^4 \neq 0$ as $n \to \infty$. The criterion for divergence implies that the given series is divergent. 5) It diverges because $\sin(n + \frac{1}{2})\pi = (-1)^n$, but the sequence $((-1)^n)$ does not converge to zero.

Section 3.5

1) Converges (CT) 3) Converges (CT) 5) Diverges (LCT) 7) Converges (CT) 9) Diverges (LCT) 11) Diverges (LCT).

Section 3.6

1) $\frac{1}{\ln(n)} \to 0$ while $\ln(n) < \ln(n + 1)$ leads to $\frac{1}{\ln(n)} > \frac{1}{\ln(n+1)}$. The alternating series test implies the convergence of the given series. 3) The series does converge due to the fact that $\frac{\sqrt{n}}{n+3} \to 0$ and $f'(x) < 0$ for $x > 3$, where $f(x) = \sqrt{x}/(x + 3)$. 5) We cannot apply the alternating series test because the sequence $(\ln(\frac{1}{n}))$ does not tend

to zero. The test for non-convergence implies that the given series is divergent since the sequence $((-1)^{n+1}\ln(\frac{1}{n}))$ does not converge to zero.

Section 3.7

1) The given series converges (due to the alternating series test) but $\Sigma_{n,1}\sin(1/n)$ diverges (apply LCT, comparing with $\Sigma_{n,1}\frac{1}{n}$). Thus, we have only conditional convergence 3) Applying the alternating series test we can conclude that the series is convergent. However, $\Sigma\frac{\sqrt{n}}{n+5}$ diverges (use LCT, comparing with the divergent series $\Sigma_{n,1}\frac{1}{\sqrt{n}}$). The answer is then that the given series is conditionally convergent 5) The series is no other than $\Sigma_{n,1}\frac{(-1)^n}{n}$, which is certainly convergent but not absolutely convergent. Hence, we have conditional convergence. 7) It converges absolutely because in section 3.5 we proved that $\Sigma_{n,1}\frac{1}{n!}$ converges 9) The series diverges because, thanks to L'Hôpital's rule, $\lim_{n\to\infty}n\sin(\frac{1}{n})=1$; thus, the sequence of terms does not converge to zero.

Section 3.8

1) Converges (ratio test) 3) Converges (root test) 5) Diverges (root test) 7) Diverges for every $x\neq 0$ (ratio test) 9) Converges (ratio test).

Section 3.9

1) Define $f(x)=\frac{\ln x}{x}$. Then $f'(x)=\frac{1-\ln x}{x^2}<0$ for $x>e$, so f is decreasing (eventually). Since $\int_2^r\frac{\ln x}{x}dx=\frac{1}{2}(\ln r)^2-\frac{1}{2}(\ln 2)^2\to\infty$ as $r\to\infty$ we can conclude that the improper integral under consideration is divergent. The integral test implies that the given series is divergent. 3) Define $f(x)=\frac{1}{x(\ln x)^p}$. We can check that $f'(x)<0$, hence f is decreasing. Moreover, $\int_2^r\frac{dx}{x(\ln x)^p}=\frac{1}{1-p}(\frac{1}{(\ln r)^{p-1}}-\frac{1}{(\ln 2)^{p-1}})\to\frac{1}{p-1}\frac{1}{(\ln 2)^{p-1}}$ as $r\to\infty$. Hence, the improper integral converges, which in turn leads to the convergence of the given series. 5) $\Sigma_{i=1}^\infty\frac{1}{i^3}-s_n\leq\int_n^\infty\frac{1}{x^3}dx$. But $\int_n^\infty\frac{1}{x^3}dx=\frac{1}{2}\frac{1}{n^2}$. So $\frac{1}{2}\frac{1}{n^2}<10^{-2}$ will be true if $n>\sqrt{50}$. Thus, we may choose $n=8$.

Section 3.10

1) Diverges (compare with $\Sigma_{n,1}\frac{1}{n}$ and use LCT, part 1). 3) Diverges because $(1-\frac{5}{n})^n\to e^{-5}\neq 0$. 5) Converges (compare with $\Sigma_{n,1}\frac{1}{n^3}$ and use LCT, part 1). 7) Converges due to the alternating series test (check that $f(x)=\frac{1}{x+3^x}$ has a negative derivative). 9) Since $\sin(n+\frac{1}{2})\pi=(-1)^n$ we have to deal with $\Sigma_{n,1}\frac{(-1)^n}{n^{3/2}}$. The alternating series test leads to the answer: it is convergent! 11) It diverges since the

sequence $\left(\frac{1}{\cos^2(n)}\right)$ does not converge to zero. 13) Use LCT (part 2), comparing with $\Sigma_{n,1}\frac{1}{n^3}$. The series converges.

Section 3.11

1) $(\frac{1}{e}, e)$ 3) Converges everywhere. 5) $(-5, 5)$ 7) $[-1, 1)$ 9) $(-1, 1)$ 11) We can see that the radius of convergence of the given series is 1. Let $s(x) = \Sigma_{n=1}^{\infty}\frac{x^n}{n}$, $-1 < x < 1$. Then $s'(x) = \Sigma_{n=1}^{\infty}x^{n-1} = \Sigma_{n=0}^{\infty}x^n = \frac{1}{1-x}$. Thus, for x in $(-1, 1)$

$$s(x) = s(x) - s(0) = \int_0^x s'(t)dt = \int_0^x \frac{dt}{1-t} = -\int_1^{1-x}\frac{du}{u} = -\ln(1-x).$$

This equality is also true at $x = -1$ since at the end of section 3.6 we proved that $\ln 2 = 1 - \frac{1}{2} + \frac{1}{3} - \frac{1}{4} + \dots$

Section 3.12

1) Integrate the given expression and use the fact that $\ln(1 + x) = \int_1^{1+x}\frac{1}{t}dt = \int_0^x \frac{du}{u+1}$ thanks to the change of variable $t = u+1$. 3) $\sqrt{1-x} = 1 - \frac{1}{2}x - \frac{1}{8}x^2 - \dots$, $-1 < x < 1$. 5) $\frac{1}{(1-x)^2} = 1 + 2x + 3x^2 + 4x^3 + \dots$, $-1 < x < 1$. 7) Observe that $c_n = \frac{f^{(n)}(0)}{n!}$. But $f(x) = 0$ for every x, so $f^{(n)}(x) = 0$ for every n and every x. In particular $f^{(n)}(0) = 0$ for every n; thus, $c_n = 0$ for every n. 9) $\frac{1}{\sqrt{1+x^2}} = 1 - \frac{1}{2}x^2 + \frac{3}{2^3}x^4 - \dots$, $-1 < x < 1$. 11) $\arcsin x = x + \frac{1}{2}\frac{x^3}{3} + \frac{3}{2^3}\frac{x^5}{5} + \dots$, $-1 < x < 1$. 13) $\frac{1}{(1-x)^3} = \frac{1}{2}\frac{d}{dx}\Sigma_{n=1}^{\infty}nx^{n-1} = \frac{1}{2}\Sigma_{n=1}^{\infty}n(n-1)x^{n-2} = \frac{1}{2}\Sigma_{n=2}^{\infty}n(n-1)x^{n-2} = \Sigma_{n=0}^{\infty}\frac{(n+2)(n+1)}{2}x^n$.

Section 3.13

1) Start with the approximation $\sin(x^2) = x^2 - \frac{x^6}{3!} + \frac{x^{10}}{5!}$. After integrating, the answer you will get is 0.000333. 3) The recursion formula happens to be $(n+2)(n+1)c_{n+2} - c_{n-1} = 0$, $n \geq 1$, and $c_0 = 0$, $c_1 = 1$. The answer is the polynomial $p(x) = x$. 5) Just notice that $\frac{1}{(y^2+r^2)^{3/2}} = \frac{1}{y^3(1+(r/y)^2)^{3/2}} \approx \frac{1}{y^3}$. Then use this approximation in the formula for the force of attraction.

Section 3.14

1) Follow the same steps that we used to prove that $\cos x$ is analytic at the origin. 3) Since

$$E_3(x) = f(x) - (f(0) + f^{(1)}(0)x + \frac{f^{(2)}(0)}{2}x^2 + \frac{f^{(3)}(0)}{3!}x^3)$$

it follows that

$$E_3^{(1)}(x) = f^{(1)}(x) - (f^{(1)}(0) + f^{(2)}(0)x + \frac{f^{(3)}(0)}{2}x^2)$$

and

$$E_3^{(2)}(x) = f^{(2)}(x) - (f^{(2)}(0) + f^{(3)}(0)x).$$

Applying L'Hôpital's rule twice we get

$$\lim_{x \to 0} \frac{E_3(x)}{x^3} = \lim_{x \to 0} \frac{E^{(1)}(x)}{3x^2} = \frac{1}{6} \lim_{x \to 0} \frac{E_3^{(2)}(x)}{x}.$$

But

$$\lim_{x \to 0} \frac{E_3^{(2)}(x)}{x} = \lim_{x \to 0} \left(\frac{f^{(2)}(x) - f^{(2)}(0)}{x - 0} - f^{(3)}(0) \right) = f^{(3)}(0) - f^{(3)}(0) = 0.$$

Consequently

$$\lim_{x \to 0} \frac{E_3(x)}{x^3} = \frac{1}{6} \times 0 = 0.$$

5) Applying integration by parts, with $u(t) = x - t$ and $v'(t) = f^{(2)}(t)$, we get

$$\int_0^x f^{(2)}(t)(x - t)dt = [(x - t)f^{(1)}(t)]_0^x + \int_0^x f^{(1)}(t)dt = -(x - 0)f^{(1)}(0) + f(x) - f(0).$$

Hence $f(x) = f(0) + xf^{(1)}(0) + \int_0^x f^{(2)}(t)(x - t)dt$. 7) We will use the method of mathematical induction. In exercise 5 we proved that the expression is valid when $n = 1$. Assume that it is valid for $n - 1$, that is,

$$f(x) = f(0) + f^{(1)}(0)x + ... + \frac{f^{(n-1)}(0)}{(n-1)!}x^n + \frac{1}{(n-1)!} \int_0^x f^{(n)}(t)(x - t)^{n-1}dt.$$

Applying integration by parts, with $u(t) = (x - t)^n$ and $v'(t) = f^{(n+1)}(t)$, we get

$$\int_0^x f^{(n+1)}(t)(x - t)^n dt = [(x - t)^n f^{(n)}(t)]_0^x + n \int_0^x (x - t)^{n-1} f^{(n)}(t)dt$$

$$= -x^n f^{(n)}(0) + n \int_0^x (x - t)^{n-1} f^{(n)}(t)dt.$$

Therefore

$$\int_0^x (x - t)^{n-1} f^{(n)}(t)dt = \frac{f^{(n)}(0)}{n}x^n + \frac{1}{n} \int_0^x f^{(n+1)}(t)(x - t)^n dt.$$

. The final step consists in replacing the inductive hypothesis in the preceding formula.

Chapter 4

1) $\hat{y} = 4028.013112x + 39.916148$, and correlation coefficient 1. Thus, $V_{max} \approx \frac{1}{4028.013112} = 2.4826 \times 10^{-4}$ and $\frac{K_m}{V_{max}} \approx 39.916148$. Consequently $K_m \approx 0.00024826 \times 39.916148 = 9.90964 \times 10^{-3}$. We can observe that the values of K_m and V_{max} are quite similar to those obtained when working with the Lineweaver-Burk plot. 3) $\hat{y} = -0.587369x + 0.602554$, and correlation coefficient -0.723707. Then $\frac{-1}{K_m} \approx -0.587369$, which in turn leads to $K_m \approx 1.7025073$, and $\frac{V_{max}}{K_m} = 0.602554$, hence $V_{max} \approx 1.7025073 \times 0.602554 = 1.0258526$. It is subject to debate whether the integrated form of Michaelis-Menten equation is a reliable tool to estimate V_{max} and K_m, using Stern's data, because the correlation coefficient is not close to -1. 5) The intersection with the vertical axis is $1/V_{max}$ while the intersection with the horizontal axis is $\frac{-1}{K_m(1+\frac{[I]_o}{K})}$. The lines corresponding to $[I]_o$, $2[I]_o$, and $3[I]_o$ pass through $(0, 1/V_{max})$, and their corresponding intersections with the horizontal axis increase (from left to right).

Chapter 5

1) $C_{in}(t) = \frac{V}{kA}\left(\frac{V_{max}C_{out}}{K_m+C_{out}} + \frac{kAC_{out}}{V}\right)(1 - e^{-kAt/V})$. When $t \to \infty$ we get $\frac{V}{kA}\frac{V_{max}C_{out}}{K_m+C_{out}} + C_{out}$. 3) We have $\frac{A}{V} = \frac{2\pi r^2 + 2\pi r^2 n}{\pi r^3 n} = \frac{2}{r}\left(\frac{1}{n}+1\right)$. But $r = \left(\frac{V}{\pi n}\right)^{1/3}$, consequently $\frac{A}{V} = \frac{2}{\left(\frac{V}{\pi n}\right)^{1/3}}\left(\frac{1}{n}+1\right) = \frac{2\pi^{1/3}}{V^{1/3}}\left(\frac{1}{n^{2/3}} + n^{1/3}\right)$. It can be seen that $\lim_{n\to\infty} \frac{A}{V} = \infty$. 5) $y(t) = \frac{b}{a}t + \frac{1}{a}\left(\frac{b}{a} - \beta\right)e^{-at} + \alpha - \frac{b}{a^2} + \frac{\beta}{a}$.

Index

www.ingramcontent.com/pod-product-compliance
Lightning Source LLC
Chambersburg PA
CBHW081429170526

45166CB00008B/2142